Advances in
Astronomy

From the Big Bang to the Solar System

THE ROYAL
SOCIETY

Royal Society Series on Advances in Science – Vol. 1

Advances in
Astronomy

From the Big Bang to the Solar System

Editor

J M T Thompson
University of Cambridge, UK

DERIVED IN PART FROM WORK ORIGINALLY PUBLISHED IN THE PHILOSOPHICAL TRANSACTIONS OF
THE ROYAL SOCIETY, SERIES A (*PHIL. TRANS. R. SOC. A, 360, 2649-3004, 2002*)

ICP Imperial College Press

Published by

Imperial College Press
57 Shelton Street
Covent Garden
London WC2H 9HE

Distributed by

World Scientific Publishing Co. Pte. Ltd.
5 Toh Tuck Link, Singapore 596224
USA office: 27 Warren Street, Suite 401-402, Hackensack, NJ 07601
UK office: 57 Shelton Street, Covent Garden, London WC2H 9HE

British Library Cataloguing-in-Publication Data
A catalogue record for this book is available from the British Library.

ADVANCES IN ASTRONOMY
From the Big Bang to the Solar System

ISBN   1-86094-577-5

Printed in Singapore by World Scientific Printers (S) Pte Ltd

FOREWORD

John D. Barrow

DAMTP, Centre for Mathematical Sciences, Cambridge University

Astronomy is currently experiencing a Golden Age. Its pace of new discovery is relentless and its interconnections with other fields continue to
multiply in unexpected ways. Once upon a time astronomy was the time-
honoured business of looking through a glass darkly, observing nearby stars
and planets, tracking the occasional visiting comet, and painstakingly developing a few reddened spectra from distant galaxies. Gradually, the as-
tronomies of other wavebands opened up by means of the technologies that
emerged from work on radar during the second world war. New energetic
phenomena were discovered that challenged theorists to understand what
went on in very strong gravitational fields where matter behaved in ways
that could never be duplicated in terrestrial laboratories. Astro-physics became a fully fledged subject. Most recently we have seen it become wedded
to the world of elementary-particle physics, as cosmologists seek to explain
many of the universe's most mysterious properties by appeal to the high-
energy physical processes in the Big Bang. And, by the same token, particle
physicists have realised that the extreme environment of the early universe
offers them a 'theoretical laboratory' in which to test their theories without
the slow and expensive construction of huge accelerators.

  In recent years we have seen the development of three strands of activity
which you will see reflected in the articles in this volume: theory and observation have been joined by the specialism of numerical simulation. The
rapid growth in computational capability has opened up the possibility of
simulating the behaviour of complicated situations, like the formation of
stars or planets, so that we can watch what transpires when equations too
complex for exact solution are allowed to run their course. What emerges
from the use of this new research tool not only challenges astronomers but

also the traditional ways of publishing the results of research. Researchers now want to publish film or huge numbers of colour images that cannot be handled by traditional avenues for scientific publication.

The steady growth in computer technology has been accompanied by spectacular achievements in engineering and electronics. This has made possible a suite of ground-based and satellite-borne astronomical detectors of exquisite sensitivity. They have enabled us to use astronomy to probe the constancy of the constants of nature with greater accuracy than was ever possible in terrestrial laboratories. They have enabled planetary astronomers to detect the tiny orbital wobbles produced by large planets orbiting more than a hundred other stars. They have made possible the detailed microwave mapping of the universal radio noise left over from the hot beginning of the universe and the partial reconstruction of the sequence of events that led to the formation of galaxies. We are also able to test the theory that in its earliest moments the universe underwent a surge of accelerated 'inflationary' expansion, and attempt to explain why more than seventy per cent of the universe is in the form of a strange 'dark energy' of unknown composition and origin. Dark energy is a mystery. We see its gravitational effects, we can measure its density, and reconstruct some of its history; we can understand its status as a sort of zero-point energy of the universe. But what is it and why is it controlling the expansion of the universe today?

Gamma-ray astronomy witnesses the fall-out from the largest explosions in the universe. Black holes are not only great cookie monsters in the sky. They can collide with one another and with neutron stars. When they do, huge bursts of high-energy gamma rays result. But we expect other evidence of events when worlds collide. Einstein predicted that gravitational waves should be produced in violent, non-spherical processes. Merging black holes and exploding stars should shake space and time enough to send ripples of gravity undulating throughout the universe. By the time they hit the Earth they will be attenuated to a level that is far smaller than the banging and crashing of earthquakes far away and passing of motorcars next door. But by ingeniously shielding detectors from identifiable forms of local noise the signal from gravitational waves can potentially be discriminated and enhanced. The first gravitational-wave detectors are now taking data, and beginning a long process of honing their sensitivity to a level where they expect the true signals of astronomical violence to be visible. These signals will play a decisive role in testing Einstein's general theory of relativity at new levels of precision.

The study of the solar system has undergone a huge resurgence in the past decade. The interior of the Sun can be probed by studying the pattern of internal vibrations that it supports. Its surface activity presents us with spectacular problems of plasma physics that may hold tantalising clues to the cycles of climate variation on Earth. From the unpredictability of huge solar storms to the predictable ups and downs of sunspot numbers, we know that an understanding of the Sun is vital if we are to unravel the effects of past and present human activity on our climate.

We used to wonder whether our Sun was unique in having a system of planets orbiting around it. Now we know that it is not. Planet formation is a generic phenomenon in the Universe. But, so far, we can only see planets larger than the Earth. And with only one exception they are found to be in orbits that are far from circular. By contrast, the planets in our solar system move in orbits that are almost circular. This may not be a coincidence. Circular orbits keep the surface temperatures fairly constant; elliptical orbits inevitably produce temperature changes that may be too extreme for early life to survive. This is a story that is just beginning. Our studies of astro-chemistry allow us to understand what types of molecular self-replication are possible, or even probable, in the Universe. In the dust of the stars we may learn something of the make-up of the material that went to make our solar system and made possible the sequence of events that led 4.6 billion years later to the studies in this volume.

John D. Barrow

John D. Barrow was born in London in 1952 and attended Ealing Grammar School. He graduated in Mathematics from Durham University in 1974, received his doctorate in Astrophysics from Oxford University in 1977 (supervised by Dennis Sciama), and held positions at the Universities of Oxford and California at Berkeley before taking up a position at the Astronomy Centre, University of Sussex in 1981. He was professor of astronomy and Director of the Astronomy Centre at the University of Sussex until 1999. He is a recipient of the Locker Prize for Astronomy and the 1999 Kelvin Medal of the Royal Glasgow Philosophical Society. He was awarded an honorary Doctor of Science degree by the University of Hertfordshire in 1999. He held a Senior 5-year Research Fellowship from the Particle Physics and Astronomy Research Council of the UK in 1994-9. He will hold the Gresham Professorship of Astronomy for the period 2003-6. He was elected a Fellow of the Royal Society in 2003.

In July 1999 he took up a new appointment as Professor of Mathematical Sciences at Cambridge University and Director of the Millennium Mathematics Project, a new initiative to improve the understanding and appreciation of mathematics and its applications amongst young people and the general public. He is also a Fellow of Clare Hall, Cambridge.

He is the author of more than 385 scientific articles and 17 books, translated into 28 languages, which explore many of the wider historical, philosophical and cultural ramifications of developments in astronomy, physics and mathematics: these include, *The Left Hand of Creation* (with Joseph Silk), *The Anthropic Cosmological Principle* (with Frank Tipler), *L'Homme et le Cosmos* (with Frank Tipler), *The World Within the World*, *Theories of Everything*, *Pi in the Sky: counting, thinking and being*, *Pérche il mondo è matematico?*, *The Origin of the Universe*, *The Artful Universe*, *Impossibility: the limits of science and the science of limits*, *Between Inner Space and Outer Space*, *The Book of Nothing* and *The Constants of Nature*. His most recent book, *The Infinite Book* has just been published by Random House. He has written a play, *Infinities*, which was performed (in Italian) at the Teatro la Scala, Milan, in the Spring of 2002 and again in 2003 under the direction of Luca Ronconi and in Spanish at the Valencia Festival. It was the winner of the Italian Premi Ubu award for best play in 2002 and the 2003 Italgas Prize for the promotion of science.

He is a frequent lecturer to audiences of all sorts in many countries. He has given many notable public lectures, including the 1989 Gifford Lectures at Glasgow University, the George Darwin and Whitrow Lectures of the Royal Astronomical Society, the Amnesty International Lecture on Science in Oxford, The Flamsteed Lecture, The Tyndall Lecture, The Brasher Lecture, The RSA Christmas Lecture for Children, and the Spinoza Lecture at the University of Amsterdam. John Barrow also has the curious distinction of having delivered lectures on cosmology at the Venice Film Festival, 10 Downing Street, Windsor Castle and the Vatican Palace.

PREFACE

1. Introduction

This book outlines the advances currently being made world-wide by researchers in the field of Astronomy. Areas covered include: cosmology and the big bang; the formation and evolution of the stars and galaxies; man's exploration of the solar system; explosive sun-spot events, and their influence on the earth's climate.

The book is the first in a new series to be published by Imperial College Press in collaboration with the Royal Society of London under the general title *Advances in Science*. The next five books will be devoted to topics within Earth Science; Mathematics; Physics; Engineering and Chemistry. All books will contain carefully selected articles by top researchers writing for a general scientific audience. Articles will be timely, topical and well illustrated with diagrams and photographs. A typical paper will describe cutting edge research, and put it into its wider context. A photograph and brief informal CV of each author will add a personal touch to each volume. This first book of the series devoted to astronomy starts with a foreword by the distinguished science writer, John Barrow, FRS. This foreword, together with the introductory article by Bruce Bassett, serves to draw together the wide-ranging topics of the contributing authors.

Many articles for this and the future books derive from invited papers published in the Philosophical Transactions of the Royal Society of London, Series A, of which I am the editor. In the present book about half the number of papers derive from this source, while the other half are carefully invited and refereed additions to make the coverage of modern astronomy more complete. Of the authors, seven hold (or have recently held) one of the prestigious Royal Society Research Fellowships.

2. Background to the Series

Founded in 1665, *Philosophical Transactions* is the world's longest running scientific journal and throughout its history has contributed hugely to the advancement of science and technology. It was used by Newton to launch his scientific career. Recent contributors include Stephen Hawking, Michael Atiyah, and Nobel Laureates Steven Weinberg and Aaron Klug. The journal is published monthly by The Royal Society, the UK academy of science, and all issues of the journal (from 1665) are archived electronically by JSTOR in the USA (www.jstor.org), accessible through most university libraries.

The three Millennium issues of *Philosophical Transactions A*, devoted to the work of young scientists, proved to be highly successful. Popular versions have now been published by Cambridge University Press as three paperback books, all carrying the generic title of *Visions of the Future*. Edited by JMT Thompson, they are devoted to *Astronomy and Earth Science, Physics and Electronics* and *Chemistry and Life Science*. These compilations capture the enthusiasm and excitement of the young scientists and provide a unique snapshot of the state of physical science at the turn of the Millennium. The series has proved to be of great interest to researchers and the general public.

To build upon this success, I am now running a rolling series of triennial Christmas issues. Replicating the Millennium pattern, I am soliciting articles from leading scientists, including holders of Royal Society Research Fellowships. For compatibility with these awards, young is interpreted as scientists with no more than ten years post-doctoral experience, and the three-year cycle will ensure that the majority of Research Fellows receive a relevant invitation. It is these Christmas issues that are being adapted to form the core content of the new book series published by Imperial College Press.

3. Invitation for Future Articles

Young researchers, worldwide, are hereby invited to submit articles for the Christmas Issues, reviewing their field of work and looking forward to new developments. Contributors are encouraged to be more speculative, and perhaps more provocative, than they would normally be in a review article. The articles should be timely and topical, and written for a general scientific audience, at about the level of *Scientific American*. They should be well illustrated with diagrams, photographs, etc, and detailed mathematics should be kept to a minimum. A paper that describes recent cutting-edge research, puts it into a wider context, and looks forward to the future is an ideal candidate. All papers will be subjected to a refereeing process which takes account of the above criteria.

4. Triennial Christmas Issues of the Journal

In selecting 46 articles for the Millennium Issues, it was apparent that three issues of the journal are needed to adequately cover the physical sciences. So for this series I am adopting the pattern of publishing one issue of *Phil. Trans.* (A) each Christmas, rolling cyclically through the physical sciences. For the second cycle the pattern will be as follows:

- 2005 to cover *Astronomy & Earth Science*
- 2006 to cover *Mathematics & Physics*
- 2007 to cover *Chemistry & Engineering*

In view of the interdisciplinary nature of much research, this classification will be applied in a flexible manner, and researchers in the biological sciences should not necessarily be deterred. Suggestions for contributions are now welcomed, according to the following timetable for a typical Journal Issue:

- 15 Oct, year before publication, Receipt of abstracts by Editor
- 30 Oct, year before publication, Acceptance decisions to authors
- 1 Feb, year of publication, Receipt of papers by the Editor
- Feb-May, year of publication, Refereeing process
- 15 Jun, year of publication, Receipt of final, refereed papers

Any young scientist who would be interested in contributing an article for the Christmas series should send an abstract to the Editor for consideration (by e-mail to jmtt2@damtp.cam.ac.uk). The abstract should be about one page in length, and a brief CV should also be attached.

In conclusion, I would like to thank some of the many people who have made this new venture possible: Cathy Brennan for her enthusiastic editorial work on the *Philosophical Transactions of the Royal Society* (A); Phil Hurst of the Royal Society and Lance Sucharov of Imperial College Press with whom this new series was planned; and Katie Lydon for her cheerful help in assembling the material at Imperial College Press.

Reference

J. M. T. Thompson (ed.), Astronomy and Earth Science, First Christmas Issue of *Phil. Trans. R. Soc. Lond.*, **A360**, 2649 (2002).

Michael Thompson, FRS
Mathematics, Cambridge

Michael Thompson

Michael Thompson was born in Yorkshire in 1937, and attended Hull Grammar School. He graduated from Cambridge with first class honours in mechanical sciences in 1958 (ScD, 1977). He was a Professor at University College London and was appointed Director of the Centre for Nonlinear Dynamics in 1991. His fourth book, *Nonlinear Dynamics and Chaos* (2nd edn., Wiley, 2002), has sold 14,000 copies. Michael was elected a Fellow of the Royal Society in 1985, and served on the Council. He won the Ewing Medal (Institution of Civil Engineers) in 1992, the IMA Gold Medal for mathematics in 2004, and was a Senior SERC Fellow.

Since 1998, Michael has been Editor of the Philosophical Transactions of the Royal Society. Special Millennium Issues, in which young scientists gave their visions of the future, were re-printed as three popular books by Cambridge University Press. A follow-up is the new *Royal Society Series on Advances in Science*, to be published by Imperial College Press.

Michael is Emeritus Professor (UCL) and a Fellow at the Dept of Applied Maths and Theoretical Physics (Cambridge). Married with two children and eight grandchildren, he enjoys astronomy with his grandchildren, wildlife photography and badminton. He is pictured above with his grandson Charlie McRobie.

CONTENTS

Cosmology and the Big Bang

CHAPTER 1

THIRTEEN BILLION YEARS IN HALF AN HOUR

Bruce A. Bassett

Department of Physics, Kyoto University, Japan & Institute of Cosmology and Gravitation, University of Portsmouth, England
E-mail: bruce.bassett@port.ac.uk

We take a high-speed tour of the approximately thirteen billion-year history of our universe focusing on unsolved mysteries and the key events that have sculpted and shaped it – from inflation in the first split second to the dark energy which is currently causing the expansion of the cosmos to accelerate.

1. Introduction

The aim of this chapter is to provide a lightning overview of the roughly thirteen billion-year history of our universe including its geography, geometry and archeology as best we know it today. The average person might read its five thousand or so words in about half an hour. Thirteen billion years in half an hour is an almost inconceivable compression.

Numerically this is akin to summarising a few hundred years worth of television adverts from around the world in a single thirty-second spot – clearly an impossible task. But fortunately, our task is possible, at least in broad-brush strokes, because there is a crucial difference – our universe has, for most of its history, been very slow in changing, allowing us to sweep over billions of years with barely a whisper.

In fact, according to our current theories, most of the evolution of our cosmos occurred in the first, tiny, fraction of its history. For example, we believe our universe cooled and expanded more in the first one millionth of a second than it has done in the rest of its history until today. In the words of Georges Lemître, a priest and one of the first cosmologists of the 20th century:

The evolution of the world can be compared to a display of fireworks that has just ended; some few red wisps, ashes and smoke. Standing on a cooled cinder, we see the slow fading of the suns, and we try to recall the vanishing brilliance of the origin of the worlds.

In writing the biography of our universe we will make an analogy with the seasons on earth, except that the order is a little different in this case: summer is followed by autumn, then by spring and finally we find ourselves now in a winter of deep discontent.

2. Summer Inside the Quantum Gravity Second

If we were to lay out the history of our universe like a roulette table and place bets where we are confident we understand what was going on at that time, then there would be a gross asymmetry in our betting.

Between the initial second after the origin of the universe as we know it – the Big Bang – and before the universe was around about five billion years old we are confident we understand the basic physics of the universe. Within the first second is a period we will call *summer* because it is the hottest period in the history of the cosmos. In contrast, the cosmic history after five billion years we will refer to as *winter*, since it is the coldest season of the Universe. During summer and winter we have very limited knowledge of the fundamental physics governing the evolution of the cosmos and the search to understand these epochs is the main driving force in modern cosmology.

It is perhaps ironic that one second, an utterly human timescale, roughly marks the first boundary of our ignorance. It is approximately the time it takes for our hearts to beat or a piece of toast to fall to the ground (buttered side down of course). Within the first second of the universe's evolution our understanding is extremely tenuous.

The main reason for this lack of understanding is that the energies and temperatures are so high throughout the whole universe that they reach far beyond what we can aspire to in particle accelerators on earth. We believe this because of some remarkable theorems proven by Stephen Hawking, Roger Penrose and others in the late 1960s and early 1970s which show that, at least within the context of Einstein's general theory of relativity, going back in time inescapably leads to increasing temperature and density of matter.

Indeed, this increase happens so rapidly that at a finite time in the past Einstein's relativity predicts that temperature and density became infinite. This is the *Big Bang*; the event from which we measure the age of the universe to be around thirteen billion years, give or take a billion years (but what's a billion years between friends!)

It is crucial to realize however, that the Big Bang – the point of infinite temperature – probably never happened. Within the first second of the Big Bang the great trust we show in Einstein must be carefully reviewed because Einstein's vision of gravity does not take into account quantum effects.

Quantum effects become crucial any time we ask a question which involves lengths scales comparable to or smaller than that of an atom. The quantum realm is a strange one indeed, divorced from everyday intuition and common sense. In it, balls can pass straight through walls and a cat can be both dead and alive at the same time.

Remarkably, the theorems proven by Hawking and Penrose show that Einstein's theory very likely breaks down at some point because the typical length scales for the Universe become smaller than the size of an atom, and so quantum effects must be included. At the same time we cannot ignore gravity because near the Big Bang gravity becomes extremely strong. As a result, the combined theory we are searching for – the holy grail of modern fundamental physics – is known as quantum gravity.

Whether or not quantum effects can smooth out and make finite the infinity density and temperature of the Big Bang is not known. Indeed it is one of the biggest unanswered questions in cosmology. Most researchers, perhaps, believe that they will – and hence that the cosmos may be much older than our current best estimate of thirteen billion years. But without a complete theory of quantum gravity we find only tantalizing hints.

The first second of cosmic history is not only mysterious because of the Big Bang. It has given us other headaches too. For example, the universe we observe is far too smooth. Gravity, the force that dominates our universe, makes things clumpy, makes things cluster. It takes matter and aggregates it into dense clumps like planets and galaxies and in so doing depletes vast regions which become essentially devoid of any matter.

But when we look out at the universe, it looks the same in all directions – the number of galaxies in any direction is basically the same all over the sky and variations in the temperature of the radiation around us – left over from the fireball in the first second – are less than one part in one hundred thousand, all over the sky.

This is difficult to understand precisely since gravity makes things clumpy, not uniform. But it gets worse. When we look out at the sky, then in standard Big Bang cosmology, points separated by more than about one degree, or about the size of the moon, could never have been in contact with each other.

And if they were never in contact, never communicated, then how is it possible that they look almost identical? It is like everyone in the world sneezing at exactly the same moment in time. It seems to be a coincidence of cosmic proportions.

Fortunately there is a solution that doesn't appeal to coincidence. If gravity is the problem, then it stands to reason that anti-gravity might provide a solution. Where gravity makes a smooth distribution of matter clumpy, anti-gravity – that is a universal *repulsive* force – will makes a clumpy distribution smooth. A short period of anti-gravity in the first second of the universe sets us up with the right conditions to give us the universe we observe by smoothing out the wrinkles in our new born universe.

In cosmology this anti-gravity has been given a special name: *inflation*, so named because it causes the universe to inflate extremely rapidly. In this view of the Universe, everything we can see in the observable cosmos inflated from a tiny embryo. The uniformity of that embryo then explains why the cosmos we observe is so uniform. The whole observable cosmos shares the same DNA.

The only problem is, how do we achieve inflation – anti-gravity is certainly not an everyday experience today – and how do we end it? Within the context of Einstein's general relativity, the only way inflation can occur is if the cosmos is dominated by some new form of matter with negative pressure. Unfortunately we have never detected any such matter!

This 'little' obstacle notwithstanding, inflation makes predictions that are in very good agreement with the precision tests we are now conducting in our cosmos. In particular, inflation provides us a wonderful bonus. In addition to giving us a universe that is very smooth, it leaves behind a subtle but vital legacy: it leaves behind quantum seeds that will later grow into the galaxies and clusters of galaxies we observe in our universe.

In summary, at present we have a successful phenomenological model of anti-gravity that appears to explain why our universe is so uniform and also explains the origin of galaxies. We still lack a fundamental theory to explain the origin of inflation, but we are working on it.

There are other problems from the initial second that remain a mystery. Most pressing perhaps of these is the question *"why are we made of matter and not anti-matter?"* Or more precisely, *"why are both author and reader made of matter?"* Anti-matter looks just like ordinary matter except that it has diametrically opposite properties (such as opposite charge and negative energy). As a result, a meeting of matter and anti-matter leads to catastrophic annihilation of both into a shower of gamma-rays – that is, very high energy light.

When we look out at the universe, then everything we see, as best we can tell, is made of matter, just as we are. We don't see great jets of gamma-rays coming from the annihilation of galaxies as they collide with anti-galaxies. So what is the problem you may ask?

The problem is that in the first fraction of a second there should have been equal amounts of matter and anti-matter in our universe. Our equations and simulations tell us that as the Universe expanded out and cooled, the matter and anti-matter would annihilate, leaving only light behind and no matter at all. So, whereas before we needed to make the universe look symmetric, now we need a mechanism to generate an asymmetry – this time between the amount of matter and anti-matter in the genesis of the universe. This is the problem of baryogenesis.

The required asymmetry is very small and getting just the right amount (and not too much) is very difficult. The universe is rather like a very fussy Goldilocks. Everything has to be *just right*. We have some candidate mechanisms to satisfy our Goldilocks but none are completely compelling just yet.

Let us end our tour of summer, by summarizing the rough properties of the cosmos after the very first second of history. The temperature is extremely high, the universe is dominated by matter rather than anti-matter and it is extraordinarily smooth, with only the finest traces of fluctuations, that will later grow into galaxies.

3. Autumn – Fading Light and Condensing Mists

Very roughly around the one-second mark, the character of our understanding of the cosmos changes dramatically. Suddenly we are in the energy range that particle physicists have been able to reach using particle accelerators on earth. Therefore we can recreate the conditions of the cosmos at that time and hence our knowledge is more secure.

At one second, the universe is dominated by the energy of light and radiation. There are also protons and electrons in equal numbers and neutrons. These particles will later join together to form the atoms from the stars, and everything we know & love, will be formed. But right now the temperature is too high for atoms to form – they are immediately ripped apart by the high-energy radiation.

But as the universe continues to expand and cool, a critical point is reached. At a certain point the temperature drops sufficiently for single protons to safely join together with single neutrons. From this point the cosmos begins manufacturing the cores of the light chemical elements such as helium and lithium. By contrast, the heavy elements – gold, silver, oxygen, carbon and so on – will only be produced much later inside the furnaces of stars.

The abundance of lithium turns out to be particularly sensitive to the density of protons and neutrons during this brief window in cosmic history. Looking out at how much lithium we see in the cosmos today therefore provides us with an exquisite barometer of what the density was only a short time after the Big Bang.

The abundance of helium in the cosmos today tells us something else. It tells us very precisely how old the Universe was at this time because of a simple fact – the neutron is unstable. Crucially, the neutron is a little bit heavier than the proton and so after about ten minutes there is a 50% chance that a free neutron will decay into a proton, an electron and a strange little particle called a neutrino.

If the Universe were very old when it became cool enough to allow helium to form then all the neutrons would have already disappeared and no helium nuclei – which is made up of two protons and two neutrons – would have been able to form. Instead we find that there is lots of helium in the cosmos.

Measuring the present abundance of helium and lithium precisely therefore gives us accurate knowledge of both the age and density of the cosmos at the time when the light elements were forged – a time known as nucleosynthesis. If we ask how old Einstein predicts the universe to be when it had that density then the result is just right. This is compelling evidence that we are on the right track at least from about one second onwards. Alternative theories, such as a steady-state universe, struggle to match the brilliant accuracy and simplicity of Big Bang cosmology that flows from Einstein's gravity.

After nucleosynthesis, the universe continues to expand and cool. Radiation – made up of light of every frequency – has up until now been the dominant energy component of the universe. But the energy in light drops quicker as the cosmos expands than does the energy of ordinary matter. The density of ordinary matter drops only because of the increasing volume of space. But light has an extra burden. The energy of each individual packet of light depends on its wavelength which is stretched by the cosmic expansion, thereby decreasing the energy of light.

As a result, there is another critical moment in the history of the universe when matter usurps the crown of the fading light as the dominant energy component in the universe. From now on, light becomes less and less important as a real player and remains only important as a messenger, bringing us information – rather like a queen who is dethroned and kept on only as a court messenger.

At this point, the universe is mainly a hot plasma soup of electrons and nuclei. While the ambient radiation has lost control of the cosmos it is still hot enough to stop the free electrons from joining the nuclei of protons and neutrons. The light has scattered very strongly from the free electrons, traveling only very short distances between consecutive scatterings.

But at a certain point, about four hundred thousands years after the Big Bang, when the universe was about one thousands times hotter and smaller than it is today, the light is weakened to such an extent that it cannot stop the unification of the electrons and nuclei, and the first atoms in the history of the Universe are produced.

Once the electrons are safely confined inside atoms the light effectively cannot reach them anymore. As a result the light decouples from matter and begins a lonely sojourn through the universe. Our queen has gone from being in a closely-knit family to being a wandering hermit, banished and unable to talk to anyone.

This is good news for us though. For when the light reaches our telescopes it has traveled vast distances, faithfully bringing us information without loss or distortion. In the mid-1960s when Arno Penzias and Robert Wilson discovered the cosmic microwave background, they did not fully realize that they were looking at a portrait of the universe when it was only about four hundred thousand years old.

This discovery has revolutionised cosmology and for the past decade we have been building ever finer and more complete versions of this portrait. The portrait has, in the last few years, become clear enough to conclusively show that space is very nearly flat! This is yet another mystery since naively we should expect space to be curved – since that is what gravity does. This is yet another conundrum that inflation can help solve.[a]

Returning to our chronological journey we have seen how, out of the hot primordial mists, first nuclei and then atoms have condensed as the universe has gradually become colder and darker. The former bright colours and high temperatures have faded in the autumn of the universe.

---

[a]See the chapter by Hiranya Peiris on the WMAP satellite and the cosmic microwave background.

From this point on, the universe simply expands and cools. The rate of expansion is slowing down, like an ageing beast which begins to feel its age and shows the passing of its youthful vitality. And as it cools, the seeds of the galaxies, perhaps laid down by inflation in the initial fraction of time, slowly begin to grow.

The temperature of the universe drops below room temperature, then below our coldest fridges. Billions of years pass. When the universe is perhaps only twenty or thirty times smaller than it is now we reach the darkest and coldest point in the universe's history. But our story is not over. In the depths of autumn the galaxy seeds are beginning to flourish.

4. Spring and the Flowering of the Cosmos

Under the nurturing hand of gravity the seeds – slight irregularities in the density of matter – have grown. Now the cold temperature allows the bigger clumps to collapse under their own gravity. As they do so the temperature and density inside them increase and at a key point the density is high enough for stars to form and for nuclear fusion to begin – nuclei of atoms are joined together — releasing large amounts of light.

The birth of the first stars leads to the production of the first heavy elements. Light once again seeps out into the universe. The spring has come to end the cosmic dark ages and there are now a plethora of spectacular events. The life and times of stars is short and violent and with their death explosions they expel the chemical elements from which we are made into the cosmos. Every atom in our bodies has come from the center of a star. We are literally made of stardust.

From this stardust planets also form. Many solar systems, harbouring new planets, have been discovered in the last ten years and their formation and detection is now an active field of research,[b] for an obvious reason – they may have incubated life in the same way that our own solar system gave rise to life on earth.

But there are even more violent events out there than the dying breathes of ordinary stars. Gamma-ray bursts[c] emit vast amounts of energy and are powered by enigmatic energy sources. Black holes probably form from overly massive stars and when they collide with stars or other black holes

---

[b]See the chapters by Andrew Markwick, Hugh Jones and Andrew Coates on planets and solar systems.
[c]See the chapter by Davide Lazzati on gamma-ray bursts.

Einstein's gravity predicts that they send out ripples in space-time – known as gravitational waves – across the universe. Detecting these waves is one of the most exciting prospects for the next ten years in astronomy[d] and would give us a new set of eyes with which to view the cosmos.

Typically the stars, gamma-ray bursts and black holes form in large regions that collapse themselves under gravity to form galaxies – collections of one hundred billion stars or more. Galaxies form, evolve, collide with each other, and somehow manage to form a wide variety of shapes, from elliptical-shaped giants and flat spirals like our own Milky Way galaxy to small, irregularly shaped dwarf galaxies. Many of these galaxies appear to have huge black holes at their centers. Why and how this is true is still unknown.[e]

When we look out at the universe, we see that the galaxies have clustered together under their mutual gravity. These clusters in turn are grouped together into vast super-clusters, megalopolises so big that it takes light hundreds of millions of years to cross them and in which each galaxy seems like a single egg in a vast swath of fish spawn.

How each galaxy forms is, however, rather mysterious. Mostly this is not because we don't understand the basic physics. Now we are in a realm of fluid dynamics, magnetic fields, turbulence and Newtonian gravity – physics whose basic properties have been understood for a century and more. The problem now is simply one of complexity – there are so many factors which are important that we struggle to deal with them all simultaneously. But with the help of massive supercomputers we are beginning to make real progress.

But there is one aspect of galaxy formation that does involve completely new physics that we don't understand yet – dark matter.[f] As early as the 1930s it was noted by Fritz Zwicky, an eccentric astronomer, that galaxies in clusters seem to move too fast for the amount of visible mass inside the cluster can support. Rather like a rock on a string that is revolving so fast that the string should break.

Zwicky's calculations, and numerous more sophisticated analyses since, have shown that the galaxies should simply fly off into space rather than move on the near-circular orbits we observe. But this problem is not only observed in clusters. The stars inside galaxies show the same mysterious

---

[d]See the chapter by Sathyaprakash on gravitational waves.
[e]See the chapter by Carole Mundell on the black holes at the centres of galaxies.
[f]See the chapter by Andrew Benson on cold dark matter.

property of moving too fast.

This strongly suggests that the force of gravity is much stronger on these large scales, than we naively imagined. There are two ways out of this: one can try to make gravity stronger, or one can add more mass. Both have the desired effect. The idea that has gained acceptance is the latter – the existence of more mass – or dark matter – which we cannot (currently) detect except through its gravitational effects.

This may sound implausible but there is, fortunately, both theoretical and experimental evidence for dark matter. Modern theories of high-energy physics such as supersymmetry and string theory naturally predict heavy particles that almost do not interact at all, except through gravity. They are ideal candidates for dark matter.

On the experimental side, astronomers have shown that not only are the dynamics of galaxies wrong, but so too is the bending of light. Zwicky suggested way back in 1937, that since light is bent by the gravity of a large mass, this would be an ideal way to sense the existence of dark matter even if we cannot see it. Since the 1980s Zwicky's ideas have been thoroughly tested – the extra bending of light has been confirmed and while we still wait for conclusive proof of dark matter, most (but not all) cosmologists accept it.

Up until 1998 the accepted cosmic dogma was that our universe was flat, dominated by dark matter, with ordinary matter in the minority. The future was therefore simple – the dark matter would continue to cluster into larger and larger aggregations and while this was going on, the expansion of the cosmos was destined to grow ever slower; grinding towards – but never quite reaching – a moment when the universe would stop expanding altogether.

But in 1998 we realized that this picture was, in all likelihood, fundamentally incorrect. With the advent of dark energy it became evidence that we have entered a cosmic winter – the second season of our profound ignorance.

5. Winter and Seventy Percent is Missing...

In 1998, sixty-nine years after Edwin Hubble had discovered that the universe is expanding, two separate groups came to a new and startling conclusion – the expansion of the cosmos is not slowing down but speeding up today! The cosmic accelerator pedal is being pressed, causing the universe to cool down very rapidly – and casting us into the grip of a cosmic winter.

The two groups based their radical conclusions on observations of distant supernovae – the death explosions of certain special stars – using them as cosmic lighthouses to measure the expansion rate of the cosmos. A supernova is an ideal beacon since for a few days it can outshine the whole galaxy of stars in which it lived.

But apart from being visible from the other side of the observable cosmos, supernovae have an absolutely crucial advantage in measuring distances. When we look out at a distant galaxy, we don't know how bright the galaxy really is. If we did, then we could estimate how far away it is simply by comparing how much the galaxy has dimmed. If space is flat, the brightness of any object decreases as the square of the distance. The same galaxy twice as far away will appear four times as dim.

The essence of the 1998 revolution was the realization, made a few years before, that a certain special supernova – given the arcane label Type Ia – carried in its light the key to determining how bright the supernova inherently was, rather like an object with a bar code with all its properties.

This supernova barcode is encoded in the realisation that there is a strong and direct connection between the inherent brightness of a supernova (which we cannot measure directly) and the time the supernova takes to dim. A typical Type Ia supernova takes one to two months to explode and then fade away again. By looking at very nearby supernovae researchers found that the longer it took for the supernova to fade away, the brighter it inherently was.

By using nearby supernovae to calibrate this connection, we can observe distant supernovae and how long they take to fade, and hence estimate how bright they were and the extent to which the light has dimmed en route to us. Hence we can estimate how far away they are and this tells us how the cosmos is expanding.

As a result, the two supernova teams were able to make the first accurate distance estimates to vastly distant objects, whose emitted light has been travelling to us for billions of years.

What the teams found was that the supernovae were too dim – and hence too far away – to be in accord with the picture of an ageing universe whose expansion was gradually slowing down towards standstill. To explain why the supernovae were so faint both groups came to the same conclusion – the universe must have started speeding up in its recent history so as to carry the supernovae further off.

This is an amazing result. Rather like a seventy-year old who suddenly takes up skydiving and full-contact karate, the cosmos isn't behaving like we expect it to. Even more strangely it implies, if we accept the picture of autumn and spring, that the cosmos began accelerating recently, roughly around the time that human life evolved on earth.

But why should this be so? Why should the cosmos begin accelerating now rather than after 10 seconds or in the future? This conundrum, yet another Goldilocks mystery, is known as the coincidence problem. Copernicus and Galileo struggled to deliver science and cosmology from the Catholic dogma that we live at a special place in the cosmos. Now it seems we find that we are living instead at a special time in the history of the cosmos. It is a mystery for which we so far have no solution and must rank as perhaps the profoundest questions in all of physics today.

The reader may suspect that the cosmos is not, in fact, really accelerating and that a simpler explanation is simply to find a way to make the supernovae fainter than expected and then all will be well. However there is rapidly growing evidence from different sources that the cosmos really is accelerating.

Key among these is the evidence from the cosmic microwave background – the radiation left over from the Big Bang - that the universe is spatially flat. However, making a flat universe requires lots of matter and when we look out at the galaxies and clusters we find that about 70% of the required amount is missing! It is important to remember that this is after we have taken into account the dark matter. In other words, about 70% of the matter in the universe does not cluster under gravity! Sound familiar? Indeed, this is rather like our earlier discussion of inflation. It appears that the beginning and end of cosmic history involved anti-gravity!

There is other evidence for cosmic acceleration too. In 2003 it was realised that the CMB shares some overlap with maps of the cosmos at the radio and X-ray wavelengths. This correlation is only expected if the universe is currently accelerating. Finding an alternative way to make supernovae dimmer would not explain this overlap.

But what could possibly be the source of this current cosmic acceleration? As with inflation in the early universe, it requires matter with negative pressure. Einstein himself had been faced with a similar need when he was trying to build a static Universe in the 1920s. He required a repulsive force to balance the attractive nature of gravity. To achieve this he introduced the cosmological constant, Λ, which is effectively anti-gravity and which he dubbed his 'greatest blunder' once the expansion of the Universe was discovered by Hubble in 1929.

Ironically Einstein not only missed the chance to predict that the universe is expanding but also missed the chance to predict that it is currently accelerating! His biggest blunder, the cosmological constant, may indeed turn out to be the source of the cosmic acceleration. On the other hand, cosmologists have other candidates known under the rubric 'quintessence' – the fifth element – or simply 'dark energy'. Whatever the source, it has recently usurped control of the universe, leaving us with two fundamental questions: *'why has it done so recently?'* and *'will it dominate the universe forever?'*

Fortunately the future looks extremely bright for hunting the dark energy. In the next decade a plethora of new experiments will be unleashed in our attempt to unravel this enigma. Two leading proposed dark energy experiments are the *Supernova Acceleration Probe* (SNAP) and the *Kilo-Aperture-Optical-Spectrograph* (KAOS).

SNAP is a satellite mission dedicated to searching out thousands of extremely distance supernovae from the quiet of space. Then there is KAOS – a next-generation galaxy survey that would hunt for dark energy in a subtle way. The distribution of galaxies in space is expected to carry a unique spatial signature, a fingerprint that gently grows with the cosmic expansion. By seeking out that fingerprint we will be able to accurately follow the expansion history of the cosmos and carefully track the properties of dark energy.

6. The End of Cosmology?

The next fifteen years will be a golden period in the history of astronomy and cosmology. New experiments will yield vast reams of information about our universe, allowing us to map it in unprecedented detail, to learn about its origin and future and to explore our place in it. In each of the next five years we can expect our knowledge of the cosmos to double.

Despite this phenomenal improvement our understanding of the very early and very late universe is extremely poor. Fundamental questions remain: what is the dark energy accelerating the universe? Why has it only come to dominate the universe recently? What is the dark matter that forms the basic cradles for galaxies, stars, planets and, ultimately, life? Is there a connection between dark matter and dark energy? What is the origin of inflation, why do we see no regions of anti-matter in the universe? What happened before the Big Bang and will the universe accelerate forever?

The coming golden period of cosmology will allow us to address most, or all, of these questions. It is possible, though extremely unlikely, that they will all be answered, but judging from past experience, the new knowledge will only lead us to new and even more fascinating questions. Fortunately as Einstein himself has reminded us – *The search for truth is more precious than its possession.*

Acknowledgments

I thank Josh Bryer and Michelle Bassett for useful suggestions on an early draft of this chapter.

Bruce Bassett

Bruce was born in Zimbabwe, received his MSc from Cape Town University and his PhD from the International School for Advanced Studies in Trieste under the supervision of Dennis Sciama. His career path took him to Oxford University as a post-doctoral researcher and then to the Institute of Cosmology and Gravitation at the University of Portsmouth as senior lecturer and Reader. He is also a Royal Society-JSPS fellow at Kyoto University in Japan.

Bruce's main research interests are inflation and dark energy. He has published a popular introduction to Einstein's gravity entitled *Introducing Relativity*. Bruce is married and spends his spare time exploring international cultures and is, for the moment at least, trying to learn Japanese.

CHAPTER 2

THE PARADIGM OF INFLATION

J. García-Bellido

Departmento de Física Teórica, Universidad Autónoma de Madrid
Cantoblanco, 28049 Madrid, Spain
E-mail: juan.garciabellido@uam.es

The standard model of cosmology is based on the hot Big Bang theory and the inflationary paradigm. Recent precise observations of the temperature and polarization anisotropies in the cosmic microwave background and the matter distribution in large scale structures like galaxies and clusters confirm the general paradigm and put severe constrains on variations of this simple idea. In this essay I will discuss the epistemological foundations of such a paradigm and speculate on its possible realisation within a more fundamental theory.

1. Introduction

Modern cosmology stands today on a firm theoretical framework, based on general relativity and the hot Big Bang theory, which describes with great precision the evolution of the universe from the first fraction of a second to our present epoch. However, this impressive framework was unable to explain the flatness and homogeneity of space, nor the origin of the matter and structures seen in the universe today. The advent of the inflationary paradigm[1,3,5] in the 1980's provided a dynamical mechanism for the generation of both matter and structure in an otherwise flat and homogeneous universe.[4] Its predictions are confirmed today with great accuracy, thanks to a true revolution in cosmological observations, coming mainly from measurements of the anisotropies[6,7] in the cosmic microwave background (CMB) and the distribution of matter in large scale structures (LSS) like galaxies and clusters of galaxies.[8,9]

Nowadays, the hot Big Bang theory and the inflationary paradigm constitute the basis of our standard model of cosmology. However, while the theory of the Big Bang is well established, there is no theory of inflation yet. At the moment, most people consider inflation as a successful idea, a paradigm, realised in specific models with concrete predictions. I will give here an overview of the present theoretical and observational status of the idea of inflation, within the most general framework, trying to be as much model-independent as possible.

I will first present an epistemological analogy that will help put into context the successes of inflation and the search for a fundamental theory of inflation. I will discuss the assumptions that are used in the construction of the most general models of inflation, with their main caveats. I will then explore its main consequences and predictions, both theoretical and phenomenological. One may think that inflation is such a basic idea that one cannot rule it out. However, I will describe ways in which one could discard not only specific models of inflation but the whole paradigm. To end this broad view of inflation, it might be useful also to state clearly some minor criticisms to the idea of inflation, as well as dispell some false claims that appear in the literature. Finally, although inflation defines a general framework in which most cosmological questions can be addressed, it is certainly not the panacea of all problems in cosmology, so it is worth mentioning those questions inflation cannot answer, at least in its present realisation. At the end, I will give a personal perspective on what I think should be the basic ingredients of a complete theory of inflation.

2. An Epistemological Analogy

The importance and outreach of a paradigm or a theory is best put into perspective by comparing it with other previous ideas whose epistemology is well known to most researchers in the field. The idea of inflation has sometimes been compared with the gauge principle of the standard model of particle physics. Like the inflationary paradigm, the gauge principle is realised in a specific model (based on a concrete choice of gauge group) with specific predictions that depend on a large number of fine-tuned parameters in order to agree with observations. Fortunately, there is a significant number of independent experiments that have fixed most of these parameters to several decimal places in a consistent framework, at least up to the energies reached with present accelerators. However, if we want to extend the standard model to higher energies, we run into problems of consistency. In order to solve those problems we need to introduce a new theory with many new ingredients. This theory is still uncertain, and will probably require new ideas based on experimental data beyond the present energy scale. Perhaps with the future particle physics colliders, at higher energies, we may find completely new and unexpected phenomenology. All of this can also be said of the inflationary paradigm and its present realisation in terms of specific models.

However, I feel this is not the best analogy for the inflationary paradigm. What I find more appropriate as an epistemological analogue is Newtonian mechanics. From our modern perspective, we can say that Newton was lucky, he proposed a force law of gravitation without a consistent theoretical framework of space and time, much to the despair of his contemporaries, Leibniz and Berkeley. However, he was smart enough to state no mechanism for this force and propose his laws of celestial mechanics with the statement: "everything works as if...". What silenced the criticisms made by both physicists and philosophers, like Mach and others, to the idea of absolute space and time was the extraordinary power of Newtonian mechanics for predicting planetary motion.

It was not until the advent of Einstein's theory of general relativity that this extraordinary coincidence was dispelled, and put Newtonian mechanics in the appropriate perspective as a concrete limit of a more fundamental theory of space and time. I believe the inflationary paradigm is nowadays in a similar situation with respect to a more fundamental theory. Present observations seem to be in complete agreement with the main predictions of inflation, but we still do not understand why it works so well. At a fundamental level, the idea of inflation is outrageously simple, and yet contains all the ingredients necessary to explain a large array of observations, many of which are independent, as well as predicting a surprisingly rich phenomenology. Moreover, most of its predictions were made much before we had any means of making contact with observations. Today, these observations are confirming the general paradigm but still have not singled out any particular model.

3. Basic Assumptions

Before describing the successes of the idea of inflation let me begin by stating clearly the basic assumptions that are used in the construction of a generic inflationary model. Variations of this generic model will be discussed below.

The inflationary paradigm assumes that gravity is described by a classical field theory, usually taken to be general relativity in 4-dimensional space-time. This assumption is not essential, it can be relaxed by including also a dilaton field, a scalar partner of the graviton field, coupled to matter like in scalar-tensor theories, and also by extending it to possibly extra compactified dimensions.

In order to explain the present homogeneity and flatness on large scales, the only requirement is an early period of accelerated expansion of the universe. In the context of general relativity, this means a negative pressure as the source of gravity. The simplest realisation is an approximately constant energy density, leading to quasi-de Sitter (exponential) expansion of the early universe. It can be parametrised in terms of an effective or fundamental scalar field, called the inflaton field, whose nature is yet unknown. Other species (matter and/or radiation) may be present, but will be diluted away very quickly by the expansion. This inflaton field may evolve slowly down its effective potential, or not. While an approximately constant energy density seems to be required, a slow-roll field is only a simplifying assumption. Non-slow-roll models of inflation exist and for the moment make predictions that are compatible with observations.

All matter fields, as well as fluctuations of the inflaton and the gravitational field, are supposed to be quantum fields evolving in a curved background. Quantum fluctuations of these fields start in the vacuum at small scales. The expansion of the universe takes these inflaton and metric fluctuations out of the de Sitter causal horizon and produces classical metric perturbations at superhorizon scales. Later on, during the radiation and matter eras, the causal horizon grows more quickly than the size of the universe and fluctuations reenter again, giving rise to radiation and matter perturbations as they fall into the potential wells of the metric.

4. General Consequences

The main consequences of inflation can be classified in two groups: those that affect the space-time background, and perturbations on this background. Due to the tremendous expansion, inflation predicts that spatial sections must be flat, that is, Euclidean, at least on our local patch. It also predicts that these spatial sections must be homogeneous and isotropic to a high degree. Both homogeneity and flatness were unstable properties under the evolution equations of the Big Bang theory and thus required highly fine-tuned initial conditions. Inflation eliminated the fine-tuning by postulating an early period of acceleration of the universe. The rest is a consequence of this simple hypothesis. For instance, if the universe had a non-trivial topology before inflation, it was redshifted away by the expansion, so that the topological cell today — the geometrical cell whose sides are nontrivially identified — should be much larger than the present Hubble volume. At the moment, this is in perfect agreement with observations of the CMB anisotropies.

Inflation not only explains in a dynamical way the observed homogeneity and flatness, but also predicts the existence of inhomogeneities and anisotropies arising from quantum fluctuations of the inflaton and the metric, and possibly other scalar fields evolving during inflation. General relativity predicts that there must be six physical degrees of freedom associated with a generic metric perturbation: two scalar, two vector (the vorticity and shear fields) and two tensor (the two polarizations of a gravitational wave) components. The perturbed Einstein equations, together with the inflaton field as matter source, imply that the two scalar metric components are linearly dependent, while the vorticity field decays very quickly during inflation. Therefore, we are left with only one scalar (giving rise to curvature perturbations) and two tensor components.

Due to quantum mechanics, superhorizon quantum fluctuations become semi-classical metric perturbations. These perturbations have an amplitude proportional to the approximately constant energy density driving inflation and, therefore, give rise to almost scale-invariant spectra of density perturbations and gravitational waves, that is, fluctuations with approximately the same amplitude on all scales. The density perturbations will later seed the structure we see in our universe, as matter falls into them and evolves through gravitational collapse. Deviations from exact scale invariance — the tilt of the spectrum — characterise a particular model of inflation and thus allow cosmologists to distinguish between them. In some cases, the tilt may also depend on scale, which puts further constrains on models of inflation. On the other hand, the tensor perturbations give rise to a spectrum of gravitational waves that may eventually be detected by laser interferometer gravitational wave observatories. Moreover, since quantum fluctuations arise from essentially free fields during inflation, both primordial spectra are expected to have a statistical distribution that is approximately gaussian.

If there were more than one scalar field evolving during inflation, their quantum fluctuations may have given rise to both energy density and entropy perturbations. The former arise from scalar curvature perturbations, while the latter is responsible for isocurvature perturbations. In some cases, this also implies a small non-gaussianity in the spectrum, and more than one spectral tilt. Unless required by observations, we will use Occam's razor and assume that inflation is driven by a single scalar field.

5. Observable Predictions

The main prediction of inflation is that spatial sections are Euclidean, i.e. with negligible spatial curvature, at least within our Hubble volume. This has been confirmed at the 2 percent level by recent measurements of CMB anisotropies, as described below. Inflation also predicts a homogeneous space-time background, with inhomogeneities or anisotropies imprinted by small amplitude quantum fluctuations giving rise to classical curvature perturbations. The amplitude is not fixed by the general paradigm, but can be computed precisely for a given model. In single-field inflation, the amplitude of the curvature perturbation remains constant outside the horizon and, therefore, its amplitude at reentry is precisely given by that at horizon exit during inflation. Thus, a generic prediction of inflation is a primordial spectrum of scale-invariant curvature perturbations on superhorizon scales, that will seed structure once they reenter the causal horizon and matter falls into them, undergoing gravitational collapse. Such a primordial spectrum was postulated in 1970, much before inflation, by Harrison and Zeldovich in order to explain the present distribution of matter, and I think it is extraordinary that such a simple idea as inflation could also explain the origin of this peculiar spectrum in terms of quantum fields in a curved background that, due to the expansion, become classical metric perturbations.

In fact, curvature perturbations are also responsible for temperature and polarization anisotropies in the CMB. As baryons fall into the potential wells of these perturbations, they induce acoustic oscillations in the plasma, with radiation pressure opposing gravitational collapse. A perturbation of a given scale can only grow after it enters the causal Hubble radius. Since all perturbations of the same wavelength entered simultaneously, they must be in the same phase of the acoustic oscillation at the time of decoupling. This gives rise to spatially coherent oscillations that stand out as peaks in the two-point angular correlation function. This pattern of acoustic oscillations in the power spectrum of temperature anisotropies is a characteristic signature of inflation, and has been clearly seen in the CMB anisotropies by Boomerang and WMAP. Any alternative theory of structure formation, like e.g. topological defects, based on an active causal origin of matter perturbations, could not explain the observed pattern of acoustic oscillations and are therefore ruled out. It is extraordinary that the surprisingly simple paradigm of inflation predicted the coherent oscillations much before they could be measured. We are now using the detailed properties of the observed spectrum to determine many cosmological parameters. For instance, the first acoustic peak in the power spectrum appears at a scale that precisely corresponds to the projected size of the acoustic horizon at decoupling, approximately one degree in the sky today, and indicates that photons have travelled since then in straight lines towards us. This implies that the universe is essentially flat, with deviations less than a few percent, again as predicted by inflation.

Furthermore, according to inflation there will also be large scale perturbations that are still outside the horizon at the time of decoupling and therefore are not enhanced by the acoustic oscillations of the plasma. These superhorizon perturbations are a specific prediction of inflation, and are responsible for the so-called Sachs-Wolfe plateau, arising from photons that are gravitationally redshifted or blueshifted as they escape from or fall into these primordial perturbations. The height of the plateau has direct information about the amplitude of the primordial perturbations and the flatness of the plateau about the tilt of the spectrum. Without an acausal mechanism like inflation that can stretch fluctuations beyond the horizon and imprint a primordial spectrum with superhorizon scales at decoupling, we could not explain the Sachs-Wolfe plateau, observed by the COsmic Background Explorer (COBE) satellite for the first time in 1992. To this plateau contributes not only the (scalar) density perturbations, but also the (tensor) gravitational wave spectrum.

However, the fact that from our particular position in the sky we cannot see more than a few realizations of the gaussian distribution of large scale fluctuations − an example of sample variance that here goes by the name of cosmic variance − means that the small tensor contribution to the Sachs-Wolfe plateau may be hidden inside the cosmic variance of the scalar component, and since gravitational waves are quickly redshifted by the universe expansion after they enter the Hubble radius, their imprint in the temperature power spectrum could be unmeasurable after all. The primordial tensor spectrum is important because it contains complementary information about the inflationary dynamics, and, in particular, its amplitude depends only on the total energy density driving inflation, which is still a great unknown.

Fortunately, the microwave background contains information not only about the energy distribution of photons, a precise blackbody spectrum, but also about their linear polarization. The polarization of the microwave background is a vector field with a gradient and a curl components (called E and B components in analogy with electromagnetism). The amplitude of the polarization spectrum is directly predicted by the size of the temperature fluctuations, since the linear polarization of CMB photons can only arise via Thomson-scattering, from quadrupolar anisotropies in orthogonal directions. This polarization is induced by photons scattering off electrons in the last scattering surface, when the universe was essentially neutral. As a consequence, the polarization spectrum is an order of magnitude smaller than the temperature spectrum, making it even more difficult to observe than the latter.

Only recently, DASI measured for the first time the polarization field at the last scattering surface. However, it was WMAP, with much bigger sky coverage and sensitivity, that measured the precise angular dependence of the polarization field, and confirmed that the temperature and the gradient (E) component of the polarization spectrum were correlated. WMAP gave the multipole expansion of the TE cross-correlation spectrum, with a pattern of peaks matching those predicted long before by the simplest models of single-field inflation. This is without any doubt one of the most important arguments in favour of inflation. While the existence of the CMB polarization spectrum is a generic prediction of Big Bang cosmology at photon decoupling, it is extremely difficult to construct *ad hoc* models of structure formation and yet be in agreement with this pattern of acoustic oscillations in both the temperature and the polarization power spectra.

Moreover, we still have the curl (B) component of the polarization spectrum. This component can only arise from the (tensor) gravitational wave primordial spectrum of perturbations coming from inflation, and has not been detected yet. It may happen that the scale of inflation is high enough for a significant amplitude of tensor perturbations to generate a B polarization field at the level needed for detection in the next generation of CMB satellites like Planck or CMBpol. At this moment, there is a large experimental and theoretical effort searching for ways to extract the rare primordial B-component of the CMB, since it contains crucial information about the scale of inflation that we may not be able to obtain otherwise.

Another interesting feature about CMB polarization is that is provides a null test of inflation. We have seen that in the absence of anisotropic stresses acting as sources during inflation, the vector component of metric perturbations is negligible. While the scalar and the tensor components are both symmetric under parity, the E- and the B-components of polarization have opposite parity. Therefore, a robust prediction of inflation is that there should be no correlation between the scalar temperature (T) and the B-component of polarization, nor between the E- and the B-components of polarization. This can be summarised as $\langle BT \rangle = \langle BE \rangle = 0$, and constitute an important test of inflation. The other four combinations, the three power spectra $\langle TT \rangle$, $\langle EE \rangle$ and $\langle BB \rangle$, and the $\langle TE \rangle$ cross-correlation are predicted by inflation to be non-zero and to possess a precise pattern of acoustic oscillations.

Furthermore, CMB anisotropies are not the only probes of inflationary predictions. The same primordial spectrum of curvature perturbations responsible for temperature anisotropies gives rise, through gravitational collapse of the primordial baryon and dark matter distributions, to the large scale structures like galaxies, clusters and superclusters. These perturbations are probed by the so-called matter power spectrum, that is, the Fourier transform of the two-point correlation function of luminous galaxies in clusters and superclusters, all the way to the horizon in our Hubble volume. If light traces matter then this power spectrum may have a precise relation to the primordial spectrum of curvature perturbations, and thus to inflation and the CMB anisotropies. The primordial adiabatic, gaussian and scale-invariant spectrum of metric perturbations seeds, through gravitational collapse, the measured matter power spectrum. Galaxy surveys both in real and redshift space have increased their covering volume at increasing rate since the first CfA catalogs in the 1970s to the present 2dF Galaxy Redshift Survey and the Sloan Digital Sky Survey, which cover a large fraction of the universe up to redshifts or order $z = 1$ for galaxies, and order $z = 3$ for quasars. These surveys have allowed cosmologists to bring into sharp focus an image of the nearby universe to unprecedented accuracy, confirming basic properties of the power spectrum, like gaussianity and scale invariance, at least on large scales, where the power spectrum has not been strongly distorted by non-linear gravitational collapse.

In the near future, with the advent of the next generation of gravitational wave observatories, it will be possible in principle to detect the approximately scale-invariant spectrum of gravitational waves generated during inflation, seen as a stochastic background in time-correlated detectors across the Earth. However, this background could remain undetectable, even by sensitive gravitational wave observatories, if the scale of inflation is too low. The detection of this new cosmic background, with specific signatures like a small negative tilt, i.e. larger amplitudes at longer wavelengths, may open the possibility of testing the idea of inflation. It happens that in all single field models of inflation, there is a unique relation between the two spectra (scalar and tensor) because the source of these metric perturbations is a single function, the effective inflaton potential. In particular, in models where the inflaton field evolves slowly towards the end of inflation and reheating, the ratio between the tensor and scalar amplitudes is uniquely related with the tilt of the gravitational wave spectrum. This is a prediction that no other theory of structure formation could have imagined a priori, and thus constitutes a test of the whole paradigm of inflation. If confirmed by observations it would have tremendous impact on inflation as a paradigm.

6. The Reheating of the Universe

The reheating epoch, in which the energy density of the inflation decayed into radiation, is also a universal feature of inflation. We know that at some point inflation must end, at least in our local patch, since today the universe contains the remnants of the radiation and matter eras. This energy conversion epoch signals the beginning of the hot Big Bang as we know it. However, the details of how it proceeds from the quasi-stationary inflationary regime to the radiation dominated era is rather model-dependent and requires some knowledge of the high energy particle physics model in which inflation is embedded. In some models it occurs in a perturbative way, as the quanta of the inflaton field decay into other field quanta to which the inflaton couples. In other models it occurs in an explosive non-perturbative way, producing a variety of interesting phenomenology like non-thermal phase transitions, with possibly the generation of topological defects, gravitational waves and/or primordial magnetic fields. These features may leave their imprint in the CMB anisotropies or in the stochastic gravitational wave background or even in the primordial intergalactic magnetic fields. Moreover, the far from equilibrium conversion of energy may help generate the matter/antimatter asymmetry at reheating, in a way that is now being explored both analytically and numerically.

Although none of these features are robust predictions of inflation, once observed they could give crucial information about the theory of high energy particle physics in which inflation is embedded, that is, the couplings and masses of the other fields to which the inflaton couples. For instance, the production of topological defects during reheating after inflation is predicted in specific models of inflation. Absence of any signatures from these cosmic defects can be used to impose stringent constraints on those models, but do not rule out other realisations of reheating.

What I think is outstanding of inflation is its ability to create such a novel and rich phenomenology from such a simple hypothesis. For inflation to work, very little has to be put in, and a lot seems to comes out, and some of its predictions have far reaching consequences.

7. Eternal Inflation

Although inflation was invented in order to alleviate the problem of initial conditions of the Big Bang, it is not in itself a theory of initial conditions. In fact, in order for inflation to proceed in the first place it is assumed that it started in a single domain that was sufficiently homogeneous across a causal horizon for the energy density to be treated as approximately constant. Once inflation started, this domain would occupy a large space-time volume due to the rapid expansion and later give rise to the whole large scale structure of space-time.[2]

Until now, I have described the observable consequences of a generic model of inflation, *i.e.* what is amenable to direct experimentation in our Hubble volume, but there is also the possibility of studying phenomena beyond our observable patch. The inflationary paradigm has a mechanism for generating metric perturbations that is completely universal, independent of the particular realisation of inflation. The spectral properties of those perturbations depends in a well known way on the precise dynamics associated with a given model, but the mechanism is a generic property of quantum field theory in curved space-times. What is less known is that backreaction of the metric plus inflaton fluctuations on the background space-time makes the inflaton field follow a Brownian motion in which half of the time the inflaton field in a given domain will jump upwards, instead of drifting down the effective potential. In those domains the rate of expansion increases. Since the amplitude of the quantum fluctuation is proportional to the local rate of expansion, the inflaton field in a few domains may continue to jump upwards, driving higher and higher rates of expansion, and therefore those domains will eventually occupy larger and larger volumes. Thus the universe becomes divided into domains in which the inflaton has drifted down the potential towards reheating and others that are still inflating, separated by distances much greater than their causal horizon. This stochastic picture leads to the self-reproduction of the universe and to eternal inflation. We may happen to be living in an island of warmth appropriate for life in an otherwise cold and eternally expanding universe. The island is probably large enough that deviations from homogeneity and flatness are negligible up to many times our present Hubble volume, so there is little chance of observing those inhomogeneous regions, at least in our lifetime as species. However, from the epistemological point of view, it is fascinating to entertain the idea that in causally disconnected regions of the universe we may have independent islands of thermalised regions in all possible stages of their evolution. Eventually all causal domains will end inflation, but at any given time the bulk of the universe is in eternal inflation. From this point of view, there is not just one Big Bang, but infinitely many, corresponding to those events where inflation ends locally and a radiation epoch ensues. This alleviates the problem of initial conditions of inflation: once a given domain allows a tiny burst of inflation, the flame will carry away expanding the rest of the universe forever. Local regions of the universe may collapse after matter domination, while others will expand for ever. Our particular patch, what we call the observable universe, is but a tiny insignificant spot in this metauniverse that encompasses all of space and time. Most probably, this grand picture of the universe will never be amenable to direct experimentation. We may have a hint of its existence from the actual pattern of CMB anisotropies, since the same quantum fluctuations that gave rise to those inhomogeneities is also responsible for the ultra large scale structure of the universe, but we may not be able to disprove it with local observations.

8. Variations on a Common Theme

Although the paradigm of inflation is based on the simple idea that the early universe went through a quasi-exponential expansion that stretched all scales, there are in principle many ways to realise this paradigm. Our first reaction is, like Newton, to make no hypothesis, but state that *everything works as if* an approximately constant energy density dominated the evolution of the early universe. Usually, this slowly changing energy density is parametrised by a single scalar degree of freedom because it is the simplest possibility, but for no other reason. This is enough to explain all observations, both in the CMB anisotropies and in the LSS distribution of matter. However, one can think of a series of variations on the same theme, for example adding another scalar field − most high energy physics models have not just one but many light scalar degrees of freedom − and induce isocurvature as well as adiabatic perturbations. They would produce specific signatures in the CMB anisotropies that have not been observed, and thus multi-field inflation is essentially ruled out, at least to some degree.

It could also happen that the inflaton field may not be treated as a free field because of its own self-coupling, because it couples to other light fields, or due to its coupling to gravity. In either case, the spectrum of quantum fluctuations is expected to have some, but small, non-gaussian statistics (only the ground state of a free field is actually a gaussian random field). For the moment, these non-gaussianities have been searched for, both in the

CMB and in LSS, without success. They would appear as deviations from the gaussian expectations in the three- and four-point correlation functions of galaxies and CMB anisotropies. A small degree of non-gaussianity is expected at galactic scales due to non-linear gravitational collapse, but for the moment what is observed is compatible with an evolved primordial gaussian spectrum. However, in the future, the information encoded in the non-gaussianities could be crucial in order to distinguish between different models.

On the other hand, a prediction that is quite robust for slow-roll inflation of the power-law type (i.e. with cubic and quartic self-interactions of the inflaton) is a precise relation between the deviations from scale invariance (the tilt of the scalar spectrum) and the duration of inflation, which makes this tilt be negative, and of order a few parts in a hundred. Another prediction of this class of models is that the tilt is also scale-dependent, but the dependence is very weak, of only a few parts in 10 000. These two predictions break down in models that are non-slow-roll, which are compatible with the rest of the inflationary phenomenology, but which predict large tilts for both scalar and tensor spectra, as well as a large scale-dependence of those tilts. Since the tilt has not been measured yet with sufficient accuracy, we cannot rule out the possibility of non-slow-roll inflation. Moreover, a large negative tilt would signal the presence of a large tensor (gravitational wave) component which would then be seen in the future CMB satellite experiments. However, this prediction is not universal. There are models of inflation, like hybrid models – that end inflation when the inflaton triggers the breaking of a symmetry associated with another field to which it couples – that can give an essentially flat spectrum, *i.e.* no tilt within the precision of planned missions, and also no scale-dendence of the tilt, nor a significant tensor component.

9. How to Rule Out Inflation

This is a tricky business since inflation is a paradigm with many realisations. Inflation makes space flat and homogeneous, but at the same time imprints space with metric fluctuations that later will seed the local structures we observe, like galaxies and clusters of galaxies. An obvious way to rule out the general idea of inflation would be to find evidence of a large spatial curvature, but the question is on what scale. Since the curvature of space depends on the duration of the period of inflation – the longer the flatter – a relatively short period might give rise to a marginally open or closed universe, depending on what sign of curvature they started with. However, curved models of inflation are somewhat fine-tuned and are nowadays ruled out at a high confidence level by the position of the first acoustic peak of the CMB temperature anisotropies.

Another way to rule inflation out is to discover that the universe is topologically non-trivial because, as far as we know from the classical theory of general relativity, the topology of space-time is invariant under the expansion of the universe; the only thing inflation does to topology is to increase the size of the topological cell. For topology to be seen today as correlated patterns in the sky, the cell should have a size smaller than the present Hubble volume. Such patterns have not been seen by the WMAP precision measurements of CMB anisotropies and, therefore, the topological cell, if it exists, must have a size much bigger than our present horizon. Yet another way to rule out the general idea of inflation would be to discover a global rotation of the universe, that is, a priviledged axis in the sky. Since inflation expands all scales isotropically, such an axis would stand out very clearly in the CMB sky today, and nothing like it has been observed by WMAP.

The previous discussion has to do with the global background properties of our observable universe. However, what characterises inflation, and allows alternative models to be distinguished, is the predicted spectrum of metric perturbations. Since the source of metric perturbations is a fundamental or an effective scalar field (it could be a condensate of some other field), there can be only two types of perturbations at first order: scalar and tensor metric perturbations. Vector perturbations only appear at second order, and therefore are suppressed, unless there are peculiar additional sources acting during inflation. Thus, a generic prediction of inflation is that the vector perturbation spectrum is absent. If someone would measure a significant primordial spectrum of vector perturbations in the CMB or LSS, one would have to question some of the assumptions of inflation, in particular single-field inflation, although it would probably not rule out the idea altogether. For the moment we have not seen any evidence of a vector perturbation in the CMB. This brings together an immediate consequence with respect to the polarization spectra, as I explained above. In the absence of any vector perturbation, parity is conserved and the cross-correlations $\langle BT \rangle$ and $\langle BE \rangle$ are expected to be zero. But this is a null test, and it may happen that detailed observations of the CMB indicate a non-zero primordial component. In that case, we would have to go back to the drawing board.

Similarly with large tensor perturbations or gravitational waves. If their amplitude is too large, compared with the scalar perturbations, it would imply an origin of inflation close to the Planckian era, and thus the classical description of space-time in terms of the general theory of relativity would not be appropriate. There are models of inflation that consider initial conditions close to the Planck boundary, but their effects are seen on scales much much larger than our Hubble volume, and by the time the fluctuations that gave rise to structure in our observable universe were produced, inflation proceeded well below Planck scale.

Concentrating now on the primordial scalar spectrum, the tilt of the spectrum is not arbitrary. It cannot be too different from scale-invariant ($n = 1$), or inflation would not have occurred in the first place. A deviation of order 100% larger or smaller than 1 is probably incompatible with most realisations of inflation, since it would require too steep potentials, although there is no general theorem, as far as I know. Typical slow-roll models of inflation predict spectral tilts that are only a few percent away from scale invariance. Some models, like hybrid inflation, are perfectly compatible with exact scale invariance, within experimental errors. In order to get significant tilts one has to go beyond slow-roll, and there are no generic analytical results for these models. One has to study them case by case.

Another key feature of primordial spectra is the scale-dependence of their spectral tilts. Again, most slow-roll models predict no significant scale-dependence, just of order a few parts in 10 000. But there could be specific models with particular features in their effective potential that change significantly at scales corresponding to those seen in the CMB and LSS, which would induce a scale-dependent tilt. In fact, there has been some debate whether WMAP has actually seen such a scale dependence in the CMB anisotropies, but there is no consensus.

In single-field models of inflation, the scalar and tensor spectra are both computed from a single analytical function, the effective scalar potential, and therefore there must a be a functional relationship between these two spectra. In slow-roll inflation, it can be cast into a relation between the ratio of the tensor to scalar amplitude and the tensor spectral tilt. Such a relation is very concrete and generic to all single-field models. We know of no other mechanism but inflation which might actually give such a peculiar relation, and for that reason it has been named "the consistency relation" of inflation. If, once the gravitational wave spectrum is discovered, such a relation did not hold, it would signal the demise of slow-roll single-field models. However, deviation from that exact relation could be due to multiple-field models, which predict a similar but different relation, with one extra parameter, that can also be measured and then checked to hold. If that doesn't happen either, then probably slow-roll models are not a viable description, and we would need to do fully non-perturbative calculations to find the exact relation between the two spectra. Eventually, whatever the relation, there must be some correlation between them that will allow cosmologists to discard all but a few realizations of inflation. That could be hard but fruitful work.

The problem, however, could be observational: the foregrounds to the CMB. In order to measure the tensor spectrum we need to detect the B-component of polarization, which may be swamped by contamination from a distorted E-component due to gravitational lensing in the intervening mass distribution between us and the last scattering surface. Moreover, on small scales it may be difficult to disentangle the primordial spectrum from the observed spectrum of perturbations. For instance, we know that non-linear gravitational collapse induces both vorticity and non-gaussianities and that gravitational lensing induces BE cross-correlation. In order to disentangle foregrounds from the CMB anisotropies, we may have to know more about the local and global distribution of matter in structures like galaxies, clusters and superclusters.

Finally, most realizations of inflation describe the universe dynamics with an effective scalar field that is essentially free during inflation, only its mass is relevant. In that case, the ground state of this quantum field gives rise to a gaussian spectrum of metric fluctuations, with only second order deviations from gaussianity. Such a prediction is very robust, at least in single-field models of inflation, and therefore, if a significant non-gaussianity were observed in the CMB anisotropies or LSS, it could signal multiple-field inflation, which could be contrasted with other predictions like isocurvature signatures; or it could indicate a significant contribution from postulated modifications of gravity on large scales. In any case, ruling out specific models of inflation leave fewer possibilities and allow cosmologists to close in on the best model, which may eventually lead to a final theory of inflation.

10. Honest Criticisms and False Claims

There are some features of inflation that are directly related with its status as a paradigm, i.e. not yet a theory. The main criticism that inflation has received over the years is that there is no unique model, but a variety of single- and multi-field models, some with fundamental descriptions, others purely phenomenological. I think this is an honest criticism; in the absence of a fundamental theory of inflation we have to deal with its effective realisation in terms of specific models. However, the models are fully predictive, they give concrete values for all observables, and can be ruled out one by one. It is possible that in the near future, with more precise cosmological observations, we may end up with a narrow range of models and eventually only one agreeing with all observations.

There are, however, some basic properties of inflation we still do not comprehend. In particular, we ignore the energy scale at which inflation occurred. It has often been claimed to be associated with the scale of grand unification, in which all fundamental forces but gravity unified into a single one, at around 10^{16} GeV, but this is still uncertain. The only observable with direct information about the scale of inflation is the amplitude of the gravitational wave spectrum, but we have not measured that yet and it may happen that it is too low for such a spectrum to be seen in the forseable future. This would be a pity, of course, but it would open the door to low scale models of inflation and a rethinking of the finetuning problems associated with them. A related criticism is that of not knowing whether the inflaton field is a low energy effective description of a more fundamental interaction or a genuine scalar field. We still have not measured any fundamental scalar field, and it could be that the inflaton is actually a geometrical degree of freedom related to the compactification of some extra dimensions, as recently proposed in the context of some string theory realizations of inflation.

Moreover, the issue of reheating of the universe remains somewhat speculative, even though a lot can be said in generic cases. For instance, there are low energy models of hybrid inflation that could have taken place − albeit with some finetuning − at the electroweak scale. In that case, it is expected that the inflaton will be some scalar degree of freedom associated with electroweak symmetry breaking and thus in the same sector as the Higgs. The gravitational wave component of the spectrum would be negligible, but at least one could explore the reheating mechanism by determining the couplings of the inflaton to other fields directly at high energy particle accelerators like the Large Hadron Collider at CERN.

On the other hand, inflation has also been for many years the target of unfounded criticisms. For instance, it has been claimed for years that inflation requires extremely small (finetuned) couplings in order to agree with CMB observations. Although this is true for some models, this is not a generic feature of inflation: hybrid inflation models agree very well with observations and do not require unnaturally small couplings and masses. Another common error is to claim that inflation is not predictive because of its many models. This is false, each especific model makes definite predictions that can be used to rule the model out. In fact, since the idea was proposed in the 1980's, the whole set of open inflation models has been ruled out, as well as models within a scalar-tensor theory of gravity. Furthermore, multi-field models of inflation have been recently contrained by observations of the temperature and polarization spectra, which do not seem to allow a significant isocurvature component to mix with the adiabatic one.

Another false claim which is heard from time to time is that inflation always predicts a red tilt, with a smaller amplitude at smaller wavelengths, but this is not true. Again, hybrid models of inflation, which are based on the idea of spontaneous symmetry breaking in particle physics, predict a slightly blue-tilted spectrum, and sometimes no tilt at all. Associated with this claim, there was the standard lore that the ratio of the tensor to scalar component of the spectrum was proportional to the tilt of the scalar spectrum. This is incorrect even in slow-roll models of inflation, being proportional to the tensor tilt, not the scalar one, which makes the prediction of the tensor amplitude more difficult.

11. What Inflation Cannot Answer

Although inflation answers many fundamental questions of modern cosmology, and also serves as the arena where further questions can be posed, it is by no means the panacea of all problems in cosmology. In particular, inflation was not invented for solving, and probably cannot solve by itself, the problem of the cosmological constant. This problem is really a fundamental one, having to do with basic notions of gravity and quantum physics. It may happen that its resolution will give us a hint to the embedding of inflation in a larger theory, but I don't think inflation should be asked responsible for not explaining the cosmological constant problem. The same applies to the present value of cosmological parameters like the baryon fraction, or the dark matter content of the universe, or the rate of expansion today. Those are parameters that depend on the matter/energy content after reheating and their evolution during the subsequent hot Big Bang eras. At the moment, the only thing we know for sure is that we are made of baryons, which contribute only 4% of the energy content of the universe. The origin of the dark matter and the cosmological constant is a mystery. In fact, the only thing we know is that dark matter collapses gravitationally and forms structure, while vacuum energy acts like a cosmic repulsion, separating these structures at an accelerated rate. They could both be related to modifications of gravity on large scales — associated with the mysterious nature of the vacuum of quantum field theories — but for the moment this is just an educated guess.

Inflation can neither predict the present age of the (observable) universe, nor its fate. If the present observation of the acceleration of the universe is due to a slowly decaying cosmological constant, the universe may eventually recollapse. Nor can inflation predict the initial conditions of the universe, which gave rise to our observable patch. Quantum gravity is probably behind the origin of the inflationary domain that started everything, but we still do not have a credible quantum measure for the distribution of those domains.

12. Conclusions and Outlook

The paradigm of inflation is based on the elegantly simple idea that the early universe went through a quasi-exponential expansion that stretched all scales, thus making space flat and homogeneous. As a consequence, microscopic quantum field fluctuations are also stretched to cosmological scales and become classical metric perturbations that seed large scale structure and can be seen as temperature anisotropies in the CMB.

The main predictions of this paradigm were drawn at least a decade before observations of the microwave background anisotropies and the matter distribution in galaxies and clusters were precise enough to confirm the general paradigm. We are at the moment improving our measurements of the primordial spectrum of metric fluctuations, through CMB temperature and polarization anisotropies, as well as the LSS distribution of matter. Eventually, we will have a direct probe of the energy scale at which inflation took place, which may open the possibility of constructing a complete theory of inflation.

For the moment, all cosmological observations are consistent with the predictions of inflation, from the background space-time to the metric and matter fluctuations. It is remarkable that such a simple idea should work so well, and we still do not know why. From the epistemological point of view, one could say that we are in a situation similar to Newtonian mechanics in the nineteenth century, before its embedding in Einstein's theory of general relativity. We can predict the outcome of most cosmological observations within the standard model of cosmology, but we do not yet have a fundamental theory of inflation.

Such a theory will probably require some knowledge of quantum gravity. This has led many theoretical cosmologists into speculations about realisations of inflation in the context of string theory or M-theory. This is certainly an avenue worth pursuing but, for the moment, their predictions cannot be distinguished from those given by other models of inflation. Eventually the higher accuracy of future CMB measurements and LSS observations will allow cosmologists to discard most models of inflation but a few, in the search for the final theory of inflation. It may require completely new ideas and perhaps even an epistemological revolution, but I am sure that, eventually, we will be able to construct a consistent theory of inflation based on quantum gravity and high energy particle physics.

Acknowledgments

It is always a pleasure to thank Andrei Linde for many enlightening discussions on the paradigm of inflation and its far reaching consequences. This work was supported by the Spanish Ministry of Science and Technology under contract FPA-2003-04597.

References

1. A.D. Linde, *Particle Physics and Inflationary Cosmology*, (Harwood Academic Press, New York, 1990).
2. A.D. Linde, "The self-reproducing inflationary universe", *Scientific Am.* November, 32 (1994).
3. A.H. Guth, *The inflationary universe*, (Perseus Books, Reading, 1997).
4. J. García-Bellido, "The origin of matter and structure in the universe", *Phil. Trans. R. Soc. lond.* **357**, 3237 (1999).
5. A.R. Liddle & D.H. Lyth, *Cosmological Inflation and Large Scale Structure*, (Cambridge U. Press, Cambridge, 2000).
6. Boomerang home page, `http://oberon.roma1.infn.it/boomerang/`
7. Wilkinson Microwave Anisotropy Probe home page, `http://map.gsfc.nasa.gov/`
8. 2dF Galaxy Redshift Survey home page, `http://www.mso.anu.edu.au/2dFGRS/`
9. Sloan Digital Sky Survey home page, `http://www.sdss.org/sdss.html`

Juan García-Bellido

Born in 1966 in Madrid, Spain, Juan García-Bellido obtained his PhD in 1992 from the Autónoma University of Madrid. He was a Postdoctoral Fellow at Stanford University (1992-94), PPARC Fellow of the Astronomy Centre at the University of Sussex (1995-96), Fellow of the TH-Division at CERN (1996-98), a Royal Society University Research Fellow at Imperial College (1998-99), and since then Professor at the Institute of Theoretical Physics in Madrid. He has published around 70 papers in theoretical physics and cosmology; attended, organised and gave lectures at international conferences and summer schools around the world, and is referee of the most prestigious journals in the field. Aged 38, he is married to a particle physicist and has two kids, a girl and a boy, aged 7 and 2, respectively. Scientific interests include the early universe, black holes and quantum gravity.

CHAPTER 3

COSMOLOGY WITH VARYING CONSTANTS

C. J. A. P. Martins

CFP, R. do Campo Alegre 687, 4169-007 Porto, Portugal,
and Department of Applied Mathematics and Theoretical Physics, CMS,
University of Cambridge, Wilberforce Road, Cambridge CB3 0WA, UK
E-mail: C.J.A.P.Martinsdamtp.cam.ac.uk

The idea of possible time or space variations of the 'fundamental' constants of nature, although not new, is only now beginning to be actively considered by large numbers of researchers in the particle physics, cosmology and astrophysics communities. This revival is mostly due to the claims of possible detection of such variations, in various different contexts and by several groups. Here, I present the current theoretical motivations and expectations for such variations, review the current observational status, and discuss the impact of a possible confirmation of these results in our views of cosmology and physics as a whole.

1. Introduction

One of the most valued guiding principles (or should one say beliefs?) in science is that there ought to be a single, immutable set of laws governing the universe, and that these laws should remain the same everywhere and at all times. In fact, this is often generalised into a belief of immutability of the universe itself—a much stronger statement which doesn't follow from the former. A striking common feature of almost all cosmological models throughout history, from ancient Babylonian models, through the model of Ptolemy and Aristotle, to the much more recent 'steady-state model', is their immutable character. Even today, a non-negligible minority of cosmologists still speaks in a dangerously mystic tone of the allegedly superior virtues of 'eternal' or 'cyclic' models of the universe.

It was Einstein (who originally introduced the cosmological constant as a 'quick-fix' to preserve a static universe) who taught us that space and time are not an immutable arena in which the cosmic drama is acted out, but are

---

in fact part of the cast—part of the physical universe. As physical entities, the properties of time and space can change as a result of gravitational processes. Interestingly enough, it was soon after the appearance of General Relativity, the Friedman models, and Hubble's discovery of the expansion of the universe—which shattered the notion of immutability of the universe— that time-varying fundamental constants first appeared in the context of a complete cosmological model,[1] though others (starting with Kelvin and Tait) had already entertained this possibility.

From here onwards, the topic remained somewhat marginal, but never disappeared completely, and even The Royal Society organised a discussion on this theme about twenty years ago. The proceedings of this discussion[2] still make very interesting reading today—even if, in the case of some of the articles, only as a reminder that concepts and assumptions that are at one point uncontroversial and taken for granted by everybody in a given field can soon afterwards be shown to be wrong, irrelevant or simply 'dead-ends' that are abandoned in favour of an altogether different approach.

Despite the best efforts of a few outstanding theorists, it took as usual some observational claims of varying fundamental constants[3] to make the alarm bells sound in the community as a whole, and start convincing previously sworn skeptics. In the past two years there has been an unprecedented explosion of interest in this area, perhaps even larger than the one caused a few years ago by the evidence for an accelerating universe provided by Type Ia supernova data. Observers and experimentalists have tried to reproduce these results and update and improve existing constraints, while a swarm of theorists has flooded scientific journals with a range of possible explanations.

I provide a summary of the current status of this topic. Rather than go through the whole zoo of possible models (which would require a considerably larger space, even if I were to try to separate the wheat from the chaff), I'll concentrate in the model-independent aspects of the problem, as well as on a few key observational results. At the end, I'll provide some reflections on the impact of a future confirmation of these time variations in our views of cosmology and physics as a whole. For more technical reviews see Refs. 4–6.

2. On the Role of the Constants of Nature

The so-called fundamental constants of nature are widely regarded as some kind of distillation of physics. Their units are intimately related to the form and structure of physical laws. Almost all physicists and engineers will have had the experience of momentarily forgetting the exact expression of a certain physical law, but quickly being able to re-derive it simply by resorting to dimensional analysis. Despite their perceived fundamental nature, there is no theory of constants as such. How do they originate? How do they relate to one another? How many are necessary to describe physics? None of these questions has a rigorous answer at present. Indeed, it is remarkable to find that different people can have so widely different views on such a basic and seemingly uncontroversial issue.

One common view of constants is as asymptotic states. For example, the speed of light c is (in special relativity) the maximum velocity of a massive particle moving in flat spacetime. The gravitational constant G defines the limiting potential for a mass that doesn't form a black hole in curved spacetime. The reduced Planck constant $\hbar \equiv h/2\pi$ is the universal quantum of action and hence defines a minimum uncertainty. Similarly in string theory there is a fundamental unit of length, the characteristic size of the strings. So for any physical theory we know of, there should be one such constant. This view is acceptable in practice, but unsatisfactory in principle, because it doesn't address the question of the constants' origin.

Another view is that they are simply necessary (or should one say convenient?) inventions: they are not really fundamental but simply ways of relating quantities of different dimensional types. In other words, they are simply conversion constants which make the equations of physics dimensionally homogeneous. This view, first clearly formulated by Eddington[7] is perhaps at the origin of the tradition of absorbing constants (or 'setting them to unity', as it is often colloquially put) in the equations of physics. This is particularly common in areas such as relativity, where the algebra can be so heavy that cutting down the number of symbols is a most welcome measure. However, it should be remembered that this procedure can not be carried arbitrarily far. For example, we can consistently set $G = h = c = 1$, but we cannot set $e = \hbar = c = 1$ (e being the electron charge) since then the fine-structure constant would have the value $\alpha \equiv e^2/(\hbar c) = 1$ whereas in the real world $\alpha \sim 1/137$.

In any case, one should also keep in mind that the possible choices of particular units are infinite and always arbitrary. For example, the metre was originally defined as the distance between two scratch marks on a bar of metal kept in Paris. Now it is defined in terms of a number of wavelengths of a certain line of the spectrum of a ^{83}Kr lamp. This may sound quite more 'high-tech' and rigorous, but it doesn't really make it any more meaningful.

Perhaps the key point is the one recently made by Veneziano in Ref. 8: there are units which are arbitrary and units which are fundamental at least in the sense that, when a quantity becomes of order unity in the latter units, dramatic new phenomena emerge. For example, if there was no fundamental length, the properties of physical systems would be invariant under an overall rescaling of their size, so atoms would not have a characteristic size, and we wouldn't even be able to agree on which unit to use as a 'metre'. With a fundamental quantum unit of length, we can meaningfully talk about short or large distances. Naturally we will do this by comparison to this fundamental length. In other words, 'fundamental' constants are fundamental only to the extent that they provide us with a way of transforming any quantity (in whatever units we have chosen to measure it) into a pure number whose physical meaning is immediately clear and unambiguous.

Still, how many really 'fundamental' constants are there? Note that some so-called fundamental units are clearly redundant: a good example is temperature, which is simply the average energy of a system. In our everyday experience, it turns out that we need three and only three: a length, a time, and an energy. However, it is possible that in higher-dimensional theories (such as string theory, see Sec. 4) only two of these may be sufficient. And maybe, if and when the 'theory of everything' is discovered, we will find that even less than two are required.

3. Standard Cosmology: Stirred, but Not Shaken?

Cosmology studies the origin and evolution of the universe, and in particular of its large-scale structures, on the basis of physical laws. By large-scale structures I mean scales of galaxies and beyond. This is the scale where interesting dynamics is happening today (anything happening on smaller scales is largely irrelevant for cosmological dynamics). The standard cosmological model, which was gradually put together during the twentieth century, is called the 'Hot Big Bang' model. Starting from some very simple assumptions, it leads to a number of predictions which have been observationally confirmed.

Three of these are particularly noteworthy. Firstly, there is Hubble's law—the fact that the universe is expanding, and galaxies are moving away from each other with a speed that is approximately proportional to the distance separating them. Secondly, Big Bang Nucleosynthesis (BBN) predicts the relative primordial abundances of the light chemical elements (which were synthesised in the first three minutes of the universe's existence): roughly 75% Hydrogen, 24% Helium and only 1% other elements. Last but not least, there is the Cosmic Microwave Background (CMB). This is a relic of the very hot and dense early universe. By measuring photons from this background coming from all directions, one finds an almost perfect black body distribution with a present temperature of only 2.725 degrees Kelvin, corresponding to a present radiation density of about 412 photons per cubic centimetre (whereas the present matter density of the universe is about 3 atoms per cubic metre).

However, despite these and many other successes, the model also has a few shortcomings. These shouldn't be seen as failures, but rather as pertinent questions to which it can provide no answer (because it wasn't built for that purpose). I'll briefly mention two of them. The first arises when instead of analysing all cosmic microwave background photons together one does the analysis for every direction of the sky. This was first done by the COBE satellite, and then confirmed (with increasing precision) by a number of other experiments. One finds a pattern of very small temperature fluctuations, of about one part in ten thousand relative to the 2.725 K. It turns out that CMB photons have ceased interacting with other particles when the universe was about 380000 years old. After that, they basically propagate freely until we receive them.

Now, temperature fluctuations correspond to density fluctuations: a region which is hotter than average will also be more dense than average. What COBE effectively saw was a map (blurred by experimental and other errors) of the universe at age 380000 years, showing a series of very small density fluctuations. We believe that these were subsequently amplified by gravity and eventually led to the structures we can observe today. The question, however, is where did these fluctuations come from? At present there are basically two theoretical paradigms (each including a range of models) which may explain this—inflation and topological defects—but they both can claim their own successes and shortcomings, so the situation is as yet not clear. On one hand, the predictions of many inflationary models seem to agree quite well with observations, but none of these successful models is well-motivated from a particle physics point of view. On the other hand, topological defect models are more deeply rooted within particle physics, but their predictions don't seem to compare so well with observations. It may happen that the final answer is some combination of the two...

The other unanswered question is, surprisingly enough, the contents of the universe. We can only see directly matter that emits light, but it turns out that most of the matter in the universe doesn't. For example, the visible parts of galaxies are thought to be surrounded by much larger 'halos' of dark matter, with a size up to 30 times that of the visible part. If all this matter were visible, the night sky would look pretty much like van Gogh's *Starry Night*.

Even though we only have indirect evidence for the existence of this dark matter, we do have a reasonable idea of what it is. About 1% of the matter of the universe is visible. Another 4% is invisible normal matter (baryons), that is mostly protons and neutrons. This is probably in the form of *MACHOS* (Massive Compact Halo Objects), such as brown dwarfs, white dwarfs, planets and possibly black holes. Roughly 25% of the matter of the universe is thought to be 'Cold Dark Matter', that is, heavy non-relativistic exotic particles, such as axions or *WIMPS* (Weakly Interacting Massive Particles). Cold Dark Matter tends to collapse (or 'clump') into the halos of galaxies, dragging along the dark baryons with it. Finally, about 70% of the contents of the universe is thought to be in the form of a 'Cosmological Constant' (that is energy of the vacuum!), or something very close to it—this is generically referred to as *dark energy*. Unlike CDM this never clumps: it tends to make the universe 'blow up' by making it expand faster and faster. In other words, it forces an accelerated expansion—which, according to the latest observations, started very recently. This has been taken somewhat skeptically by some people. In particular, a period of future acceleration of the universe, while not posing any problems for cosmologists would be very problematic for string theory (see Sec. 4) as we know it. However, this is not a basis for judgement...

These ingredients are needed for cosmological model building. One starts with a theoretical model, adds cosmological parameters such as the age, matter contents and so forth, and computes its observational consequences. Then one must compare notes with observational cosmologists and see if the model is in agreement with observation: if it doesn't one had better start again. But in the hope of eliminating some of the shortcomings of the Big Bang model, one needs to generalise the model, and yet unexplored extra dimensions are a good place to look for answers.

4. Strings and Extra Dimensions

It is believed that the unification of the known fundamental interactions of nature requires theories with additional spacetime dimensions. Indeed, the only known theory of gravity that is consistent with quantum mechanics is String Theory, which is formulated in ten dimensions.[9]

Even though there are at present no robust ideas about how one can go from these theories to our familiar low-energy spacetime cosmology in four dimensions (three spatial dimensions plus time), it is clear that such a process will necessarily involve procedures known as dimensional reduction and compactification. These concepts are mathematically very elaborate, but physically quite simple to understand. Even if the true 'theory of everything' is higher-dimensional, one must find how it would manifest itself to observers like us who can only probe four dimensions. Note that this is more general than simply obtaining the low energy limit of the theory.

But given that we only seem to be able to probe four dimensions, we must figure out why we can't see the others, that is why (and how) they are hidden. A simple solution is to make these extra dimensions compact and very small. For example, imagine that you are an equilibrist walking along a tight rope that is suspended high up in the air. For you the tight rope will be essentially one-dimensional. You can safely walk forwards or backwards, but taking a sideways step will have most unpleasant results. On the other hand, for a fly sitting on the same rope, it will be two-dimensional: apart from moving forwards and backwards, it can also safely move around it. It turns out that there are many different ways of performing such compactifications and, even more surprisingly, there are even ways to make infinite dimensions not accessible to us (more on this in Sec. 5).

A remarkable consequence of these processes is that the ordinary 4D constants become 'effective' quantities, typically being related to the true higher-dimensional fundamental constants through the characteristic length scales of the extra dimensions. It also happens that these length scales typically have a non-trivial evolution. In other words, it is extremely difficult and unnatural, within the context of higher-dimensional theories in general (and of string theory in particular) to keep their size fixed. This is technically known as the *radius stabilisation problem*. In these circumstances, one is naturally led to the expectation of time and space variations of the 4D constants we can measure. In what follows we will go through some of the possible cosmological consequences and observational signatures of these variations, focusing on the fine-structure constant ($\alpha \equiv e^2/\hbar c$, a measure of the strength of electromagnetic interactions). Before this, however, we need to make a further excursion into higher-dimensional cosmology.

5. A Cosmological Brane Scan

The so-called 'brane-world scenarios' are a topic of much recent interest in which variations of four-dimensional constants emerge in a clear and natural way. There is ample evidence that the three forces of particle physics live in $(3 + 1)$ dimensions—this has been tested on scales from 10^{-18} m to (for the case of electromagnetism) solar system scales. However, this may not be the case for gravity. Einstein's field equations have only been rigorously tested[10] in the solar system and the binary pulsar, where the gravitational field exists essentially in empty space. On smaller scales, only tests of linear gravity have been carried out, and even so only down to scales of about a fifth of a millimetre (roughly the thickness of a human hair).

Sparkled by the existence, in higher-dimensional theories, of membrane-like objects, the brane world paradigm arose. It postulates that our universe is a $(3 + 1)$ membrane that is somehow embedded in a larger space (called the 'bulk') which may or may not be compact and might even have an infinite volume. Particle physics is confined (by some mechanism that need not concern us here) to this brane, while gravity and other hypothetical non-standard model fields (such as scalar fields) can propagate everywhere. This may also provide a solution to the hierarchy problem, that is the problem of why is gravity so much weaker than any of the other three forces? The brane world paradigm's answer is simply that this is because it has to propagate over a much larger volume.

What are possible signatures of extra dimensions? In accelerator physics, some possible signatures include missing energy (due to the emission of massive quanta of gravity—gravitons—which escape into the bulk), interference with standard model processes (new Feynman diagrams with virtual graviton exchange which introduce corrections to measured properties such as cross sections), or even more exotic phenomena like strong gravity effects (a much trumpeted example being pair-creation of black holes).

For gravitation and cosmology, the most characteristic sign would be changes to the gravitational laws, either on very small or very large scales. Indeed in these models gravity will usually only look four-dimensional over a limited range of scales, and below or above this range there should be departures from the four-dimensional behaviour that would be clues for the extra dimensions. The reason why they appear on small enough scales can be understood by recalling the tight rope analogy: only something probing small enough scales (a fly as opposed to the equilibrist) will see the second dimension. The reason why they should appear on large enough scale is also easy to understand. If you lived in the Cambridgeshire fens all your life, you could perhaps be forgiven for believing that the Earth is flat and two-dimensional. However, once you travel long enough you will start to see mountains, and once you climb to the top of one of them you will realize that it is actually curved, and it must be curved into a further dimension.

Other possible clues for brane-type universes and extra dimensions include changes to the Friedman equation (for example, with terms induced by the bulk[11]), the appearance of various large-scale inhomogeneities, and variations of the fundamental constants—the main topic of this discussion. Despite this seemingly endless list of possibilities, one should keep in mind that there are strong observational tests and constraints to be faced, some of which we already discussed.

6. How to Spot a Varying Constant

We are almost ready to start looking for varying constants. But how would we recognise a varying constant, if we ever saw one? Two crucial points, already implicitly made in Sec. 2, are worth re-emphasising here: one can only measure dimensionless combinations of dimensional constants, and any such measurements are necessarily local.

For example, if I tell you that I am 1.76 metres tall, what I am really telling you is that the last time I took the ratio of my height to some other height which I arbitrarily chose to call 'one metre', that ratio came out to be 1.76. There is nothing deep about this number, since I can equally tell you my height in feet and inches! Now, if tomorrow I decide to repeat the above experiment and find a ratio of 1.95, that could be either because I've grown a bit in the meantime, or because I've unknowingly used a smaller 'metre', or due to any combination of the two possibilities. And the key point is that, even though one of these options might be quite more plausible than the others, any of them is a perfectly valid description of what's going on: there is no experimental way of verifying one and disproving the others. Similarly, as regards the point on locality, the statement that 'the speed of light here is the same as the speed of light in Andromeda' is either a definition or it's completely meaningless, since there is no experimental way of verifying it. These points are crucial and should be clearly understood.[12,13]

So it is possible (and often convenient) to build models which have varying dimensional quantities (such as the speed of light or the electron charge). However, there is nothing fundamental about such choice, in the sense that any such theory can always be re-cast in a different form, where another constant will be varying instead, but the observational consequences of the two will be exactly identical. From the observational point of view, it is meaningless to try to measure variations of dimensional constants *per se*. One can only measure dimensionless quantities, and test statements based on them. The most topical example is that of the fine-structure constant, which we have already mentioned. We are now ready to start the search. As we shall see, the current observational status is rather exciting, but also somewhat confusing.

7. Local Experiments

Laboratory measurements of the value of the fine-structure constant, and hence limits on its variation, have been carried out for a number of years. The best currently available limit is[12]

$$\frac{d}{dt}\ln\alpha < 1.2 \times 10^{-15}\,\mathrm{yr}^{-1}. \tag{1}$$

This is obtained by comparing rates between atomic clocks (based on ground state hyperfine transitions) in alkali atoms with different atomic number Z. The current best method uses ^{87}Rb vs. ^{133}Cs clocks, the effect being a relativistic correction of order $(\alpha Z)^2$. Future progress, including the use of laser-cooled, single-atom optical clocks and (more importantly) performing such experiments in space (at the International Space Station or using dedicated satellites), is expected to improve these bounds by several orders of magnitude. Note that this bound is local (it applies to the present day only).

On geophysical timescales, the best constraint comes from the analysis of Sm isotope ratios from the natural nuclear reactor at the Oklo (Gabon) uranium mine, on a timescale of 1.8×10^9 years, corresponding to a cosmological redshift of $z \sim 0.1$. The most recent analysis[13] finds two possible ranges of resonance energy shifts, corresponding to the following rates

$$\frac{\dot{\alpha}}{\alpha} = (0.4 \pm 0.5) \times 10^{-17}\,\mathrm{yr}^{-1} \tag{2}$$

$$\frac{\dot{\alpha}}{\alpha} = -(4.4 \pm 0.4) \times 10^{-17}\,\mathrm{yr}^{-1}. \tag{3}$$

The first is a null result whereas the second corresponds to a value of α that was larger in the past—compare this with the following section. There are suggestions that due to a number of nuclear physics uncertainties and model dependencies a more realistic bound might be about an order of magnitude weaker. The authors also point out that there is plausible but tentative evidence that the second result can be excluded by a further Gd sample. However, that analysis procedure is subject to more uncertainties than the one for Sm, so a more detailed analysis is required before definite conclusions can be drawn. It should also be noticed that most theories predicting variations of fundamental constants can be strongly constrained through local gravitational experiments, most notably via tests of the Equivalence Principle.[10]

8. The Recent Universe

The standard technique for this type of measurements, which have been attempted since the late 1950's, consists of observing the fine splitting of alkali doublet absorption lines in quasar spectra, and comparing these with standard laboratory spectra. A different value of α at early times would mean that electrons would be more loosely (or tightly, depending on the sign of the variation) bound to the nuclei compared to the present day, thus changing the characteristic wavelength of light emitted and absorbed by atoms. The current best result using this method is[14]

$$\frac{\Delta\alpha}{\alpha} = (-0.5 \pm 1.3) \times 10^{-5}. \qquad z \sim 2 - 3; \qquad (4)$$

Note that in comparing a rate of change at a certain epoch $(\dot{\alpha}/\alpha)$ with a relative change over a certain range $(\Delta\alpha/\alpha)$ one must choose not only a timescale (in order to fix a Hubble time) but also a full cosmological model, in particular specifying how α varies with cosmological time (or redshift). Hence any such comparison will necessarily be model-dependent.

Recent progress has focused on a new technique, commonly called the *Many Multiplet* method, which uses various multiplets from many chemical elements to improve the accuracy by about an order of magnitude. The current best result is[15]

$$\frac{\Delta\alpha}{\alpha} = (-0.543 \pm 0.116) \times 10^{-5}, \qquad z \sim 0.2 - 3.7, \qquad (5)$$

corresponding to a 4.7-sigma detection of a *smaller* α in the past. This comes from an analysis of data of 128 quasar absorption sources, obtained with the Keck/HIRES telescope, and despite extensive testing no systematic error has been found that could explain the result. On the other hand, a more recent analysis[16] using only 23 sources of VLT/UVES data and a slightly simpler analysis find the null result

$$\frac{\Delta\alpha}{\alpha} = (-0.06 \pm 0.06) \times 10^{-5}, \qquad z \sim 0.4 - 2.3. \qquad (6)$$

Note that this assumes today's isotopic abundances. If one instead assumes isotopic abundances typical of low metalicity, one does find a detection,

$$\frac{\Delta\alpha}{\alpha} = (-0.36 \pm 0.06) \times 10^{-5}, \qquad z \sim 0.4 - 2.3; \qquad (7)$$

while the true value should be somewhere between the two, these results also highlight the need for independent checks. A few other attempted measurements at much lower resolution also produce null results. Higher resolution data and further independent measurement techniques (including the use of emission lines) are needed in order to provide definitive measurements, and a number of groups—including the two above and our own—are currently working actively on this task.

A somewhat different approach consists of using radio and millimetre spectra of quasar absorption lines. Unfortunately at the moment this can only be used at lower redshifts, yielding the upper limit[17]

$$\left|\frac{\Delta\alpha}{\alpha}\right| < 0.85 \times 10^{-5}, \qquad z \sim 0.25 - 0.68. \tag{8}$$

Finally, one can also search for variations of other dimensionless constants. An example is the proton-to-electron mass ratio ($\mu = m_p/m_e$), which can be measured by measuring the wavelengths of H_2 transitions in damped Lyman-α systems (and using the fact that electron vibro-rotational lines depend on the reduced mass of the molecule, the dependence being different for different transitions). This method has produced another claimed 3-sigma detection[18]

$$\frac{\Delta\mu}{\mu} = (5.02 \pm 1.82) \times 10^{-5}, \tag{9}$$

which is based on VLT/UVES data from a single quasar at redshift $z \sim 3$.

So even though a very substantial amount of work has been put into this type of measurements, the jury is still out on whether or not we have seen the variations. The possibility of relative systematic errors between different telescopes/spectrographs is a particular cause for concern. On the other hand, if these variations existed at relatively recent times in the history of the universe, one is naturally led to the question of what was happening at earlier times—presumably the variations relative to the present day values would have been stronger then. Those measurements are in a sense harder to do, but they are crucial as an independent check, and our group has been working on them for a number of years now.[19–24]

9. The Early Universe: BBN and CMB

At much higher redshifts, two of the pillars of standard cosmology (discussed in Sec. 3) offer exciting prospects for studies of variations of constants. Firstly, nucleosynthesis has the obvious advantage of probing the highest redshifts ($z \sim 10^{10}$), but it has a strong drawback in that one is always forced to make an assumption on how the neutron to proton mass difference depends on α. No known particle physics model provides such a relation, so one usually has to resort to a phenomenological expression, which is needed to estimate the effect of a varying α on the ^4He abundance. The abundances of the other light elements depend much less strongly on this assumption, but on the other hand these abundances are much less well known observationally.

The cosmic microwave background probes intermediate redshifts ($z \sim 10^3$), but allows model-independent measurements and has the significant advantage that one has (or will soon have) highly accurate data. A varying fine-structure constant changes the ionisation history of the universe: it changes the Thomson scattering cross section for all interacting species, and also changes the recombination history of Hydrogen (by far the dominant contribution) and other species through changes in the energy levels and binding energies. This will obviously have important effects on the CMB angular power spectrum. Suppose that α was larger at the epoch of recombination. Then the position of the first Doppler peak would move smaller angular scales, its amplitude would increase due to a larger early Integrated Sachs-Wolfe (ISW) effect, and the damping at small angular scales would decrease.

We have recently carried out detailed analyses of the effects of a varying α on BBN and the CMB, and compared the results with the latest available observational results in each case. We find that although the current data has a very slight preference for a smaller value of α in the past, it is consistent with no variation (at the 2-sigma level) and, furthermore, restricts any such variation, from the epoch of recombination to the present day, to be less than a few percent. Specifically, for BBN we find[21]

$$\frac{\Delta\alpha}{\alpha} = (-7 \pm 9) \times 10^{-3}, \qquad z \sim 10^{10}, \tag{10}$$

while for the CMB we have[23]

$$0.95 < \frac{\alpha_{cmb}}{\alpha_{now}} < 1.02, \qquad z \sim 10^3. \tag{11}$$

In both cases the likelihood distribution function is skewed towards smaller values of α in the past.

One of the difficulties with these measurements is that one needs to find ways of getting around degeneracies with other cosmological parameters.[19,20,22] The recent improvements in the CMB data sets available (and particularly the availability of CMB polarisation data) will provide crucial improvements. For example, we have recently shown[24] that once data from the Planck Surveyor satellite is available, one will be able to measure α to at least 0.1% accuracy.

10. The Future: A Cosmologist's Wish List

We have seen that despite tantalising hints the currently available data is still inconclusive. What is needed to get a definitive answer? On the astrophysical side, a number of groups (including ourselves) have started programmes aiming to obtain further measurements using quasar data, with a number of different techniques, and there is reason to hope that the situation will soon be much clearer. On the cosmological side the issues to be tackled are arguably harder, and some key developments will be required. Among these, three spring to mind:

First, one would like to have a post-Planck, CMB polarisation-dedicated satellite experiment. It follows from our recent CMB analysis and forecasts[24] that the Planck Surveyor will measure CMB temperature with an accuracy that is very close to the theoretically allowed precision (technically, this is called the cosmic variance limit), but its polarisation measurements will be quite far from it—this is because the ways one optimises a detector to measure temperature or polarisation are quite different, and in some sense mutually incompatible. On the other hand, we have also shown[23,24] that CMB polarisation alone contains much more cosmological information than CMB temperature alone. Hence an experiment optimised for polarisation measurements would be most welcome.

Second, one needs further, more stringent local tests of Einstein's Equivalence Principle. This is the cornerstone of Einstein's gravity, but all theories with additional spacetime dimensions violate it—the issue is not whether or not they do, it's at what level they do it (and whether or not that level is within current experimental reach). In the coming years some technological developments are expected which should help improve these measurements, not only in the lab but also at the International Space Station and even using dedicated satellites (such as Microscope, GravityProbe-B and STEP). Either direct violations will be detected (which will be nothing short of revolutionary) or, if not, it will drastically reduce the list of viable possibilities we know of, indicating that the 'theory of everything' is very different from our current naive expectations.

Third, one needs tests of the behaviour of gravity. Because it is so much weaker than the other forces of nature, it is quite hard to test the behaviour of gravity, and surprisingly little is known about it on many interesting scales. There has been a recent surge of interest in laboratory tests on small scales (and further improvements are expected shortly), but experiments that can test it on very large (cosmological) scales are still *terra incognita*. This is relevant because, as explained above, in theories with additional spacetime dimensions gravity is non-standard on large enough and/or small enough scales. Furthermore, a non-standard large-scale behaviour could also be an alternative to the presence of dark energy (a topic of much recent interest, but beyond the scope of this article).

Apart from the experimental and observational work, there are deep theoretical issues to be clarified. Noteworthy among these is the question as to whether all dimensionless parameters in the final physical 'theory of everything' will be fixed by consistency conditions or if some of them will remain arbitrary. Today this is a question of belief—it does not have a scientific answer at present. By arbitrary I mean in this context that a given dimensionless parameter assumed its value in the process of the cosmological evolution of the universe at an early stage of it. Hence, with some probability, it could also have assumed other values, and it could possibly also change in the course of this evolution.

11. So What is Your Point?

Physics is a logical activity, and hence (unlike other intellectual pursuits), frowns on radical departures. Physicists much prefer to proceed by reinterpretation, whereby elegant new ideas provide a sounder basis for what one already knew, while also leading to further, novel results with at least a prospect of testability. However, it is often not easy to see how old concepts fit into a new framework. How would our views of the world be changed if decisive evidence is eventually found for extra dimensions and varying fundamental constants?

Theories obeying the Einstein and Strong Equivalence Principles are metric theories of gravity.[10] In such theories the spacetime is endowed with a symmetric metric, freely falling bodies follow geodesics of this metric, and in local freely falling frames the non-gravitational physics laws are written in the language of special relativity. If the Einstein Equivalence Principle holds, the effects of gravity are equivalent to the effects of living in a curved spacetime. The Strong Equivalence Principle contains the Einstein

Equivalence Principle as the special case where local gravitational forces (such as probed by Cavendish experiments or gravimeter measurements, for example) are ignored. If the Strong Equivalence Principle is strictly valid, there should be one and only one gravitational field in the universe, namely the metric.

Varying non-gravitational constants are forbidden by General Relativity and (in general) all metric theories. A varying fine-structure constant will violate the Equivalence Principle, thus signalling the breakdown of 4D gravitation as a geometric phenomenon. It will also reveal the existence of further (as yet undiscovered) gravitational fields in the universe, and may be a very strong proof for the existence of additional spacetime dimensions. As such, it will be a revolution—even more drastic than the one in which Newtonian gravity became part of Einsteinian gravity. Also, while not telling us (by itself) too much about the 'theory of everything', it will provide some strong clues about what and where to look for.

Most people (scientists and non-scientists alike) normally make a distinction between physics (studying down-to-earth things) and astronomy (studying the heavens above). This distinction is visible throughout recorded history, and is still clearly visible today. Indeed, in my own area of research, such distinction has only started to blur some thirty years or so ago, when a few cosmologists noted that the early universe should have been through a series of phase transitions (of which particle physicists knew a fair amount about). Hence it would be advisable for cosmologists to start learning particle physics. Nowadays the circle is closing, with particle physicists finally beginning to realize that, as they try to probe earlier and earlier epochs where physical conditions were more and more extreme, there is no laboratory on Earth capable of reproducing such these conditions. Indeed, the only laboratory that is fit for the job is the early universe itself. Hence it is also advisable for particle physicists to learn cosmology.

The topic of extra dimensions and varying fundamental constants is, to my mind, the perfect example of a problem at the borderline between the two areas. It will be remarkable, for example, if we can learn about String Theory through something as 'mundane' as spectroscopy. It's also one where knowledge of only one of the sides, no matter how deep, is a severe handicap. This obviously makes it somewhat difficult, but also extremely exciting.

Acknowledgments

I am grateful to Pedro Avelino, Ana Garcia and Julia Mannherz for useful comments on an earlier version of this article. This work was funded by grants FMRH/BPD/1600/2000, ESO/FNU/43753/2001 and CERN/FIS/43737/2001 from FCT (Portugal). Numerical work was performed on COSMOS, the Altix3700 owned by the UK Computational Cosmology Consortium, supported by SGI, HEFCE and PPARC.

References

1. P.A.M. Dirac, *Nature* **139**, 323 (1937).
2. W.H. McCrea and M.J. Rees (Eds.), *Phil. Trans. R. Soc. Lond.* **A310**, 209 (1983).
3. J.K. Webb *et al.*, *Phys. Rev. Lett.* **82**, 884 (1999).
4. C.J.A.P. Martins, *Phil. Trans. Roy. Soc. Lond.* **A360**, 261 (2002).
5. C.J.A.P. Martins (Ed.), *The Cosmology of Extra Dimensions and Varying Fundamental Constants* (Kluwer, 2003).
6. J.-P. Uzan, *Rev. Mod. Phys.* **75**, 403 (2003).
7. A.S. Eddington, *The Philosophy of Physical Science* (Cambridge University Press, 1939).
8. M. Duff *et al.*, *JHEP* **0203**, 023 (2002).
9. J. Polchinski, *String Theory* (Cambridge University Press, 1998).
10. C.M. Will, *Theory and Experiment in Gravitational Physics* (Cambridge University Press, 1993).
11. P.P. Avelino and C.J.A.P. Martins, *Astrophys. J.* **565**, 661 (2002).
12. H. Marion *et al.*, *Phys. Rev. Lett.* **90**, 150801 (2003).
13. Y. Fujii, *Astrophys. Space Sci.* **283**, 559 (2003).
14. M.T. Murphy *et al.*, *Mon. Not. Roy. Astron. Soc.* **327**, 1237 (2001).
15. M.T. Murphy, J.K. Webb and V.V. Flambaum, *Mon. Not. Roy. Astron. Soc.* **345**, 609 (2003).
16. H. Chand *et al.*, *Astron. Astrophys.* **417**, 853 (2004).
17. C.L. Carilli *et al.*, *Phys. Rev. Lett.* **85**, 5511 (2000).
18. A. Ivanchik *et al.*, *Astrophys. Space Sci.* **283**, 583 (2003).
19. P.P. Avelino, C.J.A.P. Martins and G. Rocha, *Phis. Lett.* **B483**, 210 (2000).
20. P.P. Avelino *et al.*, *Phys. Rev.* **D62**, 123508 (2000).
21. P.P. Avelino *et al.*, *Phys. Rev.* **D64**, 103505 (2001).
22. C.J.A.P. Martins *et al.*, *Phys. Rev.* **D66**, 023505 (2002).
23. C.J.A.P. Martins *et al.*, *Phys. Lett.* **B585**, 29 (2004).
24. G. Rocha *et al.*, *Mon. Not. Roy. Astron. Soc.* **352**, 20 (2004).

Carlos Martins

Born in Viana do Castelo (Portugal), Carlos Martins graduated from the University of Porto in 1993 with an interdisciplinary Physics and Applied Mathematics degree (specialising in astrophysics), and obtained his PhD from the University of Cambridge in 1997. Currently 34, he divides his time between the Centre for Physics of the University of Porto, DAMTP in Cambridge, and the Institut d'Astrophysique in Paris.

He works at the interface between astrophysics, cosmology and particle physics, focusing on testing theories of the early universe observationally. Specific recent interests include the various roles of scalar fields in astrophysics and cosmology. Outside (and sometimes during) office hours he can be found on basketball courts, attending classical music concerts (especially if they feature Bach, Mahler or Monteverdi) or going to movies. He is also an avid reader, has a slight addiction to chocolate, and is making an effort to improve his cooking and to learn a fifth language.

CHAPTER 4

SMALL SCALES, BIG ISSUES FOR COLD DARK MATTER

Andrew J. Benson

Department of Physics, Keble Road, Oxford, OX1 3RH, United Kingdom
E-mail: abenson@astro.ox.ac.uk

One of the theoretical cornerstones of modern cosmology is the concept of cold dark matter. This theory supposes that almost 90% of the mass in the Universe is made up of some unknown, invisible (hence "dark") particle which interacts with the rest of the Universe only through gravity and which is cold in the sense that the particles move slowly (like molecules in a cold gas). One of the grand achievements of observational cosmology in recent years has been to uncover strong evidence in favour of this idea, through means as diverse as gravitational lensing, surveys of millions of galaxies and studies of the cosmic microwave background. This observational evidence however, applies almost entirely on rather large scales (comparable to the sizes of galaxy clusters and larger). On smaller scales there is an uncomfortable disagreement between the theory and observations. Many galaxies seem to contain too little dark matter in their centres, while galaxies such as the Milky Way ought to have hundreds of small companion galaxies, but apparently have only a handful of them. I will describe these recent confirmations and refutations of the cold dark matter hypothesis and discuss their possible consequences. In particular, I will highlight some of the theoretical work which has been prompted by this discussion and describe how a debate over dark matter is leading to improvements in our understanding of how galaxies form.

1. Introduction

There now exists a broad consensus as to the contents of our Universe. It has been known for a long time that the matter which we can observe directly with our telescopes (mostly stars in galaxies, but also gaseous material emitting at radio wavelengths) is but a small fraction of the total matter in the Universe. Several different techniques have been used to determine the total mass present in different regions of the Universe. For example, gravitational lensing of light allows an estimate of the total mass present in a galaxy cluster to be estimated, while the speed at which a galaxy rotates depends upon the total mass of material it contains. All of these measurements imply the presence of much more mass than can be accounted for by the matter that we can observe. The remainder, the so-called "missing-matter", is dark—it does not emit any photons that we can detect. A long history of observational measurements, culminating in the current generation of cosmic microwave background (CMB) experiments,[3,33] confirm that our Universe contains large quantities of this dark matter. Furthermore, we now know that this dark matter cannot be made of the usual stuff (electrons, protons, neutrons etc.—known as "baryonic matter" on account of most of its mass being in the form of baryons), and instead must be made of some exotic particle. The mass of our Universe consists of about 84% dark matter and 15% baryonic matter.[3] It is interesting to note that only a small fraction (around 8%) of the baryonic matter is in the form of visible stars,[11] so when studying galaxies we are really only probing the very tip of the mass iceberg. Experiments which measure the rate of expansion of our Universe (by observing distant supernovae) show that a mysterious "dark energy" contributes over twice as much to the energy of our Universe as does dark matter, and is causing the expansion to accelerate.[35,38,45] The origins of this dark energy are even less clear than those of dark matter, although theories (ranging from the presence of a cosmological constant to the exotic physics of extra dimensions) are abundant.

How do we know that the dark matter cannot be normal, baryonic material? Seconds after the Big Bang, nuclear reactions occurring in the high temperature early Universe produced a variety of light elements and isotopes such as deuterium and lithium.[37] The amount of each species produced depends upon just how much baryonic material was available to drive the reactions. Highly accurate measurements of the amount of deuterium in our Universe[28] now show that there cannot be enough baryonic material present in the Universe to account for all of the missing mass—thus we require something else. We also know that dark matter must interact with itself and other particles only weakly; in general it is considered to interact only through the gravitational force (although other interactions may have been important under the very different conditions found in the early Universe). The candidates for the dark matter particle can be split into two broad categories: hot and cold. Hot dark matter would consist of particles which were moving at close to the speed of light when they last interacted with other particles, while cold dark matter particles, being much more massive, would move at much lower speeds at this point. Hot dark matter models have been effectively ruled out by studies of the large scale distribution of mass in our Universe.[51,52] Galaxies only form in the densest regions of the Universe. However, the large velocities of hot dark matter particles allow them to quickly stream out of any dense region in which they may find themselves. As a consequence, those dense regions tend to get smoothed away. Only the largest gravitational pulls can confine hot dark matter with the consequence that the only objects to form are superclusters—there would be no isolated galaxies such as we observe. Cold dark matter does not suffer from this problem, and in fact gravitationally bound cold dark matter objects, or "halos[a]", form in a hierarchical way, growing through the merging together of smaller halos.[18]

Thus, the cold dark matter theory has become the standard theory for the bulk of the matter content of the Universe. Without having to specify exactly what the dark matter is, merely that it is cold and interacts only through the force of gravity, this theory has met with tremendous success in explaining many observed phenomena in our Universe. Over the past twenty years or so, observers and theorists alike have attempted to test this theory. In the remainder of this article I will review some of the successes of cold dark matter (hereafter CDM) theory and then concentrate on the current challenges that it is facing.

2. Large Scales

The easiest way to compare the predictions of CDM theory with observations is by looking at very large scales in the Universe—scales ranging from the size of the largest of galaxy clusters to the size of the Universe itself. There is good reason to believe that in order to understand the properties of CDM on these truly cosmological scales all we need to know about is how the force of gravity works and what the initial distribution of CDM was shortly after the Big Bang. We certainly believe that we have a complete theory of gravity (at least in the classical regime—a quantum theory of gravity still eludes us), and there is a well defined, and therefore testable, theory for the initial conditions (the "inflationary" theory proposed by Guth,[20] which suggests that small variations in the dark matter density at very early times arose from the quantum fluctuations of a decaying particle which drove a period of rapid, accelerated expansion in the

---

[a]This name arises because they often surround galaxies, forming a "halo" of material around the galaxy.

very early Universe). Consequently, we can make firm predictions about the distribution of dark matter on large scales. Amazingly, on these scales the predictions have so far been confirmed by every observational test applied. I will briefly describe some of these tests below.

The classical way to study the distribution of dark matter is to map out the positions of galaxies. On scales much larger than galaxies and clusters of galaxies gravity is always the dominant force: other forces, such as electromagnetism, are always effectively short-range, since positive and negative charges can cancel out. With gravity, we only have positive mass, so there is no cancelling out and the force is long range. Consequently, on large scales, the positions of galaxies should have been determined purely by gravity, and that is primarily determined by the distribution of dark matter. To put this simply, we expect the distribution of galaxies to trace that of the dark matter, perhaps with some simple offset (or "bias") reflecting the fact that galaxies like to form in the densest regions of the Universe.[25] Mapping the positions of galaxies therefore allows us to infer the distribution of dark matter. Furthermore, galaxies are known to have velocities in addition to that due to the Hubble expansion of the Universe.[21] The only plausible origin for these velocities is the gravitational pull of large amounts of dark matter. Mapping the velocities of galaxies can therefore tell us about how dark matter in the Universe is arranged.[31]

Maps of galaxies have been made for many decades, but in the past few years two large surveys, the Two-degree Galaxy Redshift Survey[13] (2dFGRS) and the Sloan Digital Sky Survey[53] (SDSS) have mapped the positions of around one million galaxies, more than all previous surveys combined. The 2dFGRS has shown that the distribution of galaxies on scales of around 10Mpc to 1000Mpc is exactly what one would expect for the standard CDM theory, lending strong support to this model. An analysis of galaxy velocities in this survey further strengthens the case for CDM.[21]

A more direct way to probe the distribution of dark matter exploits one of the remarkable consequences of general relativity, namely that light will follow a curved path in the vicinity of massive objects. One astrophysical consequence of this is that light coming from different regions of a single galaxy will follow slightly different curved paths on its way towards us due to the slightly different distributions of mass along those paths. This leads to the image of the galaxy that we observe being distorted—a perfectly circular galaxy would appear slightly elliptical because of this gravitational lensing. For a single galaxy, this lensing-induced ellipticity cannot be distinguished from the intrinsic ellipticity of the galaxy. However by mapping the ellipticities of thousands of galaxies in a patch of the sky the small distortions induced by lensing (typically only a few percent in magnitude) can be separated out statistically. This allows the distribution of matter which caused them to be inferred. While measuring the shapes of galaxies to such accuracy is highly challenging (telescope optics also have a tendency to make circular objects appear elliptical) careful analysis has revealed that the distribution of dark matter is statistically compatible with the CDM predictions.[22]

Finally, the CDM model can be used to predict the properties of temperature variations in the cosmic microwave background (the light left over from the Big Bang that we observe at microwave wavelengths today). These temperature variations are caused by the slight differences in dark matter density in the very early Universe (about 380,000 years after the Big Bang). Experiments such as WMAP[3] have now mapped the distribution of these temperatures to an accuracy of a few μK across the whole sky, and have confirmed that they meet the expectations from CDM theory. (These experiments also confirm some of the major predictions of the inflation theory.[34])

3. Small Scales

Since everything seems to work out well for CDM on large scales the next logical step is to study its behaviour on small scales. Two important things happen when we look at small scales—firstly we lose our ability to make accurate predictions about dark matter and secondly the observations no longer agree with predictions. We can examine these two, perhaps not unrelated events one at a time.

We lose our ability to make predictions about dark matter for the simple reason that, on small scales, dark matter is no longer the only important actor on the stage. The baryonic material (mostly hydrogen and helium) which makes up about 15% of the mass of the Universe is often concentrated into high density lumps (also known as galaxies) where its density frequently exceeds that of the dark matter in the same region. It is exactly these regions which we can probe observationally because they tend to form stars and so emit light. Consequently, while on large scales it was only gravity which had any influence on the distribution of mass, on small scales there is much more to consider. The baryonic material is affected not only by gravity, but by hydrodynamical and electromagnetic forces. Atomic and molecular physics greatly influence the distribution of baryonic material since they allow it to cool and lose energy, while nuclear physics is crucial in star formation which is believed to have a profound influence on the structure of entire galaxies. Note that it is not that we do not understand some of this additional physics—there are well defined theories for all of it—merely that we cannot solve the equations. The largest supercomputers in the world are capable only of solving the equations describing the gravitational forces in the Universe (or, in a much cruder way, the hydrodynamical forces also). So far, no computer is able to simultaneously solve all of the above physics, but with ever increasing computer power it may indeed be possible to achieve this in the not so distant future.

Despite these difficulties, theorists have made predictions about the behaviour of CDM on small scales, and observers have tested these predictions. It is here that we find a number of instances where the CDM theory fails to agree with the observational data. The first problem becomes very obvious when we compare our Local Group of galaxies with what CDM theory predicts the distribution of dark matter around of Milky Way to be. Our Local Group consists of our own Galaxy, the Milky Way, the Andromeda galaxy and a few tens of other, smaller galaxies which are, in general, orbiting around the Milky Way or Andromeda. These satellites include the Magellanic Clouds and several smaller dwarf galaxies. CDM theory predicts there to be several hundred small concentrations of dark matter in orbit around our Milky Way (see Figure 1). Some of these concentrations, or substructures, are presumably the abode of the Milky Way's satellite galaxies. This is all very well, but, if this is the case, why do we observe there to be only slightly more than ten such galaxies in orbit around the Milky Way? Why not one for every one of the hundreds of substructures predicted by CDM? This problem, first pointed out by Kauffmann, White & Guiderdoni,[27] has attracted a lot of attention in recent years. Initially proposed as a severe challenge for CDM by Moore *et al.*,[30] several possible solutions have now been proposed.

There are, broadly speaking, two possible ways out of this problem. Either CDM theory is wrong, or most of the substructures fail to make galaxies.[b] Since we know that modelling galaxy formation is tricky, we should certainly first check that there isn't some straightforward way to prevent galaxies forming in substructures before ruling out the CDM model. In

---

[b]Another possibility is to simply change the initial distribution of cold dark matter in the early Universe in order to remove some of the small scale density enhancements which will grow into the substructures.[26] I will consider "CDM theory" here to imply the standard initial conditions, so such alternatives will fall into the "CDM is wrong" category.

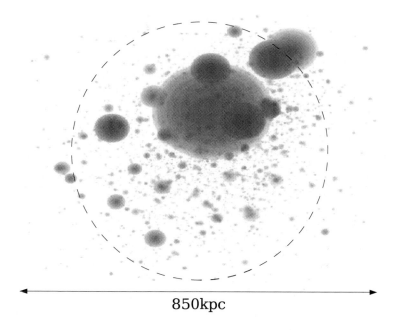

850kpc

Fig. 1.   Substructure in a cold dark matter halo. Each grey sphere represents a single dark matter substructure embedded within a larger halo (not shown for clarity, although the dashed line indicates the extent of the larger halo). The distribution of substructure positions and masses was taken from the work of Benson *et al.* (2004).

fact, a plausible solution, first suggested by Rees[36] and developed recently by several workers including this author,[5,10,43] has been uncovered, and relies on the fact that the Universe became ionized long in the past by some of the first galaxies to form. Massive stars in galaxies emit photons that are sufficiently energetic that they can ionize hydrogen atoms. If these photons escape from the galaxy in which they originate they find themselves in the intergalactic medium (the space in between galaxies), and will ionize the hydrogen that exists there. As a side effect, some of the photon's energy will be transformed into thermal energy in the ionized hydrogen, causing its temperature to rise. Rising temperature means rising pressure, and soon after the hydrogen in the Universe becomes completely ionized its pressure becomes sufficient to resist the pull of gravity of the small dark matter substructures. Consequently, these substructures are unable to acquire any gas from which to make stars and so they remain dark and invisible. It is only those few substructures which happened to form before the Universe became ionized that were able to make stars, and it is these select few that we observe in orbit around the Milky Way today. We have recently completed the most detailed calculations of this process to date, exploiting an analytic model of galaxy formation to determine how many satellite galaxies would be expected in CDM theory once we account for the above effects. The conclusions of this work are that the reionization of the Universe *can* sufficiently reduce the predicted number of satellites such that theory and observations agree (although there are differences in some of the details that still need to be understood). We would of course, like to count satellites around galaxies beyond our Local Group (since it is never good to base a conclusion on observations of a single system), but for now there appears to be no problem in producing only small numbers of visible satellites in CDM theory.

This leads to an interesting question—if the substructures are there but we can't see them, is there any other way in which we can detect their presence? The only possibilities rely on the gravitational effects of the substructures. One such effect is, once more, gravitational lensing. The substructures should occasionally pass in front of a distant quasar (a massive black hole swallowing gas) making it appear brighter. How can we tell if a quasar is brighter than it ought to be? We have to examine very special systems where the quasar has already been lensed once, by a distant galaxy, producing two images of the quasar which, in the absense of any substructure, should have almost identical brightnesses. The presence of a substructure near one of these images would then make it appear brighter than the other (see Figure 2), and we can conclude that a substructure is nearby. These special quasars are rare, but analysis of the few that do exist shows that they are indeed occasionally lensed by a substructure. The frequency with which this happens is close to that predicted by CDM theory.[14]

A second way that we can test for the presence of these substructures is by their influence on the disks of galaxies. Disks of galaxies are remarkably thin—the thickness of the disk is much smaller than the radius as can be seen in Figure 3. Occasionally, some of the dark matter substructures must pass near by the disk of a galaxy as their orbit carries them into the central regions of the dark matter halo which they inhabit. The gravitational pull of the substructure will tug on stars in the galaxy disk. The stars tend to gain energy through this interaction and the more energy they have the further they can travel away from the mid-plane of the disk. The result is that the disk starts to become thicker. If there are too many substructures disks would become so thickened that they would no longer look like disks, too few substructures and disks might be much thinner than we observe.

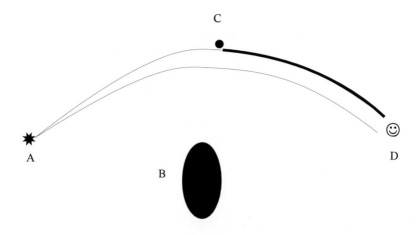

Fig. 2. The geometry of a lensing event which may be used to infer the presence of substructure. Light from the quasar, A, is lensed by an intervening galaxy, B, before being observed by the observer, D. The lensing galaxy creates two images of the quasar which, under certain conditions, will be magnified by the same amount and so will appear equally as bright to the observer. One of the images happens to pass close to a substructure, C, which lenses it further, making it appear brighter (as indicated by the thicker line type).

So, does CDM predict too many, too few or just the right number of substructures to thicken disks to the measured degree? This seemingly simple question is difficult to answer quantitatively, due to the complexity of the interaction between substructures and galaxy disks. Nevertheless, theorists have attempted to answer it. The problem has been tackled using both numerical N-body simulations[24,47] and analytic techniques.[46] Recently, we have developed a much more detailed analytical model of this thickening process, allowing us to make a careful assessment of the amount of heating which occurs. By following the heating due to all of the substructures which interact with a disk over its several billion year lifetime we are able to determine its thickness at the present day. Furthermore, we can repeat this analysis for many model galaxies and so determine the distribution of disk thicknesses. While many uncertainties remain the current results suggest that the CDM prediction for the number of substructures is not only consistent with the thickness of the disk of the Milky Way and other galaxies, but in fact agrees remarkably well with them.

There is one further problem with CDM substructure. According to numerical simulations the substructures act as sites of gas cooling in the early Universe. This gas is later used to build galaxies such as the Milky Way.

A. J. Benson

Fig. 3.   The galaxy NGC 4013 as seen by the Hubble Space Telescope. The galaxy is seen almost exactly edge-on from our vantage point. Image courtesy of NASA and The Hubble Heritage Team (STScI/AURA).

However, by cooling into the substructures early on this gas loses most of its angular momentum to the surrounding dark matter. As a consequence, the galaxies made in numerical simulations tend to possess much less angular momentum than real galaxies—they therefore spin too slowly and are forced to contract to very small sizes before their rotation can support them against gravity, making them much smaller than the sizes of observed galaxies. The way to solve this problem is reasonably straightforward—stop the gas cooling into the substructures at early times.[48] If this is what really happens in the Universe we have to ask what is missing from the numerical simulations. The answer, it seems, is feedback. When stars form in a galaxy the most massive ones quickly reach the end of their lives and become supernovae. These supernovae produce huge amounts of energy which could easily cause all the gas currently cooling into a substructure to be blown back out. This "feedback" has been an important component of models of galaxy formation for many years, but is difficult to implement in numerical simulations. State of the art numerical calculations have been able to include simple models of feedback, and they do seem to solve this problem of disk sizes. As a result the new simulations are able to form galaxies of about the right size and angular momentum.[1,39]

The final challenge to CDM on small scales is perhaps the most serious. Some of the first evidence for dark matter came from galaxy rotation curves[40]—a measure of the speed at which a galaxy rotates as a function of the distance from the centre of that galaxy. The faster a galaxy rotates the more gravitating mass must exist inside of it to prevent the rotation causing it to fly apart. The measured rotation curves of galaxies show that galaxies rotate much too fast to be held together by just the mass of their luminous stars, thus the need for some dark mass to hold them together.

So far, so good, but what does CDM theory predict that these rotation curves should look like? The rotation curve is directly related to the distribution of mass in a halo and therefore to the density at each radius within the halo. The question of how density should change with radius has been addressed using numerical simulations. Typically, these simulations have included only dark matter, and so cannot form galaxies. Nevertheless, we can measure from the simulation what the dark matter would contribute to the rotation of a galaxy which did live in such a halo.

The first in depth study, carried out by Navarro, Frenk & White,[32] indicated that all dark matter halos had a density profile of roughly the same form—in the outer regions of the halo the density was inversely proportional to the radius cubed, while in the inner regions it was inversely proportional to the radius, with a smooth transition in between. There has been considerable debate about just how correct this result is on small scales in recent years. While everyone agrees on the behaviour of the density on large scales, on small scales there has been much debate about the "central slope", which simply measures the way in which density depends on radius in the central regions through the relation $\rho \propto r^{-\alpha}$. The original Navarro, Frenk & White prediction says $\alpha = 1$. Others[19] have argued for $\alpha = 1.5$, while more recent studies perhaps indicate that α keeps on changing as we consider smaller and smaller radii. Nevertheless, there is general agreement that the density must keep on rising as we move to smaller scales.

How does this compare with what is observed? Before we make such a comparison we must keep in mind the important fact that these predictions are for systems of pure dark matter, whereas we can only measure

A. J. Benson

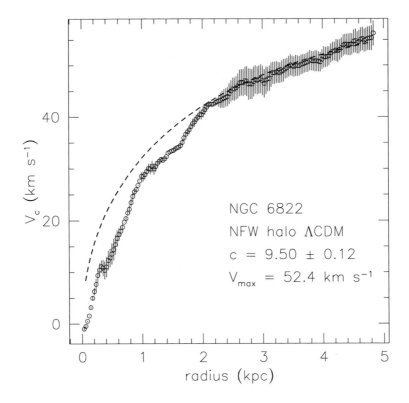

Fig. 4. The rotation curve of galaxy NGC6822. The circles with error bars show the measured rotation curve,[50] while the best fitting NFW model is shown by the dashed line. V_{max} gives the maximum velocity of the NFW model, while c is the concentration of the NFW model which determines the radius at which the slope of the NFW density profiles changes from -3 to -1. Figure courtesy of Erwin de Blok.

rotation curves for galaxies, where by definition there is some non-dark matter around. The logical place to test the prediction is in galaxies where we believe the total mass is dominated by dark matter, as these might well be the most similar to our dark matter simulations. Low surface brightness galaxies—galaxies which are rather more diffuse and extended than the norm—are a good example of such systems. Considerable observational effort has been expended to measure rotation curves in these systems, and the results don't look good for CDM.[50] Figure 4 shows the measured rotation curve for the galaxy NGC 6822, together with an attempt to reproduce this measurement using the predictions of CDM theory. Clearly the agreement is not very good.

Is this the end of CDM? We have learnt that our understanding of galaxy formation is poor enough that there may be reasons why the measurement should not agree with the simplest theory. We know that the theoretical predictions were made from simulations containing only dark matter. What would have happened had we included some gas as well, and allowed this gas to form a galaxy? The answer to that question remains unclear, but there are several possibilities. Although the density of gas in an LSB galaxy is much less than that of the dark matter at the present day, it may not have been so in the past. Could this gas have altered the distribution of dark matter away from the simple predictions? The answer is "yes, maybe." Stars and gas in galaxies frequently form bars—linear arrangements of material in the central regions of galaxy disks. The energy and angular momentum of these bars may be efficiently tranferred to the halo dark matter through gravity. The dark matter particles that gain this energy will be thrown out of the central regions of the halo, reducing the density dark matter there. This may help bring the rotation curve into agreement with the observations,[49] although this remains a matter of debate.[42] Alternative solutions abound: expelling dark matter from the centres of forming galaxies by coupling it to an outflow from the galaxy;[7] throwing out dark matter from the centre by having it gain energy from substructures through dynamical friction;[17] changing the nature of the dark matter particle, making it warm or self interacting;[2,9,12,16,23] changing the initial conditions to make them smoother on small scales;[26] and even changing the law of gravity.[15] Many of these require a better understanding of the process of galaxy formation before we can decided whether or not they really occur.

There is always a need to be cautious of taking the observations at face value as well. Although they have been made with tremendous care and attention to detail, we can nevertheless ask if something was misinterpretted. For example, the measurements are usually made from the motions of gas in the galaxy, which is assumed to move on simple, circular paths around the galaxy centre. What if it moves in a more complicated way around the centre (as will certainly be the case if a bar exists in the galaxy)? The observational estimates of the rotation curve would be incorrect in this case. In some cases it is not at all easy to determine just where the centre of the galaxy is anyway, making it hard to determine a rotation curve. These, and other uncertainties lead to systematic uncertainties in the determinations of rotation curves,[44] and so the problem may yet turn out to be not a real problem at all.

Nevertheless, this problem of cuspy halos is probably one of the most pressing for CDM. It does not just show up in LSB galaxies, recent work has failed to find evidence for sufficient dark matter in the central regions of galaxy clusters as well,[41] although once again the measurement can only be made in a region dominated by luminous matter which may have influenced the distribution of dark matter. There is clearly much more work to be done before we have any definitive evidence as to the validity or otherwise of the cold dark matter hypothesis.

4. Conclusions

The cold dark matter theory is remarkably simple—it assumes only that the dark matter moved slowly when it last interacted with other particles and that it now interacts only via gravity. Yet, it has been extremely successful at predicting the distribution of matter in our Universe on large scales. Our role as scientists is to remain sceptical of any theory and seek new ways to test it. There has therefore been a huge effort in recent years to test the predictions of the cold dark matter theory in very different regimes, such as on both large and small scales. The results are far from conclusive—while on large scales everything seems to fit together, many observational tests of dark matter on small scales seem, at first, to contradict the theoretical predictions. I have shown in this article how we must be very cautious of taking these results at face value, for the simple reason that, on small scales, the presence of luminous matter and its complicated physics, makes any prediction highly approximate. In most cases there are plausible ways in which the presence of luminous matter can be used to explain the apparent differences between theory and observations. To progress we need to develop these plausibility arguments into stronger, water-tight cases. To do so we will need to advance not our understanding of cold dark matter, but our understaning of how galaxies form and interact with their environment.

How then do we make progress? In a recent article, Binney[8] has suggested that the only way to move forward in our astrophysical understanding of CDM is to treat it as an established theory, in the same category as electromagnetism or general relativity. In comparison, a recent article by Merrifield[29] adopts a different view. While reaching the same basic conclusion that we need to understand galaxy formation better before we can really test CDM, Merrifield believes that CDM should be considered to be a mere hypothesis until more definite proof is established. It is perhaps indicative of the false dichotomy in astronomy that a theorist and observer have such different philosophies about how we should treat the CDM theory, even though in the end they reach the same basic conclusions. What is clear is that we must now focus on understanding galaxy formation in sufficient detail that we can make sufficiently accurate predictions regarding CDM on small scales—only by doing this will we be able to test the validity of the CDM hypothesis. Of course, a full understanding of galaxy formation will be extremely valuable in its own right. Observational data are already providing extremely detailed insights into the galaxy formation process, and proposed telescopes (such as JWST, OWL, TMT etc.) should provide much more in the next few decades. Galaxy formation theory is currently lagging behind somewhat, its level of detail being somewhat too poor to fully exploit these data. If this situation can be rectified then we should expect to learn much about galaxies and dark matter in the near future.

Acknowledgments

I would like to thank Erwin de Blok for kindly providing the rotation curve plot. The GALFORM semi-analytic model of galaxy formation, without which much of the worked described here would not have been possible, was developed by Carlton Baugh, Shaun Cole, Carlos Frenk and Cedric Lacey.

References

1. M. G. Abadi, J. F. Navarro, M. Steinmetz, V. R. Eke, *ApJ*, **591**, 499 (2003).
2. V. Avila-Reese, P. Colin, O. Valenzuela, E. D'Onghia, C. Firmani, *ApJ*, **559**, 516 (2001).
3. C. L. Bennett *et al.*, *ApJS*, **148**, 1 (2003).
4. A. J. Benson, S. Cole, C. S. Frenk, C. M. Baugh, C. G. Lacey, *MNRAS*, **311**, 793 (2000).
5. A. J. Benson, C. S. Frenk, C. G. Lacey, C. M. Baugh, S. Cole, *MNRAS*, **333**, 177 (2002).
6. A. J. Benson, C. G. Lacey, C. S. Frenk, C. M. Baugh, S. Cole, *MNRAS* in press (astro-ph/0307298).
7. J. J. Binney, O. Gerhard, J. Silk, *MNRAS*, **321**, 471 (2001).
8. J. J. Binney, in S. Ryder, D. J. Pisano, M. Walker and K. Freeman eds. in the proceedings of IAU Symposium 220 "Dark Matter in Galaxies".
9. P. Bode, J. P. Ostriker, N. Turok, *ApJ*, **556**, 93 (2001).
10. J. S. Bullock, A. V. Kravtsov, D. H. Weinberg, *ApJ*, **539**, 517 (2000).
11. S. Cole *et al.*, *MNRAS*, **326**, 255 (2001).
12. P. Colin, V. Avila-Reese, O. Valenzuela, *ApJ*, **542**, 622 (2000).
13. M. Colless *et al.*, *MNRAS*, **328**, 1039 (2001).
14. N. Dalal, C. S. Kochanek, *ApJ*, **572**, 25 (2002).

15. W. J. G. de Blok, S. S. McGaugh, *ApJ*, **508**, 132 (1998).
16. V. R. Eke, L. Shlosman, Y. Hoffman, *ApJ*, **560**, 636 (2001).
17. A. El-Zant, L. Shlosman, Y. Hoffman, *ApJ*, **560**, 636 (2001).
18. C. S. Frenk, S. D. M. White, M. Davis, G. Efstathiou, *ApJ*, **327**, 507, (1998).
19. S. Ghigna, B. Moore, F. Governato, G. Lake, T. Quinn, J. Stadel, *ApJ*, **544**, 616 (2000).
20. A. H. Guth, *Phys. Rev. D*, **23**, 347 (1981).
21. E. Hawkins *et al.*, *MNRAS*, **346**, 78 (2003).
22. H. Hoekstra, H. K. C. Yee, M. D. Gladders, *NewAR*, **46**, 767 (2002).
23. C. Hogan, J. Dalcanton, *PhRvD*, **62**, 817 (2000).
24. S. Huang, R. G. Carlberg, *ApJ*, **480**, 503 (1997).
25. N. Kaiser, *ApJ*, **284**, 9 (1984).
26. M. Kamionkowski, A. Liddle, *PRL*, **84**, 4525 (2000).
27. G. Kauffmann, S. D. M. White, B. Guiderdoni, *MNRAS*, **264**, 201 (1993).
28. D. Kirkman, D. Tytler, N. Suzuki, J. M. O'Meara, D. Lubin, *ApJS*, **149** 1 (2003).
29. M. R. Merrifield, in S. Ryder, D. J. Pisano, M. Walker and K. Freeman eds. in the proceedings of IAU Symposium 220 "Dark Matter in Galaxies".
30. B. Moore, S. Ghigna, F. Governato, G. Lake, T. Quinn, J. Stadel, P. Tozzi, *ApJ*, **524**, 19 (1999).
31. V. K. Narayan, D. H. Weinberg, E. Branchini, C. S. Frenk, S. Maddox, S. Oliver, M. Rowan-Robinson, W. Saunders, *ApJS*, **136**, 1 (2001).
32. J. F. Navarro, C. S. Frenk, S. D. M. White, *ApJ*, **490**, 493 (1997).
33. C. B. Netterfield *et al.*, *ApJ*, **571**, 604 (2002).
34. H. V. Peiris *et al.*, *ApJS*, **148**, 213 (2003).
35. S. Perlmutter *et al.*, *ApJ*, **517**, 565 (1999).
36. M. J. Rees, *MNRAS*, **218**, 25 (1986).
37. H. Reeves, J. Audouze, W. Fowler, D. N. Schramm, *ApJ*, **179**, 909 (1973).
38. A. G. Reiss *et al.*, *AJ*, **116**, 1009 (1998).
39. B. E. Robertson, N. Yoshida, V. Springel, L. Hernquist, astro-ph/0401252 (2004).
40. V. C. Rubin, *IAUS*, **84**, 211 (1979).
41. D. J. Sand, T. Treu, R. S. Ellis, *ApJ*, **574**, 129 (2002).
42. J. A. Sellwood, *ApJ*, **587**, 638 (2003).
43. R. S. Somerville, *MNRAS*, **572**, 23 (2002).
44. R. A. Swaters, B. F. Madore, F. C.van den Bosch, M. Balcells, *ApJ*, **583**, 732 (2003).
45. J. L. Tonry *et al.*, *ApJ*, **594**, 1 (2003).
46. G. Tóth, J. P. Ostriker, *ApJ*, **389**, 5 (1992).
47. H. Velázquez, S. D. M. White, *MNRAS*, **304**, 254 (1999).
48. M. L. Weil, V. R. Eke, G. Efstathiou, *MNRAS*, **300**, 773 (1998).
49. M. D. Weinberg, N. Katz, *ApJ*, **580**, 627 (2002).
50. D. T. F. Weldrake, W. J. G. de Blok, F. Walter, *MNRAS*, **340**, 657 (2003).
51. S. D. M. White, C. S. Frenk, M. Davis, *ApJ*, **274**, 1 (1983).
52. S. D. M. White, M. Davis, C. S. Frenk, *MNRAS*, **209**, 27 (1984).
53. http://www.sdss.org/

Andrew Benson

Andrew Benson is a Royal Society University Research Fellow studying the physics of galaxy formation at the University of Oxford. Before joining Oxford, Andrew studied for his undergraduate degree at the University of Leicester before pursuing a Ph.D. at the University of Durham where he worked with Carlos Frenk. After graduating from Durham, Andrew took up a Prize Fellowship in Astronomy at the California Institute of Technology, where he regularly engaged in the old Caltech tradition of rubbing Millikan's nose for good luck.

CHAPTER 5

VIOLENCE AND BLACK HOLES IN THE HEARTS OF GALAXIES

Carole G. Mundell

Astrophysics Research Institute, Liverpool John Moores University,
Twelve Quays House, Egerton Wharf,
Birkenhead, CH41 1LD, U.K.
E-mail:cgmastro.livjm.ac.uk

Violent activity in the nuclei of galaxies has long been considered a curiosity in its own right; manifestations of this phenomenon include distant quasars in the early Universe and nearby Seyfert galaxies, both thought to be powered by the release of gravitational potential energy from material being accreted by a central supermassive black hole (SMBH). The study of galaxy formation, structure and evolution has largely excluded active galactic nuclei (AGN), but recently this situation has changed, with the realisation that the growth of SMBHs, the origin and development of galaxies and nuclear activity at different epochs in the Universe may be intimately related. The era of greatest quasar activity seems to coincide with turbulent dynamics at the epoch of galaxy formation in the young, gas-rich Universe; ubiquitous black holes are then a legacy of this violent era. Closer to home, a fraction of ordinary galaxies have re-ignited their central engines. These galaxies are more established than their distant cousins, so their activity is more puzzling. I review the evidence for causal links between SMBHs, nuclear activity and the formation and evolution of galaxies, and describe opportunities for testing these relationships using future astronomical facilities.

1. Introduction

Over the last 50 years, astronomers have been intrigued by enormously energetic objects called Active Galactic Nuclei (AGN), a violent phenomenon occurring in the nuclei, or central regions, of some galaxies with intensities and durations which cannot easily be explained by stars, thus providing some of the first circumstantial evidence for theoretically-predicted supermassive black holes. Despite their intriguing properties they were largely

---

The material in this chapter is taken from *Phil. Trans. R. Soc. Lond. A*, Copyright 2002.

viewed as interesting but unimportant freaks in the broader study of galaxy formation and evolution, leading astronomers studying the properties of galaxies to exclude the small fraction of galaxies with active centres as irritating aberrations. Here I describe the discovery of AGN and the variety of classifications that followed; I describe some features of unifying models of the central engine that attempt to explain the varied properties of different AGN classes that give rise to the classification. The search for supermassive black holes in AGN and non-active galaxies is discussed along with the developing realisation that all galaxies with significant bulge components might harbour dormant supermassive black holes as remnants of a past adolescent period of quasar activity and therefore possess the potential to be re-triggered into activity under the right conditions, making nuclear activity an integral part of galaxy formation and evolution.

2. The Early Studies of Active Galactic Nuclei

The discovery of AGN began with the development of radio astronomy after World War II when hundreds of sources of radio waves on the sky were detected and catalogued (e.g. Third Cambridge Catalogue (3C)[1] and its revision (3CR)[2]), but the nature of these strong radio emitters was unknown. Astronomers at Palomar attempted to optically identify some of the catalogued radio sources; Baum & Minkowski[3] discovered optical emission from a faint galaxy at the position of the radio source 3C295 and, on studying the galaxy's spectrum, or cosmic bar-code, measured its redshift and inferred a distance of 5000 million light years, making it the most distance object known at that time. Distances can be inferred from Hubble's law whereby the more distant an object, the faster it appears to be receding from us, due to the expansion of the Universe. Chemical elements present in these objects emit or absorb radiation at known characteristic frequencies and when observed in a receding object, the observed frequency is reduced or *redshifted* due to the Doppler effect; the same physical process that causes a receding ambulance siren to be lowered in pitch after it passes the observer.

Attempts to find visible galaxies associated with other strong radio sources such as 3C48, 3C196 and 3C286 failed and only a faint blue, star-like object at the position of each radio source was found - thus leading to their name 'quasi-stellar radio sources', or 'quasars' for short. The spectrum of these quasars resembled nothing that had previously been seen for stars in our Galaxy and these blue points remained a mystery until Maarten

Schmidt[4] concentrated on 3C273, for which an accurate radio position was known.[5] The optical spectrum of the blue source associated with the radio emitter seemed unidentifiable until Schmidt realised that the spectrum could be clearly identified with spectral lines emitted from hydrogen, oxygen and magnesium atoms if a redshift corresponding to 16% of the speed of light was applied. The same technique was applied successfully to 3C48[6] and demonstrated that these objects are not members of our own galaxy but lie at vast distances and are super-luminous. Indeed, the radiation emitted from a quasar (L $\gtrsim 10^{13}$ L$_\odot$, where the Sun's luminosity is L$_\odot = 3.8 \times 10^{26}$ Watt) is bright enough to outshine all the stars in its host galaxy. Such energies cannot be produced by stars alone and it was quickly realised that the release of gravitational potential energy from material falling towards, or being accreted by, a *supermassive black hole* at the galaxy centre, ~ 100 times more energy efficient than nuclear fusion, was the only effective way to power such prodigious outputs.[7]

A black hole is a region of space inside which the pull of gravity is so strong that nothing can escape, not even light. Two main kinds of black holes are thought to exist in the Universe. *Stellar-mass* black holes arise from the the collapsed innards of a massive star after its violent death when it blows off its outer layers in a supernova explosion; these black holes have mass slightly greater than the Sun, compressed into a region only a few kilometres across. In contrast, *supermassive black holes*, which lurk at the centres of galaxies, are 10 million to 1000 million times more massive than the Sun and contained in a region about the size of the Solar System. The emission of radiation from a supermassive black hole appears at first to be contradictory; however, the energy generating processes take place outside the black hole's point-of-no-return, or *event horizon*. The mechanism involved is the conversion of gravitational potential energy into heat and light by frictional forces within a disc of accreting material, which forms from infalling matter that still possesses some orbital energy, or angular momentum, and so cannot fall directly into the black hole.

Radiation from AGN is detected across the electromagnetic spectrum and today, nuclear activity in galaxies has been detected over a wide range of luminosities, from the most distant and energetic quasars, to the weaker AGN seen in nearby galaxies, such as Seyferts,[8] and even the nucleus of our own Milky Way.

C. G. Mundell

3. AGN Orientation - Looking at it from All Angles

After the initial discovery of radio-loud AGN, the advent of radio inter-ferometry soon led to detailed images of these strong radio emitters[9,10] which revealed remarkable long thin jets of plasma emanating from a central compact nucleus and feeding extended lobes, often at considerable distances from the AGN, millions of light years in the most extreme cases. The radio emission is synchrotron radiation produced by electrons spiraling around magnetic fields in the ejected plasma; Figure 1 shows a radio image of the classic radio galaxy Cygnus A in which the nucleus, jets and lobes are visible. These dramatic jets and clouds of radio-emitting plasma were interpreted as exhaust material from the powerful central engine.[11,12]

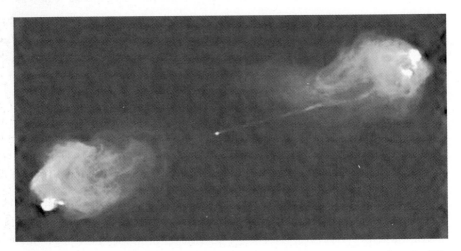

Fig. 1.   6-cm radio image of the classic radio galaxy Cygnus A (courtesy Chris Carilli).

3.1. *Too fast to believe - remarkable jets in radio-loud AGN*

The sharpest radio images, made repeatedly over many years using networks of radio telescopes spanning the globe, resulted in 'movies' of the motion of material in the jets. The blobs of plasma in these jets were apparently being ejected at many times the speed of light, c, appearing to violate fundamental laws of physics. It was quickly realised that such *superluminal* motion, was an optical illusion caused by the plasma moving at *relativistic* speeds, i.e. $\gtrsim 0.7c$, and being ejected towards us at an angle close to our line of sight.[13] Relativistic motion appears to be present for jet matter over hundreds of thousands of light years and the detailed physical driving mechanisms remain an area of active study. The relativistic motion of jet matter has an enormous impact on the appearance of these objects and is possibly the single-most important contributor to the variety of observed morphological types.

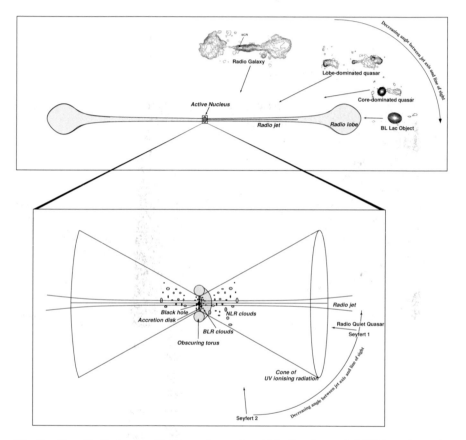

Fig. 2.   Top: Radio-loud unification scheme, in which the observed AGN type depends on the observer's viewing angle to the ejection axis of the radio jet. Bottom: Sketch of an AGN central engine with a central black hole surrounded by (a) an accretion disc that emits cones of ultraviolet ionising radiation and defines the radio jet launch direction, (b) a torus of dust a gas that accounts for the different observed kinds of radio-quiet AGN by blocking our view of the accretion disc and dense, rapidly-moving ionised gas clouds in the broad-line region (BLR) when viewed edge-on (type 2 objects). Less-dense ionised clouds in the narrow-line region (NLR) lie above the plane of the torus and are visible from all angles.

The fast motion of jet material also causes extreme apparent brightening, or Doppler boosting, of the radiation and greatly amplifies any flickering, or variability, in the light levels. Today, the wide range of observed radio structures, brightnesses and levels of variability can be understood in terms of the angle at which we view the high-speed plasma jet. *Radio galaxies* like Cygnus A are orientated perpendicular to our line of sight, lying in the plane of the sky, appear rather symmetrical, and as expected, show no variability or superluminal motion. At the other extreme are bright, compact and highly variable *BL Lac objects*, which are being observed head-on. Figure 2 shows a sketch of this model in which a jet of plasma is ejected from either side of the central engine at relativistic speeds; object classification depends on the angle of the jet to our line of sight. Objects viewed at intermediate angles are seen as either extended, 'lobe-dominated' quasars or relatively compact, 'core-dominated' quasars.[14]

3.2. *Obscuring-Doughnuts in Radio-Quiet AGN*

Radio-quiet quasars and Seyferts are ~10 times more common, but 100 to 1000 times weaker at radio wavelengths and significantly less extended than their *radio-loud* cousins,[15] but orientation still has important effects, this time on the optical properties. Optical spectroscopy provides a powerful diagnostic tool for probing the physical conditions in astronomical objects; as described earlier, chemical elements have a characteristic spectral signature and physical conditions within a gas can be inferred from distortions of this chemical bar-code. In particular, broadening of the spectral lines indicates a spread in gas-cloud velocities; the relative brightnesses of spectral lines indicate the intensity of ultraviolet radiation shining on the gas.

Measurements of the optical spectra of Seyfert nuclei show spectral lines from gas ionised (i.e. gas in which atoms have been stripped of one or more electrons) by strong ultraviolet radiation that is too intense to be produced by a collection of stars and is thought to originate from a hot accretion disc. All Seyfert nuclei contain a region of moderate-density ionised gas, the Narrow Line Region (NLR), extending over several hundred light years where the spectral line-widths correspond to gas velocties of a few hundred $km\,s^{-1}$. Closer in, within ~0.1 light year of the black hole, is the Broad-Line Region (BLR), a much denser region of gas velocities up to 10,000 $km\,s^{-1}$. Seyferts were originally classified into two types; type-1 that show evidence for both a BLR and an NLR, and type-2 that show only an NLR.[16,17]

The mystery of the missing BLRs in type-2 Seyferts was solved elegantly in 1985 when Antonucci & Miller[18] discovered a hidden BLR in the *scattered* light spectrum of the archetypal Seyfert 2 galaxy NGC 1068, which closely resembled that of a Seyfert type 1. This discovery led to the idea that the BLR exists in all Seyferts and is located inside a doughnut, or torus, of molecular gas and dust; our viewing angle with respect to the torus then explains the observed differences between the unobscured, broad-line Seyfert 1s, viewed pole-on, and the obscured, narrow-line Seyfert 2s, viewed edge-on. Hidden Seyfert 1 nuclei can then be seen in reflected light as light photons are scattered into the line of sight by particles above and below the torus acting like a "dentist's mirror".[18-21] The lower panel of Figure 2 shows a sketch of a Seyfert nucleus with the different types of AGN observed as angle between line of sight and torus axis increases. Figure 3 shows an image of the molecular torus in NGC 4151, surrounding the mini, quasar-like radio jet emanating from the centre of the galaxy, as predicted by the unification scheme.

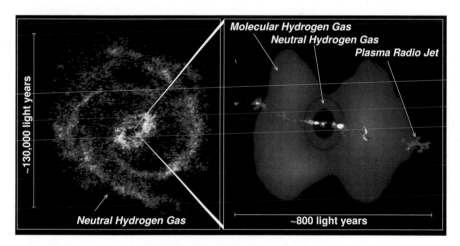

Fig. 3.   Left: Radio image of neutral hydrogen gas in the spiral Seyfert host galaxy NGC 4151[22]; Right: composite image of the central regions of NGC 4151 showing a 1.4-GHz radio image of the well-collimated plasma jet surrounded by an obscuring torus of molecular hydrogen imaged at 2.2μm[23] and an inner ring of neutral hydrogen inferred from absorption measurements.[24]

Radio quiet quasars also have broad and narrow lines and are considered to be the high luminosity equivalents of Seyfert type 1 galaxies. A population of narrow-line quasars, high luminosity equivalents to obscured Seyfert 2s, are predicted by the unification scheme but, until now, have remained elusive. New optical and infrared sky surveys are beginning to reveal a previously undetected population of red AGN[25] with quasar type 2 spectra[26] and weak radio emission.[27] A significant population of highly obscured but intrinsically luminous AGN would alter measures of AGN evolution, the ionisation state of the Universe and might contribute substantially to the diffuse infrared and X-ray backgrounds.

3.3. *Further unification?*

The presence of gas emitting broad and narrow optical lines in radio-loud AGN and the discovery of mini radio jets in Seyferts[28] led to further consistency between the two unification schemes. Nevertheless, the complete unification of radio-loud and radio-quiet objects remains problematic, particularly in explaining the vast range in radio power and jet extents, and might ultimately involve the combination of black hole properties, such as accretion rate, black hole mass and spin, and orientation.[29,30]

4. Searching for Supermassive Black Holes

Although incontrovertible observational proof of the existence of supermassive black holes (SMBHs) has yet not been found, evidence is mounting to suggest the presence of massive dark objects, or large mass concentrations at the centres of galaxies. Black holes, by definition, cannot be 'seen' and instead one must look for the consequences of their presence. The presence of SMBHs has been inferred indirectly from the energetics of accretion required to power luminous AGN and explain rapid flux variability and, more directly, from kinematic studies of the influence of the black hole's gravitational pull on stars and gas orbiting close to it in the central regions of both active and non-active galaxies. Theoretical models rule out alternatives to a supermassive black holes such as collections of brown or white dwarf stars, neutron stars or stellar-mass black holes which would merge and shine or evaporate too quickly.[32–35]

4.1. *Quasar lifetimes and the black hole legacy*

Soon after the discovery of quasars it became clear that they were most common when the Universe was relatively young with the peak of the quasar epoch at redshift z∼2.5 or a *look-back time* of 65% of the age of the Universe (See Figure 4); today bright quasars are rare and weaker Seyferts dominate instead. The number of dead quasars or relic, dormant black holes left today can estimated by applying some simple arguments to the quasar observations. Soltan[36] integrated the observed light emitted by quasars, and, assuming the power source for quasar light is accretion of material by a supermassive black hole with a mass-to-energy conversion efficiency of 10% and that the black hole grows during the active phase, predicted the total mass in relic black holes today. Knowing the number of galaxies per unit volume of space,[37] if one assumes that all galaxies went through a quasar phase at some time in their lives, then each galaxy should, on average, contain a $\sim 10^8$ M$_\odot$ black hole as a legacy of this violent, but short-lived period ($\sim 10^7$ to $\sim 10^8$ years). Alternatively, if only a small fraction of galaxies went through a quasar phase, the active phase would have lasted lasted longer ($>10^9$ years) and the remnant SMBHs would be relatively rare, but unacceptably massive ($>10^9$ M$_\odot$).[38-40]

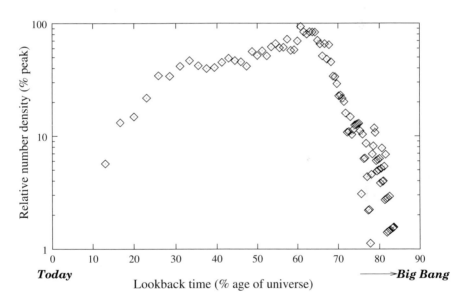

Fig. 4.   Relative number of galaxies per unit volume of space (as a fraction of the peak value) detected in the SLOAN Digital Sky Survey, as a function of *look-back time* i.e. time running backwards from now to the Big Bang [using data from Schneider[31]].

More complex models including quasar evolution[41,42] and the effects of galaxy growth[43] favour short-lived periods of activity in many generations of quasars, or a mixture of continuous and recurrent activity.[44-46] The complex physics of accretion and black hole growth, however, remain an area of active study.[47,48] Nevertheless, the range of black hole mass of interest is thought to be $M_\bullet \sim 10^6$ to $10^{9.5} M_\odot$, with the lower mass holes being ubiquitous.[49]

4.2. Irresistible black holes - motion of gas and stars

Although the prodigious energy outputs from powerful quasars offer strong circumstantial evidence that supermassive black holes exist, most notably in driving the ejection and acceleration of long, powerful jets of plasma close to the speed of light,[50] it has not, until recently, been possible to make more direct kinematic measurements of the black hole's gravitational influence. The mass of a central object, the circular velocity of an orbiting star and the radius of the orbit are related by Newton's Laws of motion and gravity. Precise measurements of the velocities of stars and gas close to the centre of a galaxy are then used to determine the mass of the central object.

The strongest dynamical evidence for black holes comes from studies of centre of our own Galaxy and a nearby Seyfert, NGC 4258; a decade of painstaking observations of a cluster of stars orbiting around the mildly active centre of the Milky Way, within a radius of 0.07 light years of the central radio source Sgr A*, suggest a central mass of $M_\bullet = (2.6 \pm 0.2) \times 10^6$ $M_\odot$.[34,35,51,52] Discovery of strong radio spectral lines, or megamasers, emitted from water molecules in a rapidly rotating nuclear gas disc at the centre of NGC 4258 implies a centre mass $M_\bullet = (4 \pm 0.1) \times 10^7$ $M_\odot$ concentrated in a region smaller than 0.7 light years,[53] again small enough to rule out anything other than a black hole.[32,33] Precision measurements of black hole masses in other galaxies using a variety of techniques, although challenging and still model dependent, have become increasingly common[54,55,56] and now more than 60 active and non-active galaxies have black hole estimates.

5. Black Hole Demographics - the Host Galaxy Connection

In general, galaxies consist of two main visible components - a central ellipsoidal bulge and a flat disc structure commonly containing spiral arms - together making a structure resembling two fried eggs back-to-back. Elliptical galaxies have no discs and are dominated by their bulges, maintaining their shapes by the random motions of their stars; spiral galaxies, like our own Galaxy and nearby Andromeda have prominent discs and are supported mainly by rotation, with rotation speeds between 200 km s^{-1} and 300 km s^{-1}. Some spiral galaxies contain a bar-like structure that crosses

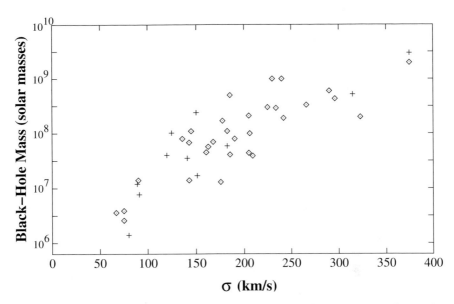

Fig. 5.   Black hole mass against host galaxy bulge stellar velocity dispersion for nearby AGN (open diamonds) and non-active galaxies (+ symbols) [using data from Kormendy & Gebhardt[49] and Ferrarese *et al.*[57]].

the nucleus; the spiral arms then begin at the ends of the bar and wind outwards. If the bar is narrow and straight it is classed as a 'strong' bar and if oval-shaped (essentially an elongated bulge) it is 'weak'. Dynamical simulations have revealed that in the region of the bar, stars do not travel on circular orbits as they do in the disc, but instead follow more elongated elliptical, or 'non-circular' paths.

With the great progress made recently in measuring the mass of central supermassive black holes in a significant number of active and non-active galaxies, correlations with their host galaxy properties are now possible. Maggorian *et al.*[54] confirmed the correlation between the brightness of a galaxy bulge (and hence stellar mass) and the mass of its central black hole[58] establishing a best fit to the linear relation of $M_{\bullet}=0.006M_{bulge}$, despite a large scatter. A much tighter correlation was subsequently discovered between the velocity dispersion (σ) of stars in the host galaxy bulge and the central black hole mass.[56,59] The velocity dispersion is a measure of the range of random speeds present in star motions and is potentially a more reliable galaxy mass indicator than total starlight; the greater the spread in speeds, the more massive the galaxy bulge. The tightness of the correlation points to a connection between the formation mechanism of the galaxy bulge and central black hole although the physics involved are not yet known. The $M_\bullet - \sigma$ relation for a mixture of nearby active and non-active galaxies (Figure 5), measured using a variety of techniques, shows the relationship between bulge and black hole is very similar for both, although investigations continue to establish the precise form of the correlation and whether it is universal for active and non-active galaxies. If universal, this relationship would provide exciting confirmation that non-active galaxies contain dormant versions of the same kind of black holes that power AGN.

No correlation exists between galaxy disc properties and black hole mass, and disc galaxies without bulges do not appear to contain supermassive black holes,[60] suggesting discs form later and are not involved in the process that intimately links the black hole and bulge.

6. AGN and Their Environment

6.1. *The violent early Universe*

The relationships between black holes and their host galaxies are increasingly compelling but unanswered questions remain concerning the relationship between star formation, galaxy formation, quasar activity and black hole creation in the early Universe. Observations of faint galaxies in the Hubble Deep Field suggest a peak in star formation that matches that of the quasar epoch[61] implying a close link between star formation and quasar activity. More recent measurements, however, suggest that the star formation activity may be constant for redshifts greater than 1 with the onset of substantial star formation occurring at even earlier epochs, at redshifts beyond 4.5.[62] An increasing number of new quasars are also being found at redshifts greater than 4[31,63] providing constraints for cosmological models of galaxy formation and continuing the debate on the relationship between quasar activity, star formation and the creation of the first black holes.[64]

The life cycle of an AGN involves a mechanism to trigger the infall of gas to create an accretion disc and continued fuelling, or replenishment, of this brightly-shining accretion disc. A number of models have suggested that at intermediate to high redshifts it may be moderately easy to trigger and fuel AGN, where galaxies might be more gas rich, star formation is vigorous and collisions between galaxies are common.[43] Kauffman & Haehnelt[65] suggest a model in which galaxy and quasar evolution at early times was driven by mergers of gas-rich disc galaxies, which drove the formation and fuelling of black holes and created today's elliptical galaxies, thereby tying together host galaxy and black hole properties. As the Universe ages, a decreasing galaxy merger rate and available gas supply and increasing accretion timescales produce the decline in bright quasars.

An alternative hypothesis, linking black hole and bulge growth with quasar activity, involves strong bars in early galaxies;[66] early disc galaxies developed strong bars which were highly efficient at removing angular momentum from disc gas and funnelling it towards the centre to feed and grow a black hole. This represents the bright quasar phase in which the black hole grows rapidly, but on reaching only a few percent of the mass of the host disc, the central mass concentration soon destroys the bar due to an increasing number of stars that follow random and chaotic paths, thereby choking off the fuel supply and quenching the quasar. In addition, the increase in random motion in the disc leads to the creation of a bulge. A disc might be re-built some time later if the galaxy receives a new supply of cold gas, perhaps from a 'minor merger' whereby a small gaseous galaxy or gas-cloud falls into the main disc and is consumed by the disc without causing significant disruption, and without significantly affecting the black hole mass. This scenario nicely accounts for the relationship between black hole masses and bulge properties and lack of correlation with disc properties.

An important unknown parameter in these models is the amount of cold gas in progenitor disc galaxies and how it evolves with time; it is expected that the Universe was more gas-rich in the past,[67] but observations of neutral hydrogen (HI) and molecular gas such as carbon monoxide (CO) with new generation facilities, such as Atacama Large Millimeter Array (ALMA), the Giant Metrewave Radio Telescope (GMRT), the Extended Very Large Array (EVLA) and the proposed Square Kilometre Array (SKA), will offer exciting opportunities to measure the gaseous properties of distant galaxies directly to further our understanding of galaxy formation and evolution and its relationship to quasar and star-formation activity.

6.2. *Re-activating dormant black holes in nearby galaxies*

While the most luminous AGN might coincide with violent dynamics in the gas-rich universe at the epoch of galaxy formation,[43] nuclear activity in nearby galaxies is more problematic since major galaxy mergers, the collision of two equal-mass disc galaxies, are less common and galaxy discs are well established; reactivation of ubiquitous 'old' black holes is therefore likely to dominate. Host-galaxy gas represents a reservoir of potential fuel and, given the ubiquity of supermassive black holes, the degree of nuclear activity exhibited by a galaxy must be related to the nature of the fuelling rather than the presence of a black hole.[68,69] Gravitational, or tidal forces exterted when two galaxies pass close to one another may play a role in this process, either directly, when gas from the companion, or outer regions of the host galaxy, is tidally removed and deposited onto the nucleus, or by causing disturbances to stars orbiting in the disc and leading to the growth of structures such as bars, in which stars travel on elliptical paths and drive inflows of galactic gas.[70-75]

Numerous optical and IR surveys of Seyfert hosts have been conducted but as yet show no conclusive links between nuclear activity and host galaxy environment. Neutral hydrogen (HI) is an important tracer of galactic structure and dynamics and may be a better probe of environment than the stellar component. HI is often the most spatially extended component of a galaxy's disc so is easily disrupted by passing companions, making it a sensitive tracer of tidal disruption.[73] In addition, because gas can dissipate energy and momentum through shock waves,[75] whereas collisions between stars are rare, the observable consequences of perturbating the HI in galactic bars are easily detectable. However, despite the diagnostic power of HI, until recently few detailed studies of HI in Seyferts have been performed.[76,77]

The strength of a galaxy collision, which depends on initial galaxy properties such as mass, separation and direction of closest approach, ranges from the most violent mergers between equal mass, gas-rich disc galaxies, to the weakest interaction in which a low mass companion on a fly-by path interacts with a massive primary. In this minor-merger case the primary disc is perturbed but not significantly disrupted or destroyed. Indeed, Seyfert nuclei are rare in strongly interacting systems, late-type spirals and elliptical galaxies[78,79] and sometimes show surprisingly undisturbed galactic discs despite the presence of HI tidal features.[73] Seyfert activity may therefore involve weaker interactions or minor mergers between a primary galaxy and a smaller companion or satellite galaxy, rather than violent major mergers.[80] A key question is whether the gaseous properties of normal galaxies differ from those with Seyfert nuclei and a deep, systematic HI imaging survey of a sample of Seyfert and normal galaxies is now required.

7. Unanswered Questions and Prospects for the Future

Studies of galaxies and AGN are being revolutionised by impressive new sky surveys, such as SLOAN and 2DF, which have already significantly increased the number of known galaxies and quasars in the Universe. In the next decade and beyond, prospects for understanding AGN and their role in galaxy formation and evolution are extremely promising given the number of planned new instruments spanning the electromagnetic spectrum.

• We do not know whether galaxies grow black holes or are seeded by them; the James Webb Space Telescope will find the smallest black holes at the earliest times and allow us to relate them to the first galaxies and stars.

• The amount of cold gas in galaxies through cosmic history is a key ingredient in star-formation, quasar activity and galaxy evolution models but is still unknown. The study of gas at high redshifts with ALMA, the GMRT and the EVLA will revolutionise our understanding its role in these important phenomena and provide powerful constraints for cosmological models.

• Current models of AGN physics - fuelling, accretion discs and the acceleration of powerful radio jets - remain speculative; detailed studies of X-ray emitting gas, e.g with the highly ambitious X-ray space interferometer MAXIM, might offer valuable new insight into the energetics and physical structure of this extreme region.

• Finally, the detection and detailed study of gravitational waves, using the space-based detector LISA, from massive black holes living in black-hole binary systems or in the very act of merging will prove the existence of SMBHs and perhaps provide insight into the origin of the difference between radio-loud and radio-quiet AGN.

Fig. 6. The author preparing instrumentation on the 4.2-m William Herschel Telescope on the island of La Palma.

References

1. D.O. Edge, J.R. Shakeshaft, W.B. McAdam, J.E. Baldwin and S. Archer, *Mem. R. Astr. Soc.* **68**, 37 (1959).
2. A.S. Bennett, *Mem. R. Astr. Soc.* , **68**, 163 (1961).
3. W.A. Baum and R. Minkowski, *Astron. J.* **65**, 483 (1960).
4. M. Schmidt, *Nature* **197**, 1040 (1963).
5. C. Hazard, M.B. Mackey and A.J. Shimmins, *Nature* **197**, 1037 (1963).
6. J.L Greenstein and T.A. Matthews, *Astron. J.* **68**, 279 (1963).
7. D. Lynden-Bell, *Nature* **223** 690 (1969).
8. C.K. Seyfert, *Astrophys. J.* **97**, 28 (1943).
9. A.H. Bridle and R.A. Perley, *Annu. Rev. Astron. Astrophys.* **22**, 319 (1984).
10. A.H. Bridle, D.H. Hough, C.J. Lonsdale, J.O. Burns and R.A. Laing, *Astron. J.* **108**, 766 (1994).
11. P.A. Scheuer, *Mon. Not. R. Astr. Soc.* **166**, 51 (1974).
12. R.D. Blandford and M.J. Rees, *Mon. Not. R. Astr. Soc.* **169**, 395 (1974).
13. R.D. Blandford and M.J. Rees, In *Pittsburg Conference on BL Lac Objects*, (ed. A.M. Wolfe), p. 328. University of Pittsburg (1978).
14. M.C. Urry and P. Padovani, *Publ. Astron. Soc. Pac.* **107**, 803 (1995).
15. P. Goldschmidt, M.J. Kukula, L. Miller and J.S. Dunlop, *Astrophys. J.* **511**, 612 (1999).
16. E.Y. Khachikian and D.W. Weedman, *Astrofizica* **7**, 389 (1971).
17. E.Y. Khachikian and D.W. Weedman, *Astrophys. J.* **192**, 581 (1974).
18. R.R.J. Antonucci and J.S. Miller, *Astrophys. J.* **297**, 621 (1985).
19. H.D. Tran, *Astrophys. J.* **440**, 565 (1995).
20. R.R.J. Antonucci, *Annu. Rev. Astron. Astrophys.* **31**, 473 (1993).
21. B.J. Wills, In *Quasars and Cosmology*, (ed. G. Ferland & J. Baldwin) ASP Conf. Ser. **162**, p.101. San Francisco: ASP.(1999).
22. C.G. Mundell, A. Pedlar, D.L. Shone and A. Robinson, *Mon. Not. R. Astr. Soc.* **304**, 481 (1999).
23. B.R. Fernandez, A.J. Holloway, J. Meaburn, A. Pedlar and C.G. Mundell, *Mon. Not. R. Astr. Soc.* **305**, 319 (1999).
24. C.G. Mundell, J.M. Wrobel, A. Pedlar and J.F. Gallimore, *Astrophys. J.* **583**, 192 (2003).
25. R.M. Cutri, B.O. Nelson, J.D. Kirkpatrick, J.P. Huchra and P.S. Smith, *Bull. Am. Astron. Soc.* **198**, 3317 (2001).
26. S.G. Djorgovski, R. Brunner, F. Harrison, R.R. Gal, S. Odewahn, R.R. de Carvalho, A. Mahabal,and S. Castro, *Bull. Am. Astron. Soc.* **195**, 6502 (1999).
27. J.S. Ulvestad, S.G. Djorgovski, R.J. Brunner and A. Mahabal, *Bull. Am. Astron. Soc.* **197**, 10804 (2000).
28. A.S. Wilson and J.S. Ulvestad, *Astrophys. J.* **263**, 576 (1982).
29. A.S. Wilson and E.J.M. Colbert, *Astrophys. J.* **438**, 62 (1995).
30. T.A. Boroson, *Astrophys. J.* **565** 78 (2002).
31. Schneider, D.P. *et al.*, *Astron. J.* **123**, 567 (2002).
32. E. Maoz, *Astrophys. J.* **447**, L91 (1995).

33. E. Maoz, *Astrophys. J.* **494**, L181 (1998).
34. R. Genzel, A. Eckart, T. Ott and F. Eisenhauer, *Mon. Not. R. Astr. Soc.* **291**, 219 (1997).
35. R. Genzel *et al.*, *Mon. Not. R. Astr. Soc.* **317**, 348 (2000).
36. A. Soltan, *Mon. Not. R. Astr. Soc.* **200**, 115 (1982).
37. J. Loveday, B.A. Peterson, G. Efstathiou and S.J. Maddox, *Astrophys. J.* **390**, 338 (1992).
38. A. Cavaliere, E. Giallongo, A. Messina and F. Vagnetti, *Astrophys. J.* **269**, 57 (1983).
39. A. Cavaliere and A.S. Szalay, *Astrophys. J.* **311**, 589 (1986).
40. A. Cavaliere and P. Padovani, *Astrophys. J.* **333**, L33 (1988).
41. S. Tremaine, In *Unsolved Problems in Astrophysics*, p.137. Princeton University Press (1996).
42. S.M. Faber *et al.*, *Astron. J.* **114**, 1771 (1997).
43. M.G. Haehnelt and M.J. Rees, *Mon. Not. R. Astr. Soc.* **263**, 168 (1993).
44. T.A. Small and R.D. Blandford, *Mon. Not. R. Astr. Soc.* **259**, 725 (1992).
45. R. Cen, *Astrophys. J.* **533**, L1 (2000).
46. Y-Y Choi, J. Yang and I. Yi, *Astrophys. J.* **555**, 673 (2001).
47. R.D. Blandford and M.C. Begelman, *Mon. Not. R. Astr. Soc.* **303**, L1 (1999).
48. A.C. Fabian, *Mon. Not. R. Astr. Soc.* **308**, L39 (1999).
49. J. Kormendy and K. Gebhardt, In *The 20th Texas Symposium on Relativistic Astrophysics*, (ed. H. Martel & J.C. Wheeler), AIP. (2001)
50. M.J. Rees, E.S. Phinney, M.C. Begelman and R.D. Blandford, *Nature* **295**, 17 (1982).
51. A. Eckart, and R.Genzel, *Mon. Not. R. Astr. Soc.* **284**, 576 (1997).
52. A. Ghez *et al.*, *Nature* **407**, 349 (2000).
53. M. Miyoshi, J. Moran, J. Herrnstein, L. Greenhill, N. Nakai, P. Diamond and M. Inoue, *Nature* **373**, 127 (1995).
54. J. Magorrian *et al.*, *Astron. J.* **115**, 2285 (1998).
55. G.A. Bower *et al.*, *Astrophys. J.* **492**, L111 (1998).
56. K. Gebhardt *et al.*, *Astrophys. J.* **539**, L13 (2000).
57. L. Ferrarese, R.W. Pogge, B.M. Peterson, D. Merritt, A. Wandel and C.L. Joseph, *Astrophys. J.* **555**, L79 (2001).
58. J. Kormendy and D. Richstone, *Annu. Rev. Astron. Astrophys.* **33**, 581 (1995).
59. L. Ferrarese and D. Merrit, *Astrophys. J.* **539**, L9 (2000).
60. K. Gebhardt *et al.*, *Astron. J.* **122**, 2469 (2001).
61. P. Madau, H. Ferguson, M. Dickinson, M. Giavalisco, C. Steidel and A. Fruchter, *Mon. Not. R. Astr. Soc.* **283**, 1388 (1996).
62. C.C. Steidel, K.L. Adelberger, M. Giavalisco, M. Dickinson and M. Pettini, *Astrophys. J.* **519**, 1 (1999).
63. X. Fan *et al.*, *Astron. J.* **122**, 2833 (2001).
64. Z. Haiman and A. Loeb, *Astrophys. J.* **552**, 459 (2001).
65. G. Kauffman and M.G. Haehnelt, *Mon. Not. R. Astr. Soc.* **311**, 576 (2000).
66. J.A. Sellwood, In *Galaxy Dynamics*, (ed. D.R. Merritt, M. Valluri & J.A. Sellwood) ASP Conf. Ser. **182**, p.96. San Francisco: ASP. (1999).

67. A.J. Barger, L.L. Cowie, R.F. Mushotzky and E.A. Richards, *Astron. J.* **121**, 662 (2001).
68. I. Shlosman and M. Noguchi, *Astrophys. J.* **414**, 474 (1993).
69. J.A. Sellwood and E.M. Moore, *Astrophys. J.* **510**, 125 (1999).
70. A. Toomre and J. Toomre, *Astrophys. J.* **178**, 623 (1972).
71. S.M. Simkin, H.J. Su and M.P. Schwarz, *Astrophys. J.* **237**, 404 (1980).
72. I. Shlosman, J. Frank and M.C. Begelman, *Nature* **338**, 45 (1989).
73. C.G. Mundell, A. Pedlar, D.J. Axon, J. Meaburn and S.W. Unger, *Mon. Not. R. Astr. Soc.* **227**, 641 (1995).
74. E. Athanassoula, *Mon. Not. R. Astr. Soc.* **259**, 328 (1992).
75. C.G. Mundell and D.L. Shone, *Mon. Not. R. Astr. Soc.* **304**, 475 (1999).
76. E. Brinks and C.G. Mundell, In '*Extragalactic Neutral Hydrogen* (ed. E.D. Skillman) ASP Conf. Ser. **106**, p.268. San Francisco: ASP. (1997).
77. C.G. Mundell, In *Gas and Galaxy Evolution*, (ed. J.E. Hibbard, M. Rupen, J.H. van Gorkom) ASP Conf. Ser. **240**, p.411. San Francisco: ASP. (1999).
78. W.C. Keel, R.C. Kennicutt, E. Hummel and J.M. van der Hulst, *Astron. J.* **90**, 708 (1985).
79. H.A. Bushouse, *Astron. J.* **91**, 255 (1986).
80. M.M. De Robertis, H.K.C. Yee and K. Hayhoe, *Astrophys. J.* **496**, 93 (1998).

Carole G. Mundell

Born in Sheffield, Yorks, Carole Mundell studied at Glasgow University where she graduated with a double first class honours in Physics and Astronomy in 1992. She obtained her PhD from the University of Manchester in 1995, then held a PPARC Research Fellowship at Jodrell Bank until 1997. She spent two years as a Postdoctoral Research Associate at the University of Maryland in College Park before returning to the UK, in 1999, to take up a Royal Society University Research Fellowship at the Astrophysics Research Institute of Liverpool John Moores University. Currently a Royal Society University Research Fellow, Carole's scientific interests include galaxy dynamics, active galaxies and gamma-ray bursts. She enjoys Ballroom and Latin American dancing, playing the piano, hiking and spending time with her young son.

Probing the Universe

CHAPTER 6

FIRST YEAR WILKINSON MICROWAVE ANISOTROPY PROBE RESULTS: IMPLICATIONS FOR COSMOLOGY AND INFLATION

Hiranya Peiris

Department of Astrophysical Sciences
Princeton University
Peyton Hall, Ivy Lane, Princeton, NJ 08544, U.S.A.

The *Wilkinson Microwave Anisotropy Probe* (*WMAP*) is currently mapping the temperature and polarization anisotropies of the cosmic microwave background (CMB) radiation on the full sky at five microwave frequencies. We summarize the major scientific results from the analysis of the first-year *WMAP* data, released on February 11, 2003: (1) the *WMAP* data alone fits a standard cosmological model, giving precise determinations of its six parameters; (2) the universe was reionized very early on in its history; (3) a joint analysis of *WMAP* data with smaller scale ground-based CMB experiments and large-scale structure data yields improved constraints on the geometry of the universe, the dark energy equation of state, the energy density in stable neutrino species, and the Inflationary paradigm.

1. Introduction

The *Wilkinson Microwave Anisotropy Probe* (*WMAP*) experiment[1] has completed one year of observing the Cosmic Microwave Background (CMB), helping us to reclaim a little more land in the country of the Unknown. The CMB is a primary tool for determining the global characteristics, constituents, history and eventual fate of the universe. *WMAP* was launched on June 30, 2001 from Cape Canaveral. It observes the CMB sky from an orbit about the second Lagrange point of the Earth–Sun system, L_2, a highly thermally-stable environment where it is unaffected by the Earth's emission and magnetic field.

Figure 1 shows the key components of the *WMAP* spacecraft. The central design philosophy of the *WMAP* mission was to minimize sources of systematic measurement errors. To achieve this goal, *WMAP* utilizes a dif-

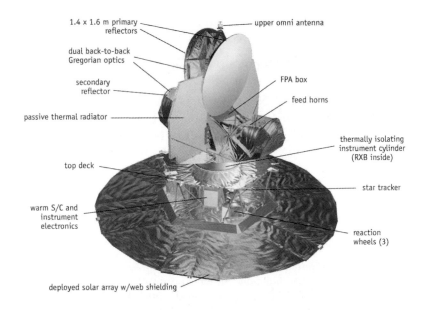

Fig. 1.   A schematic diagram of the *WMAP* spacecraft (Illustration by NASA/WMAP Science Team).

ferential design; it observes the temperature differences between two directions in the sky, using a back-to-back set of nearly identical optics.[2,3] These optics focus the radiation into horns[4] that feed differential microwave radiometers.[5,6] Calibration errors are $< 0.5\%$, and the low level of systematic error is fully characterized in Refs. 3 and 6–8. The differential measurements, combined with a scan strategy whereby each pixel is scanned many times with different orientations of the *WMAP* beam, significantly suppress undesirable frequency-dependent noise and noise correlations between pixels. An accurate characterization of the *WMAP* in-flight beam pattern is essential to the goal of precision measurements of the CMB. This is accomplished by using in-flight observations of Jupiter to quantify the shapes of the main beam and side-lobes.

A CMB map is the most compact representation of CMB anisotropy without loss of information. Full-sky maps in five frequency bands from 23–94 GHz are produced from the radiometer data of temperature differences measured over the full sky. The angular resolution of the *WMAP* satellite is 30 times greater than the previous full sky map as produced by the *COBE* satellite[9] in 1992 (Fig. 2). The multi-frequency data enables the separation of the CMB signal from the foreground dust, synchrotron

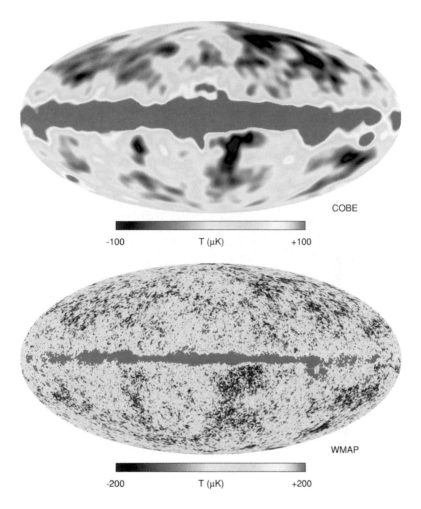

Fig. 2.   Top[11]: The all-sky CMB temperature anisotropy map produced by the *COBE* satellite c. 1996. Bottom[11]: The all-sky temperature anisotropy seen at 61 GHz by the *WMAP* satellite c. 2003. The large continuous band in the middle consists of foreground microwave emission from our own Galaxy. The light and dark spots represent regions which are hotter and colder than average, respectively. The primordial anisotropy is at the level of 1 part in 100,000.

and free-free emission from our Galaxy,[10] since the CMB signal does not vary with frequency but the foreground emission does. Furthermore, pixels containing known point sources are excised from the maps before they are used for cosmological analysis.

The maps, their properties, and a synopsis of the basic results of *WMAP*'s first year of operation are presented in Ref. 11. The CMB fluctuations are consistent with obeying Gaussian statistics,[12] and therefore the information in the CMB maps can be represented by an angular power spectrum. *WMAP* has measured the temperature (*TT*) power spectrum to $2 \leq \ell \leq 900$, cosmic-variance-limited for $\ell < 354$[13] (the cosmic variance error is a fundamental limit on measuring the CMB, arising from the fact that one can only measure $(2\ell + 1)$ modes on the sky corresponding to each ℓ-mode.). The temperature-polarization (*TE*) cross-power spectrum has also been measured to $\ell \leq 450$.[14] The methodology used to analyze this wealth of new information is presented in Refs. 15 and 16–18 discuss some of the cosmological implications of the *WMAP* results, which will be summarized in this contribution.

2. The Importance of CMB Measurements to Cosmology

The CMB is the thermal afterglow of the Big Bang. If we rewind the expansion of the universe, the further we look back into the past, the hotter and denser the universe becomes. Eventually we come to a time when distances in the universe were only 0.1% their present size, when the temperature was high enough to ionize the universe. Back then, the universe was filled with a plasma of protons, electrons and photons (plus a few He nuclei and trace amounts of "metals", as astrophysicists call anything higher up in the periodic table). CMB photons are emitted at the era of *recombination*, the so-called "last scattering surface" (LSS) when the primordial plasma coalesced into neutral atoms. Before this time, the number density of free electrons was so high that the mean free path of photons Thompson scattering off electrons was extremely short. As the universe expands, it cools, and the wavelengths of the photons in the plasma get stretched out, lowering their energy. Eventually the photon energies became low enough that, at a temperature of $\sim 3000°$ K, the photon energies become too low to keep the universe ionized. At this point, the protons and electrons in the plasma (re)combined to form neutral hydrogen, and the mean free path of the photons increased to essentially the size of the observable universe. They have since traveled almost unhindered through the universe.

When we observe this radiation, we see the universe when it was only 380,000 years old. Now, 13.7 billion years later, it has cooled to microwave frequencies. The CMB is nearly uniform, with a mean temperature of $T_0 = 2.725$ K and a nearly-perfect black-body spectrum, as measured by the

COBE satellite.[19,20] The discovery of the CMB[21] together with the black-body nature of its spectrum was of fundamental importance to cosmology because it proved the idea of a *hot Big Bang* — the universe was hot and dense in the past.[22] Equally important is the the fact that the CMB has slight variations of 1 part in 100,000 in its temperature.[23] This *temperature anisotropy* reflects the primordial inhomogeneities in the underlying density field that provided the seeds for cosmological structure formation through gravitational instability.

2.1. *Temperature anisotropy*

The physics governing the evolution of the temperature anisotropy is very well understood. The basic ideas were worked out independently on both sides of the Iron Curtain in 1970.[24,25] The power spectrum of the CMB temperature anisotropies (see Sec. 3) contains three distinct regimes — (1) at angular separations larger than $\sim 1°$, there is a relatively flat plateau; (2) at intermediate angular scales, between $\sim 1°$ and a few arcminutes, there is a series of peaks and troughs which are (3) exponentially damped at subarcminute scales. These features can be understood in terms of a simple physical analogy.

Just before recombination, the universe contained a tightly-coupled baryon-photon fluid as described above, along with dark matter which was not coupled to the baryon-photon fluid since it does not participate in electromagnetic interactions. There were tiny perturbations in the density, and hence gravitational potentials, over a wide range of length-scales. The perturbations in the dark matter grow continuously as the universe expands, but the gravitationally-driven collapse of perturbations in the photon-baryon fluid is resisted by the pressure of the photons in the fluid, which acts as a restoring force. This means that an overdensity in the photon-baryon fluid falls into a potential well, is compressed till the collapse is halted by the radiation pressure of the photons, and then rebounds till the expansion is halted by the weight of the fluid and the gravity of the dark matter potential well, causing the mode to recollapse once more. In short, the tug of war between gravity and pressure sets up a sound wave in the photon-baryon fluid, and the physics is essentially described by a forced harmonic oscillator,[26] with the overall amplitude of the wave-form being fixed by the initial amplitude of the perturbations. On sub-degree scales, smaller than the sound horizon at the LSS, what we observe are modes which have had time to undergo this oscillation. Since last scattering is nearly instantaneous, modes with different wavelengths are caught at different phases in their oscillation. Modes that were at maximum compression or rarefaction correspond to *peaks* of the CMB power spectrum (which is the *square* of the amplitude) and the *troughs* correspond to velocity maxima (where the densities correspond to neutral compression, i.e. the velocity is out of phase by $\pi/2$ with respect to the density). In terms of a map of the CMB temperature anisotropy, a compression (overdensity) makes a cold spot, since a photon loses energy and redshifts in climbing out of the extra deep potential well; and a rarefaction (underdensity) makes a hot spot compared to the average CMB temperature, since the photon does not lose as much energy in climbing out of a shallower potential well.

This perfect fluid approximation becomes invalid when photon diffusion starts to become important. On scales smaller than the photon mean free path, photons free-stream out of overdensities, erasing the perturbation. The equation becomes a forced harmonic oscillator with a friction term representing viscous damping, and the temperature fluctuations are exponentially damped on sub-arcminute scales.[27]

In addition to these hydrodynamical effects, temperature anisotropy arises at the very largest scales, from modes with wavelengths larger than the sound-crossing distance at last scattering. These modes are essentially unaffected by causal physics, frozen in their initial configurations. Purely general-relativistic effects create temperature anisotropy at these large scales; for example, the so-called Integrated Sachs–Wolfe (ISW) effect[28] changes the temperature of the CMB as CMB photons traverse time-varying gravitational potentials. This happens when gravitational potentials decay at early times when radiation dominates and at late times when dark energy dominates.

We do not directly observe the sound waves at the LSS; what we measure is their projection on the sky. Since the sound crossing scale at the LSS is a well defined quantity, a "standard rod", the angle it subtends is related to the geometry of the space between the observer (*WMAP* satellite at L2, today) and the last scattering surface (at redshift $z \sim 1100$); the sound crossing scale would subtend an angle of $\theta_s \simeq 230$ Mpc/d_A, where d_A is the (comoving angular diameter) distance to the LSS. d_A is dependent on the geometry of space-time. The photons from the LSS follows a geodesic line. Thus, if the spatial geometry of the universe were flat (Euclidean), the photons would follow a simple straight-line path from the LSS to WMAP. If it were positively curved, such as the surface of a sphere, the photons follow a curved trajectory (such as the lines of longitude on the surface of the Earth) and d_A is smaller than in a flat universe. If it were negatively curved (saddle-shaped), d_A is larger. This is the leading order effect.

The first order effect is sensitive to the equation of state of the universe. Since, according to general relativity, it is the matter-energy content that shapes the geometry of space-time, a precise measurement θ_s tells us about the total matter-energy density content, because the expansion rate of the universe, and hence d_A, depends on the latter. Let's consider this dependence in the simplest case of a flat universe. The expansion rate H, also known as the Hubble parameter, is given by

$$H^2 = \left(\frac{\dot{a}}{a}\right)^2 = \frac{8\pi G}{3}\rho, \tag{1}$$

where ρ is the total energy density of the universe. a is the *scale factor* by which the universe has expanded since the Big Bang — the physical distance is proportional to the comoving distance times the scale factor. As the universe expands, matter-energy within it evolves as $\rho \propto a^{-3(1+w)}$, where w is the *equation of state* given by $w = P/\rho$, the ratio of pressure to density. The universe starts out dominated by radiation (which has pressure, and an equation of state $w = 1/3$), then becomes matter-dominated (pressure-less, with equation of state $w = 0$) and finally becomes dominated by *dark energy*, characterized by an unusual equation of state $w < 0$: the pressure of dark energy is *negative*. One possible candidate for the dark energy is Einstein's "cosmological constant", the energy of empty space resulting from the quantum uncertainty principle. This is a spatially homogeneous component whose density is constant in time. Another possibility is "quintessence"; a dynamic, time-evolving, spatially varying form of energy created by a oscillating quantum field with a very long wavelength. For a cosmological constant, $w = -1$, whereas for stable quintessence models $-1 < w < 0$. Further, w does not vary with time for the former and may vary for the latter. The equation of state changes the distance as $d_A = \int da(a^2 H)^{-1}$, illustrating why a precise measurement of θ_s is sensitive to w.

2.2. *Polarization anisotropy*

The CMB anisotropies not only have temperature, but polarization as well.[29,30] The Thompson scattering cross-section as a function of solid angle depends on polarization. If a free electron "sees" an incident radiation pattern that is isotropic, then the outgoing radiation remains unpolarized because orthogonal polarization states cancel out. However, if the incoming radiation field is a quadrupolar anisotropy, a net linear polarization is generated via Thompson scattering. Since quadrupolar temperature anisotropy is generated at last scattering when the tight-coupling approximation is broken, linear polarization results from the relative velocities of electrons and photons on scales smaller than the photon diffusion length-scale. Since both the velocity field and the temperature anisotropies are created by density fluctuations, a component of the polarization should be correlated with the temperature anisotropy. An important corollary to this argument is that no additional polarization anisotropy is generated after last scattering, since there are no free electrons to scatter the CMB photons. However, when the first generation of stars form, their strong UV light reionizes the universe; free electrons scatter CMB photons, introducing some optical depth and uniformly suppressing the power spectrum of the temperature fluctuations by $\sim 30\%$. Free electrons see the local CMB quadrupole at the redshift of star-formation and polarize the CMB at large scales where no other mechanism of polarization operates.[31] The CMB linear polarization anisotropy is small - in the standard picture of the thermal history, it is only a few percent of the temperature anisotropy. Consequently it is much harder to measure than the former.

The CMB polarization anisotropy also has the potential to reveal the signature of primordial gravity waves.[32,33] The CMB temperature anisotropy, being a scalar quantity, cannot uniquely differentiate between contributions from density perturbations (a scalar quantity) and gravity waves (a tensor quantity). However, polarization has "handedness", and thus can discriminate between the two. The polarization anisotropy can be decomposed into two orthogonal modes. *E-mode*: the curl-free mode (giving polarization vectors that are radial or tangential around a point on the sky) is generated by both density and gravity wave perturbations. *B mode*: the divergence-free mode (giving polarization vectors with vorticity around a point on the sky) can only, to leading order, be produced by gravity waves. This appears at large angular scales. On smaller scales, another effect which was sub-dominant on large scales kicks in, and eventually becomes dominant: weak gravitational lensing of the E-mode (as CMB photons are deflected by large scale structure that they traverse through to reach us) converts it to B-mode polarization.[34] The primordial B-mode anisotropy, if it exists, is an order of magnitude smaller than the E-mode polarization. This, combined with the difficulty of separating primordial B-modes from these "foreground" B-modes, makes the measurement of a potentially-existing primordial gravity wave contribution a great challenge — the "holy grail" of CMB measurements.

The presence of polarization potentially gives four types of measurements of the CMB anisotropy (temperature, linear polarization, tensor polarization, and the cross-correlation between the linear polarization and the temperature). The dependence on cosmological parameters of each of these components differs, and hence a combined measurement of all of them greatly improves the constraints on cosmological parameters by giving increased statistical power, removing degeneracies between fitted parameters, and aiding in discriminating between cosmological models.

2.3. *Initial conditions and connection to the early universe*

Since the CMB is a snapshot of the conditions when the universe was only 380,000 years old, plus some small processing which occurred when the CMB photons were en route to us, it preserves a more pristine picture of the very early universe than any other probe to which we have access. It is believed that, in the initial moments of the Big Bang, a period of exponential expansion called inflation caused the universe to expand by at least a factor of 10^{26} in an infinitesimal time (from when the universe was about 10^{-35} seconds old to when it was about 10^{-33} seconds old).[35-38] The expansion is driven by a hypothetical quantum field called the *inflaton*, which has negative pressure (in this respect, it is similar to the dark energy component discussed above). The actual size of the universe grows so much that it becomes much larger than the distance that light could have traveled since the Big Bang (i.e. our observable horizon). Any inhomogeneities that preceded inflation are erased and the universe becomes flat and smooth throughout our observable patch, in the same way that the surface of the earth looks flat when viewed from a small aircraft, even though its global shape is spherical. However, the oscillations of the inflaton also create tiny quantum mechanical fluctuations, and such scenarios predict that these fluctuations correspond to the perturbations imprinted on the CMB and the large scale distribution of galaxies. This is the currently dominant theory of generation of the initial inhomogeneities.

Inflationary models make detailed predictions about the statistics of the hot and cold spots on the CMB sky: how many patches there should be of each angular size, and how much hotter or colder than the average they should be, and so forth. The simplest inflationary scenarios predict several general consequences for the form of CMB perturbations: (1) Flat geometry, i.e. the observable universe should have no spatial curvature. (2) Gaussianity, i.e. the primordial perturbations should correspond to Gaussian random variables to a very high precision. (3) Scale invariance, i.e. to a first approximation, there should be equal power at all length-scales in the perturbation spectrum, without being skewed towards high or low frequencies. (4) Adiabaticity, i.e. the temperature and matter density perturbations should satisfy the condition $\delta T/T = \frac{1}{3}\delta\rho/\rho$. This is the same correlation found in the adiabatic compression of a gas — regions of high density are also hotter. It implies that a positive fluctuation in the number density of one species is a positive fluctuation in all the other species. (5) *Superhorizon fluctuations*, i.e. there exist correlations between anisotropies on scales larger than the causal horizon, beyond which two points could not have exchanged information at light-speed during the history of the universe. This corresponds to angular separations on the sky larger than $\sim 2°$. In addition, particular models of inflation predict significant amounts of (6) primordial gravity waves, which gives rise to temperature and polarization anisotropies as described above.

The CMB is the cosmological data set where it is easiest to characterize/remove systematics and to interpret, because it is produced by relatively simple physical processes and observed with mature technologies.[39] With this knowledge of the underlying physics in hand, one can make precise predictions for the observed temperature anisotropy for a given set of initial conditions and background cosmological model, and compare them with measurements of the CMB temperature and polarization power spectra. In this way, one can deduce the constituents and the initial conditions of the universe, reconstruct its history and evolution, and potentially obtain a window into the epoch when all the structure we see today originated.

3. Interpretation of *WMAP* Data

How would one extract cosmological information from a CMB map in practice? We can express the temperature fluctuations in the CMB sky ($\delta T(\theta, \phi)$ where (θ, ϕ) denotes the position angle) by expanding it in spherical harmonics:

$$\delta T(\theta, \phi) = \sum_{\ell,m} a_{\ell m} Y_{\ell m}(\theta, \phi) \ . \tag{2}$$

If the anisotropies form a Gaussian random field,[a] all the statistical information in the CMB map is contained in the *angular power spectrum*:

$$C_\ell = \frac{1}{2\ell + 1} \sum_m |a_{\ell m}|^2 \ . \tag{3}$$

---

[a]We find no evidence for deviation from Gaussianity.[12]

Thus we can describe the cosmological information contained in the millions of pixels of a CMB map in terms of a much more compact data representation.

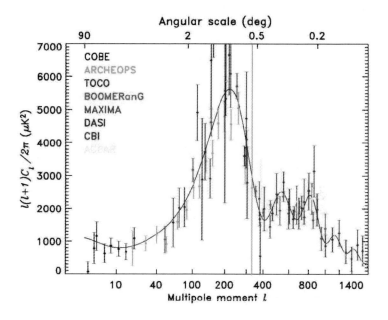

Fig. 3.   The angular power spectrum of the CMB temperature (*TT*) power spectrum from a compilation of pre-*WMAP* data.[13] The solid line is the best fit model for *WMAP* data.

Figure 3 shows a compilation of pre-*WMAP* CMB temperature (*TT*) angular power spectrum measurements. One may see evidence by eye for the existence of the projected acoustic waveform in this data, and a χ^2 analysis certainly tells us that the expected waveform fits the data well (*chi*^{2} is a non-parametric test of statistical significance that gives the degree of confidence one can have in accepting or rejecting a hypothesis). However, the errors are still large enough that the power of the data to discriminate amongst cosmological models and to measure cosmological parameters precisely is impaired.

Figure 4 shows the CMB temperature anisotropy (*TT*) angular power spectrum (top) and the temperature-polarization (*TE*) cross-power spectrum (bottom) measured by *WMAP* after one year of operation. The new *TT* measurement leaves no doubt that we are seeing projected acoustic

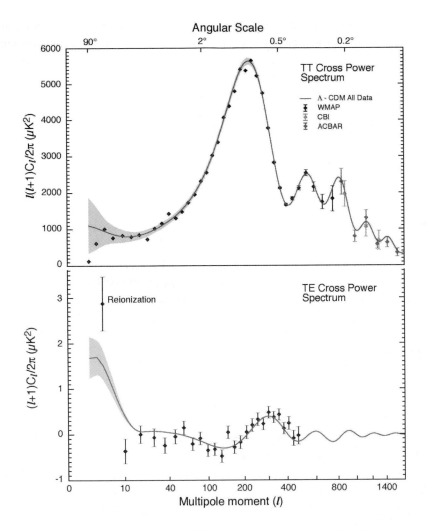

Fig. 4. Top[13]: *WMAP* temperature (*TT*) power spectrum: the gray band is the cosmic variance error-bar while the error-bars on the data points are the noise-error bars. The cosmic variance error is a fundamental limit on measuring the CMB, arising from the fact that one can only measure $(2\ell+1)$ modes on the sky corresponding to each ℓ-mode. The first year *WMAP* data is cosmic-variance limited to $\ell = 354$. The solid line is the best fit model for *WMAP* data. Bottom[14]: The temperature-polarization (*TE*) cross-power spectrum. The solid line is the best fit model for *WMAP* data. The excess power at $\ell < 10$ is due to early star formation; the excess is shown as a binned error bar from 10 multipoles.

waves on the sky, and the errors are dramatically reduced. Last year, the first detection of the CMB polarization (*EE*) signal was announced;[40] the *WMAP TE* measurement is the first measurement of the cross-correlation of temperature with polarization. It represents not just a detection but an actual power-spectrum that can be used to learn about cosmology.

These power spectra constitute our primary dataset. The technical details of obtaining them and their errors, and the likelihood analysis techniques we used to recover cosmological parameters from them are described in Refs. 13–15. Now we will use these data to constrain the standard LCDM cosmological model.

3.1. *Implications for the standard concordance cosmological model*

The "concordance" cosmological model (the so-called LCDM model) is a flat, low density universe with adiabatic initial conditions, composed of radiation, baryons, dark matter and a cosmological constant. Its six parameters are: the dark matter density (parameterized by Ω_c), the physical baryon density ($\Omega_b h^2$), the Hubble parameter ($H_0 = 100h$ km/s/Mpc), the spectral slope of the primordial power spectrum (n_s), the amplitude of fluctuations (parameterized by the present-day RMS fluctuations smoothed on 8 Mpc/h spheres σ_8), and the optical depth to the last scattering surface (τ). For physicists, these six parameters are equivalent to a different set of more familiar parameters.

While it is beyond the scope of this article to give a detailed explanation of how the C_ℓ determine these cosmological parameters,[b] Fig. 5 gives a summary of how specific features in the CMB power spectrum yield information about different parameters. The best fit values for these cosmological parameters and their 1σ error bars[16] are reported in Table 1. These results yield two important conclusions.

3.1.1. *Discovery of the early reionization of the universe*

Polarization at large angular scales is a distinctive signature of reionization (see Sec. 2.2) and *WMAP* has detected a significant *TE* correlation at $\ell < 7$. This extra power corresponds to an optical depth due to Thompson scattering of $\tau = 0.17 \pm 0.04$.[14] The optical depth is proportional to the

---

[b] For a physically-motivated and lucid explanation for the beginner, please see the website created by W. Hu at http://background.uchicago.edu/

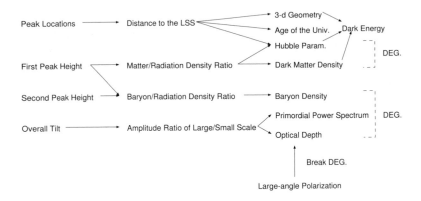

Fig. 5.   The relation between specific features in the CMB angular power spectrum and the cosmological parameters. "DEG" means that one can move along a *degeneracy direction* between the two parameters while keeping the C_ℓ essentially invariant (Illustration by E. Komatsu).

Table 1.   LCDM best fit parameters to *WMAP* data. Left: format for physicists; Right: format for astrophysicists. All errors are 1σ.

Dark matter	$(2.25 \pm 0.38) \times 10^{-27} \mathrm{kg/m^3}$	Ω_c	0.26 ± 0.07
atomic density	$(2.7 \pm 0.1) \times 10^{-7} \ \mathrm{cm^{-3}}$	$\Omega_b h^2$	0.024 ± 0.001
Age	13.4 ± 0.3 Gyr	h	0.72 ± 0.05
σ_8	0.9 ± 0.1	σ_8	0.9 ± 0.1
n_s	0.99 ± 0.04	n_s	0.99 ± 0.04
z_{reion}	17	τ	0.17 ± 0.04

mean column density N_e of free electrons along the photon line-of-sight; $\tau \sim \sigma_T N_e$. Since we know the mean *number* density of electrons in the universe (from the baryon density) we can estimate the distance to the epoch of reionization (presumably also the epoch of formation of the first stars) as $d_e \sim N_e/n_e$. The new measurement says that $\sim 15\%$ of the CMB light was re-scattered since the universe reionized. Assuming reionization happened instantaneously, the estimated reionization redshift is $z \sim 20$ or 200 million years after the Big Bang.[14]

This means that the universe reionized significantly earlier than expected by most researchers. Previous observations of Lyman–α absorption systems of quasars due to intervening clouds of neutral hydrogen suggested that the universe was completely ionized from redshift $z = 0$ to $z \sim 6$ but that the ionization fraction begins to drop beyond $z \sim 6$.[41] This was interpreted as the universe being instantaneously reionized at $z \sim 6$, when it was 1 Gyr old. Reconciling these two measurements by assuming a long-duration reionization epoch is difficult, and the relevant physical processes are still uncertain. Thus, this new measurement sheds new light on the end of the "dark ages" when the first stars formed in the universe, and once again reveals that the universe has a more complicated history than we expected.

3.1.2. *We now have a phenomenological "standard cosmological model"*

The flat LCDM model with adiabatic initial conditions fits this new set of precision observations remarkably well: only six parameters fit 1346 data points with acceptable χ^2.[16] The best fit LCDM model is consistent not only with *WMAP* data and other CMB experiments, but also a host of other cosmological observations at different scales and redshifts, obtained using different techniques. For example, the H_0 determination is in agreement with the *Hubble Space Telescope* measurement obtained from observations of stars of known intrinsic luminosity in nearby galaxies; and the age determination is in good agreement with other measurements based on the ages of the oldest stars. The model also fits large scale structure survey data (galaxy power spectrum from the 2dFGRS and SDSS surveys, Lyman–α forest power spectrum).

When using *WMAP* data alone, some degeneracies among cosmological parameters remain; the main degeneracy is between n_s and τ. Degeneracies among cosmological parameters increase the error-bars on marginalized quantities (i.e. integrated probabilities over all other parameters), but this will significantly improve with more years of *WMAP* operation.

There are two assumptions in the standard LCDM model which should be verified.

- **Flatness:** We assumed that the universe was at the critical density, $\Omega_{tot} = 1$; in other words, the cosmological constant energy density was fixed by $\Omega_\Lambda = 1 - \Omega_m$ where Ω_m is the total density in cold dark matter and baryons. We can test this assumption by adopting a 7-parameter model where Ω_Λ and Ω_m are allowed to vary independently. Introducing an extra parameter introduces new degeneracies among the cosmological parameters which can be lifted by adding external data sets (see Sec. 4). By doing this we obtain a constraint on the flatness of the universe, $\Omega_{tot} = 1.02 \pm 0.02$ (68% confidence level), consistent with our assumption. From the *WMAP* data alone, the constraint is only slightly worse; $\Omega_{tot} = 1.04 \pm 0.04$, again consistent with flatness.

- **Adiabatic Initial Conditions:** This assumption is supported by the internal consistency between the *WMAP TT* and *TE* power spectra. For primordial adiabatic initial conditions, the temperature power spectrum gives a precise prediction for the temperature-polarization cross-correlation power spectrum (Fig. 6), which is different from the predictions for other types of initial conditions (e.g. isocurvature, causal seed models). Note that there is an anti-correlation peak in C_ℓ^{TE} at $\ell \sim 140$ and a positive correlation peak at $\ell \sim 300$, but no correlations at $\ell \sim 220$ at which the temperature power spectrum shows its highest peak. As explained in Sec.2.2, polarization is being generated by the relative velocity of the photons and electrons at the LSS, and the velocity mode is off in phase by $\pi/2$ from the density mode which generates the first peak of the CMB temperature power spectrum. This adiabatic prediction is directly verified by the *TE* data.

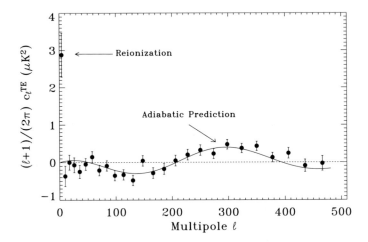

Fig. 6. The temperature-polarization (*TE*) cross-power spectrum.[14] The solid line is the prediction from the temperature data for adiabatic initial conditions.

4. Combining *WMAP* Data with Other Data Sets

WMAP data can be combined with complementary cosmological data sets: (1) most importantly, to test the self-consistency of the standard cosmological model; (2) *if* consistency is demonstrated, the additional information in the external data can be used to lift degeneracies among cosmological parameters present in the *WMAP* data alone. The compilation of complementary data sets used in this work are shown in Fig. 7. Every effort has been made to model the systematics of the external data sets conservatively, as described in Ref. 15; However it is understood that the non-CMB data come from physical processes that are much more complex than the physics of the CMB. Hence they will never be as "clean" as the CMB. Armed with the statistical power of these external data sets, we will now test models beyond the simple six-parameter LCDM model.

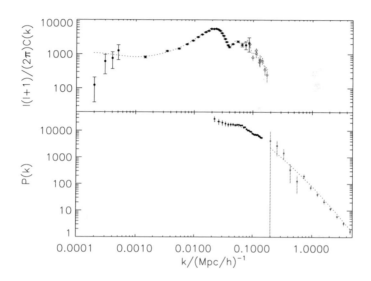

Fig. 7. Top panel[42]: *WMAP*, smaller-scale CMB experiments (CBI[43,44] and ACBAR[45]); the multipole ℓ has been converted to physical wavenumber k. Bottom panel[42]: 2dFGRS[46,47] galaxy clustering data and Lyman–α forest.[48,49] The data are complementary in scale and in redshift. The solid line is the best fit LCDM model to *WMAP* data. The extrapolation of this model to low redshift and smaller scales fit the data remarkably well. The results presented here have been obtained without the Lyman–α data, which contain the greatest level of systematic errors.

4.1. Dark energy equation of state

Dark energy properties can be parameterized by the equation of state $w = P/\rho$ where P denotes pressure and ρ the density. For a cosmological constant $w = -1$ while for the alternative candidate, "quintessence", a dynamical, time-evolving and spatially varying energy component, $-1 < w < 0$; there are unstable models that give $w < -1$. Although, in principle, w can vary in time for a quintessence-type model, we only consider here models with constant w; hence what we obtain is an "effective" value for w, a weighted average of the w values from $z = 0$ to $z = 1100$. By combining different data sets we find no evidence for the dark energy being different from a cosmological constant: $w = -0.98 \pm 0.12$.[16] Figure 8 presents the constraints on the dark energy equation of state.

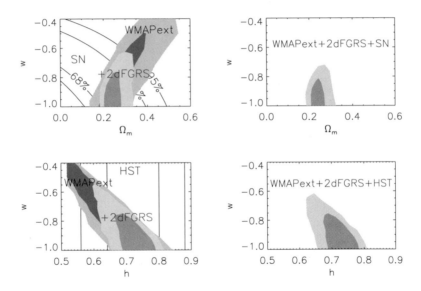

Fig. 8. Constraints on Dark Energy Properties.[16] These panels show the degeneracy directions between the dark energy equation of state w, the fractional matter density Ω_m and the local expansion rate h of the Universe. The dark areas in these plots contain 68% of the probability and the light areas contain 95% of the probability. The upper left panel shows the error contours for a combination of *WMAP* with other CMB experiments (marked WMAPext) and for a combination of the CMB data with the Two Degree Field Galaxy Redshift Survey (marked 2dFGRS). The solid lines in the figure show the 68% and 95% error contours for supernova data (SN). One can see the long degeneracy directions in the individual data sets. The lower left panel shows similar constraints this time using the measurement of h by the *Hubble Space Telescope*, as an alternative to supernova data. In the right panels, we show how the degeneracy is broken by combining all the data, leading to a much smaller error region. Note that, by measuring any one data set better, one merely narrows the width of its degeneracy, and does not shorten its length.

4.2. *Neutrino mass*

So far we have assumed that neutrinos are massless, but we can try to constrain the neutrino mass in the context of a flat LCDM model. Neutrinos stream freely out of the potential wells; thus, if they have mass, they tend to erase fluctuations on small scales, and suppress the growth of cosmic structures on those scales. From current observations, the CMB alone is not sensitive to the neutrino mass but we can improve the constraint by including information on the shape and amplitude of the matter power spectrum (the amplitude of clustering of matter) from large scale structure surveys (see Fig. 9). We obtain a constraint on the physical energy density of neutrinos: $\Omega_\nu h^2 < 0.0067$ at the 95% confidence level when combining the CMB data (WMAP+CBI+ACBAR) with 2dFGRS.[16] For three degenerate neutrino species, this implies $m_\nu < 0.23$ eV. If we add the Lyman–α data this constraint remains virtually unchanged.

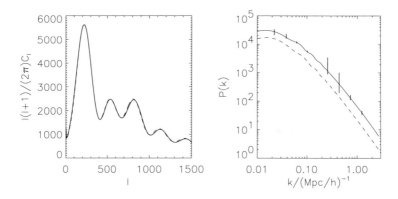

Fig. 9. Two models[15]: one with massless neutrinos the other with three degenerate neutrino species each with a mass of 0.6 eV. These two models are virtually indistinguishable from the CMB power spectrum (left) but the matter power spectra are very different on scales probed by large-scale structure surveys (right). It is clear that the two models can be easily distinguished if the matter power spectrum shape and amplitude are known.

5. Implications for Inflation

WMAP has made several key observations that are of importance in constraining inflationary models: (a) The universe is consistent with being flat.[16] (b) The primordial fluctuations are described by random Gaussian fields.[12] (c) The *WMAP* detection of an anti-correlation between CMB temperature and polarization fluctuations at angular separations $\theta > 2°$ (corresponding to the *TE* anti-correlation seen on scales $\ell \sim 100 - 150$ in Fig. 6) is a distinctive signature of adiabatic fluctuations on superhorizon scales at the epoch of decoupling, confirming a fundamental prediction of the inflationary paradigm.[17] (d) Both by itself and in combination with complementary data, *WMAP* has been used to put interesting constraints on the "primordial" observables characterizing the primordial scalar and tensor power spectra predicted by single field inflationary models: the scalar power spectrum slope n_s, its running $dn_s/d\ln k$, and the tensor-to-scalar ratio r[17] (Fig. 10). These constraints in turn give limits on the first three derivatives of the inflationary potential $V(\phi)$. We can already put tight constraints on models which give rise to a large tensor contribution and a red tilt ($n_s < 1$): for example, minimally-coupled monomial potentials of the type $V(\phi) \propto \phi^n$ have already been ruled out for $n \geq 4$. (e) Two-field inflationary models with an admixture of adiabatic and CDM isocurvature components are tested; the initial conditions are consistent with being purely adiabatic.[17]

We note that an exact scale-invariant spectrum ($n_s = 1$ and $dn_s/d\ln k = 0$) is not yet excluded at more than 2σ level. This prevents us from ruling out broad classes of inflation models with different physical motivations. Excluding this point would have profound implications in support of inflation, as physical single field inflationary models predict non-zero deviation from exact scale-invariance.

6. Conclusions

We have presented the highlights of the implications of the *WMAP* first year data release for cosmology and inflation. Cosmology now has a phenomenological "standard model", in a similar state as particle physics was three decades ago. We can make measurements of the six parameters of this model precisely, in some cases to percent accuracy (see Table 1). However, fundamental questions like the origin of inflation (the initial conditions of our universe) and the nature of dark matter and dark energy (which constitute 95% of the energy density of the universe) are totally unknown. Over the coming years, improving CMB, large-scale structure, gravitational lensing and supernovae data etc. will provide ever more rigorous tests of the model. But answering the difficult question of why the universe obeys this bizarre model is a great challenge for theoretical physics.

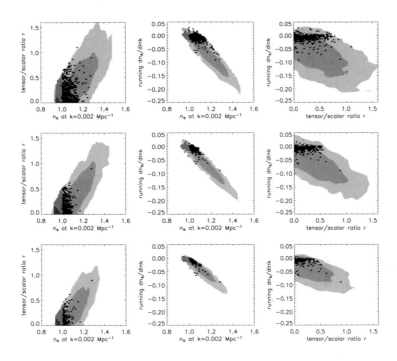

Fig. 10. Constraining inflationary models[17]: Left to right, r vs. n_s, $dn_s/d\ln k$ vs. n_s, and $dn_s/d\ln k$ vs.r. 68% and 95% joint constraints from the *WMAP* data alone (top row), plus the small-scale CMB data from CBI and ACBAR, and galaxy survey data from the 2dFGRS (middle row), plus the Lyman–α forest data (bottom row). Different colored dots represent predictions from different classes of inflationary models.

Acknowledgments

The *WMAP* mission is made possible by the support of the Office of Space Sciences at NASA Headquarters and by the hard and capable work of scores of scientists, engineers, technicians, machinists, data analysts, budget analysts, managers, administrative staff, and reviewers. *WMAP* is named in honor of Prof. David Wilkinson (1935-2002). The *WMAP* Science Team responsible for the scientific analysis of the *WMAP* first year results are: C. Barnes, C. Bennett (PI), M. Halpern, R. Hill, G. Hinshaw, N. Jarosik, A. Kogut, E. Komatsu, M. Limon, S. Meyer, N. Odegard, L. Page, H. Peiris, D. Spergel, G. Tucker, L. Verde, J. Weiland, E. Wollack, E. Wright. Fig. 5 has been adapted from E. Komatsu, to appear in Proceedings of the Fourth International Conference on Physics Beyond the Standard Model "Beyond the Desert 03", Castle Ringberg, Germany, June 9-14 (2003).

The *Legacy Archive for Microwave Background Data Analysis* (*LAMBDA*)[c] provides detailed information about the *WMAP* in-flight operations, as well as the data products and software for the first year data release. If using these data sets please refer to Ref. 13 for the *TT* power spectrum and Ref. 14 for the *TE* power spectrum. We also make available a subroutine that computes the likelihood for *WMAP* data given a set of C_ℓ; please refer to Ref. 15 if using this routine.

References

1. C. L. Bennett *et al.*, Astrophys. J. **583**, 1 (2003).
2. L. Page *et al.*, Astrophys. J. **585**, 566 (2003).
3. L. Page *et al.*, Astrophys. J. Suppl. **148**, 39 (2003).
4. C. Barnes *et al.*, Astrophys. J. **143**, 567 (2002).
5. N. Jarosik *et al.*, Astrophys. J. **145**, 413 (2003).
6. N. Jarosik *et al.*, Astrophys. J. Suppl. **148**, 29 (2003).
7. G. F. Hinshaw *et al.*, Astrophys. J. Suppl. **148**, 63 (2003).
8. C. Barnes *et al.*, Astrophys. J. Suppl. **148**, 51 (2003).
9. C. L. Bennett *et al.*, Astrophys. J. Lett., **464**, L1 (1996).
10. C. L. Bennett *et al.*, Astrophys. J. Suppl. **148**, 97 (2003).
11. C. L. Bennett *et al.*, Astrophys. J. Suppl. **148**, 1 (2003).
12. E. Komatsu *et al.*, Astrophys. J. Suppl. **148**, 119 (2003).
13. G. F. Hinshaw *et al.*, Astrophys. J. Suppl. **148**, 135 (2003).
14. A. Kogut *et al.*, Astrophys. J. Suppl. **148**, 161 (2003).
15. L. Verde *et al.*, Astrophys. J. Suppl. **148**, 195 (2003).
16. D. N. Spergel *et al.*, Astrophys. J. Suppl. **148**, 175 (2003).
17. H. V. Peiris *et al.*, Astrophys. J. Suppl. **148**, 213 (2003).
18. L. Page *et al.*, Astrophys. J. Suppl. **148**, 233 (2003).
19. J. C. Mather *et al.*, Astrophys. J. Lett. **354**, L37 (1990).
20. J. C. Mather *et al.*, Astrophys. J. **512**, 511 (1999).
21. A. A. Penzias and R. W. Wilson, Astrophys. J. **142**, 1149 (1965).
22. R. H. Dicke *et al.*, Astrophys. J. **142**, 414 (1965).
23. G. F. Smoot *et al.*, Astrophys. J. Lett. **396**, L1 (1992).
24. P. J. E. Peebles and J. T. Yu, Astrophys. J. **162**, 815 (1970).
25. R. A. Sunyaev and Ya. B. Zel'dovich, Astrophys. Space Sci. **7**, 3 (1970).
26. W. Hu and N. Sugiyama, Astrophys. J. **444**, 489 (1995).
27. J. Silk, Astrophys. J. **151**, 459 (1968).
28. R. K. Sachs and A. M. Wolfe, Astrophys. J. **147**, 73 (1967).
29. M. J. Rees, Astrophys. J. Lett. **153**, L1 (1968).
30. N. Kaiser, Mon. Not. R. Astron. Soc. **202**, 1169 (1983).
31. M. Zaldarriaga, Phy. Rev. D. **55**, 1822 (1997).
32. U. Seljak and M. Zaldarriaga, Phys. Rev. Lett. **78**, 2054 (1997).

---

[c]http://lambda.gsfc.nasa.gov

33. M. Kamionkowski, A. Kosowsky and A. Stebbins, Phys. Rev. Lett. **78**, 2058 (1997).
34. M. Zaldarriaga and U. Seljak, Phys. Rev. D. **58**, 023003 (1998).
35. A. H. Guth, Phys. Rev. D. **23**, 347 (1981).
36. V. F. Mukhanov, H. A. Feldman, and R. H. Brandenberger, Phys. Rept. **215**, 203 (1992).
37. A. A. Starobinsky, JETP Lett. **30**, 682 (1979).
38. A. A. Starobinsky, Sov. Astron. Lett. **11**, 133 (1985).
39. M. Kamionkowski, D. N. Spergel, and N. Sugiyama, Astrophys. J. Lett. **426**, 57 (1994).
40. J. Kovak *et al.*, submitted to Astrophys. J., astro-ph/0209478 (2002).
41. R. H. Becker *et al.*, Astron. J. **122**, 2850 (2001).
42. L. Verde *et al.*, N. Astr. Rev. in press (2003).
43. B. S. Mason *et al.*, Astrophys. J. **591**, 540 (2003).
44. T. J. Pearson *et al.*, Astrophys. J. **591**, 556 (2003).
45. C. L. Kuo *et al.*, submitted to Astrophys. J., astro-ph/0212289 (2003).
46. M. Colless *et al.*, Mon. Not. R. Astron. Soc. **328**, 1039 (2001).
47. W. J. Percival *et al.*, Mon. Not. R. Astron. Soc. **327**, 1297 (2001).
48. R. Croft *et al.*, Astron. J. **581**, 20 (2002).
49. N. Gnedin and A. J. Hamilton, Mon. Not. R. Astron. Soc. **334**, 107 (2002).

Hiranya Peiris

Born in Sri Lanka, Hiranya studied at Cambridge University, where she graduated with first class honours in Natural Sciences (Physics) in 1998. She obtained her PhD in Astrophysics from Princeton University in 2003. Her primary research interests are in cosmology, observational signatures of physics beyond the Standard Model, and the interface between cosmology and theoretical physics. In her spare time she enjoys music, Argentine tango, outdoor pursuits, and is an avid practitioner of yoga. At present, Hiranya is a Hubble Fellow at the University of Chicago.

CHAPTER 7

QUEST FOR GRAVITATIONAL WAVES

B. S. Sathyaprakash

School of Physics and Astronomy, Cardiff University
P.O. Box 913, Cardiff, United Kingdom
E-mail: B.Sathyaprakashastro.cf.ac.uk

Nearly a century has passed since gravitational waves were first conceived by Poisson in 1908 and shown by Einstein to be an essential consequence of general relativity in 1916. The quest for gravitational waves began with the famous Weber's bar experiment. To date we have no conclusive detection of these elusive waves but many believe we are at the verge of a breakthrough. In this article I will highlight the worldwide quest for gravitational waves and what fundamental physics, astrophysics and cosmology we might learn by opening this new window of observation.

1. Introduction

Gravitational radiation arises in any relativistic theory of gravity: it is the inescapable consequence of reconciling gravitational interaction with special relativity. Gravitational waves are described by a second-rank tensor field with as many as six independent components in the most general case but just two independent degrees of freedom in Einstein's theory. In general relativity, gravity is synonymous with the geometry of spacetime, with Einstein's equations providing a way to compute the metric of the background spacetime given the distribution of matter and energy. A change in the distribution alters the background metric and a consequent evolution in the background curvature of spacetime. The non-stationary part of the evolution of spacetime geometry is what we refer to as gravitational wave. For weak gravitational fields there is no need to make a formal connection between gravitational waves (GW) and the geometry of spacetime: One can imagine the waves (in Einstein's gravity) as massless, spin-2 fields, propagating on a light-cone in the flat background spacetime of special relativity. However, in the presence of strong gravitational fields it is not easy to disentangle the background from the waves and poses a challenge to numerical relativists who are required to address such problems when dealing with the interaction of strong fields as in the case, for example, of colliding black holes. Because of the difficulties associated with the theoretical understanding of the nature of GW many doubted if the waves really exist, but these thoughts were laid to rest thanks to the pioneering work of Bondi and others in the 50s and 60s.

Although GW have not been directly detected yet, the back-reaction force caused by their emission has been observed and shown to agree[36] with general relativity to better than 0.5% in an astronomical binary system discovered by Hulse and Taylor. The binary comprises of two neutron stars each of mass roughly 1.4 $M_\odot$ and an orbital period of 7.75 Hrs. The acceleration in the system leads to the emission of GW and the loss of orbital energy and angular momentum. The emitted radiation causes the two bodies to fall towards each other and hence a reduction in their orbital period. The tiny rate at which the period is decaying, about 70 micro seconds per year, has been measured thanks to the continuous monitoring of the system by Taylor and co-workers. The Hulse-Taylor binary, and two others that have been discovered since then, provide a test bed for different theories of gravity.

I will begin with an overview of gravitational waves in Sec. 2. A brief summary of the GW interferometers around the world and a list of upcoming projects is given in Sec. 3. The main focus of this article will be the astronomical sources of gravitational radiation and astrophysical and cosmological measurements afforded by GW observations which will be dealt with in Sec. 4. We refer to recent reviews[16,19,35] for further reading.

2. Nature of Gravitational Waves

Gravitational waves are produced by the non-axisymmetric motion of bulk matter and represented by a second rank tensor but with just two independent degrees of freedom which constitute the the *plus* and *cross* polarisations. Plus and cross waves incident perpendicular to a plane containing a circular ring of *free* particles deform the ring as shown in Fig. 1. Monitoring the distance from the centre of the ring to "free" mirrors placed at the ends of two orthogonal radial directions will enable the detection of a passing wave. This is the principal behind a laser interferometer GW antenna wherein the differential change in the arm lengths will be recorded

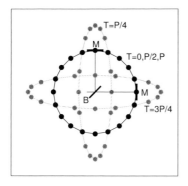 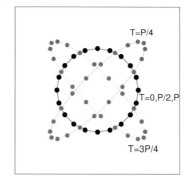

Fig. 1. The response of a circular ring of *free* particles to waves of period P with either plus (left) and cross (right) polarisation. The ring (radius C) is continuously deformed into an ellipse with axes $(1 + h/2)C$ and $(1 - h/2)C$. Also depicted are the locations of the beam-splitter and mirrors that are freely suspended inside a vacuum tube in an interferometric detector.

as a change in the fringe pattern of the laser exiting the interferometer. Heuristically, the influence of the radiation on matter can be characterized by a dimensionless amplitude which is a measure of the deformation resulting from the waves passing through the matter. For instance, if ℓ is the distance between two free masses (as in Fig. 1) then a wave of amplitude h causes a change in the distance between the masses by $\delta\ell = h\ell/2$.

We conclude this introduction by giving order-of-magnitude estimate of the gravitational wave amplitude h, luminosity $\mathcal{L}$ and frequency ω for a self-gravitating system at a distance r from Earth, whose physical size is R and total mass M[a]

$$h \sim \frac{M}{r}\frac{M}{R}, \quad \mathcal{L} \sim \left(\frac{M}{R}\right)^5 \sim v^{10}, \quad \omega^2 \sim \frac{M}{R^3}, \tag{1}$$

where $v = \sqrt{M/R}$ is the dynamical velocity in the system. Let us note the salient features of the waves: firstly, the amplitude of the radiation is a product of the internal and external gravitational potentials of the source[34] and is greater the more compact a source is. Secondly, the luminosity is also

---

[a]This article chiefly deals with compact objects, namely neutron stars (NS) and black holes (BH). Unless specified otherwise we shall assume that a NS has a mass of $M = 1.4M_\odot$ and radius $R = 10$ km, and by a stellar mass BH we shall mean a black hole of mass $M = 10M_\odot$. We shall assume a flat Universe with a cold dark matter density of $\Omega_M = 0.27$, dark energy of $\Omega_\Lambda = 0.7$, and a Hubble constant of $H_0 = 65\,\mathrm{km\,s^{-1}\,Mpc^{-1}}$ and use a system of units in which $c = G = 1$, which means $1\,M_\odot \simeq 5 \times 10^{-6}\,\mathrm{s} \simeq 1.5$ km, 1 Mpc $\simeq 10^{14}$ s.

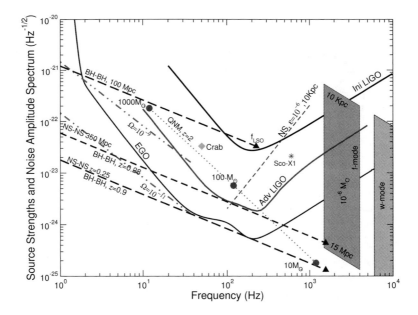

Fig. 2.   The sensitivity of the various working/planned instruments and the strength ($\sqrt{f}\bar{h}(f)$, where $\bar{h}(f)$ is the Fourier transform of the signal) of sources. See text for details.

higher for more compact and relativistic sources, raising very steeply as a function of velocity. The factor to covert the luminosity from $G = c = 1$ units to conventional units is $\mathcal{L}_0 \equiv c^5/G \simeq= 3.6 \times 10^{59}$ erg s^{-1}. Since $v < 1$, $\mathcal{L}_0$ denotes the largest luminosity a source could ever have and generally $\mathcal{L} \ll \mathcal{L}_0$. Finally, GW frequency is the same (to within a factor of 2) as the dynamical frequency in the system.

For a stellar mass ($M = 1\,M_\odot$) source of size $R = 10M$ at a distance of 100 Mpc, the above formulas give $h \sim 5 \times 10^{-23}$ and $f_{\rm GW} \sim 1\,kHz$, while for a supermassive ($M = 10^6\,M_\odot$) source of size $R = 10M$, at $z = 1$ (luminosity distance 7 Gpc), we have $h \sim 7 \times 10^{-18}$ and $f_{\rm GW} \sim 1\,{\rm mHz}$. These amplitudes and frequencies are typical for the sources to be detected with the ground- and space-based antennas, respectively, and indicate the kind of sources expected to be observed in them.

3. Interferometric GW Detectors

There are currently six long baseline detectors in operation: the American Laser Interferometer Gravitational-Wave Observatory (LIGO)[2] – a network of three detectors, two with 4 km arms and one with 2 km arms, at two sites (Hanford, Washington and Livingstone, Louisiana) – the French-Italian VIRGO detector with 3 km arms at Pisa,[13] the British-German GEO 600[26] with 600 m arms at Hannover and the Japanese TAMA with 100 m arms in Tokyo.[38] Plans are well underway both in Europe and the USA to build, by 2008, the next generation of interferometers that are 10–15 times more sensitive than the initial interferometers. On a longer time-scale of 2015 technology will have improved to build antennas that are a factor 50 better in GW amplitude, and detection rates a factor 1.25×10^5 higher, than the initial interferometers. The European Gravitational-Wave Observatory, or EGO, which is currently being studied[7] is an example of such an antenna.

ESA and NASA have plans to build, by 2013, the Laser Interferometer Space Antenna (LISA)[8] which consists of three spacecraft in heliocentric orbit, at the vertices of an equilateral triangle of size 5 million km and located with respect to Sun 60 degrees behind the Earth. LISA will operate in the frequency range of 0.1–10 mHz. There have been proposals to build an antenna in the frequency gap 0.01–1 Hz of LISA and ground-based detectors; the *Deci-Hertz Interferometer Gravitational-Wave Observatory* (DECIGO)[37] by the Japanese team and the *Big-Bang Observer* (BBO) have been conceived to detect the primordial GW background and to answer cosmological questions on the expansion rate of the Universe and dark energy.

In ground-based interferometric antennas highly reflective mirrors (losses $\lesssim 10^{-5}$) are freely suspended (quality factors $\gtrsim 10^6$) at the ends of two orthogonal arms (length $\ell \sim$ km) inside vacuum tanks (pressure $\sim 10^{-8}$ mbar) and high power lasers (effective power of 10 kW) are used to measure extremely tiny strains ($\delta\ell/\ell \gtrsim 10^{-21}$–$10^{-23}$ for transient bursts and 10^{-25}–10^{-27} for continuous wave sources), in a wide range (10–1000 Hz) of frequency. Space-based detectors are passive interferometers that work by tracking distance between what are truly *free* test-masses housed inside each spacecraft. The technology required to shield the test-masses from external disturbances (cosmic rays, solar radiation and particles) will be demonstrated in an ESA mission called LISA Pathfinder, scheduled for launch in 2008.

Gravitational wave interferometers are quadrupole detectors with half the peak sensitivity over a third of the sky. A single antenna cannot determine the polarisation state of a *transient* wave or the direction to the source that emits the radiation. However, a network of three antennas are able to measure five numbers (three amplitudes and two independent time-delays in the arrival of the signal) and thereby determine the distance and

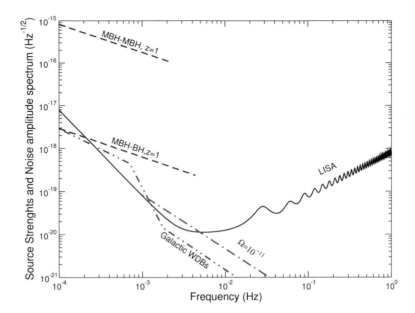

Fig. 3.   The sensitivity of LISA and the strength $\sqrt{f}\tilde{h}(f)$, where $\tilde{h}(f)$ is the Fourier transform of the signal, of sources. See text for details.

direction to the source, the polarisation angle and the orientation of the source. For long-lived sources the Doppler modulation of the signal caused by the motion of the detector relative to the source can be used to resolve the source on the sky.

Figs. 2 and 3 plot the expected sensitivity of the various ground-based antennas and LISA, respectively. The plots show the noise amplitude spectrum of the detectors as a function of frequency. Also plotted are the GW amplitude of various sources (integrated over a frequency band equal to the frequency for transient signals, integrated for four months in the case of continuous waves and a combination of both for stochastic signals). Source strengths have been scaled down by a suitable threshold so that a source point/curve lying on or above a noise curve can be detected with a confidence of 99% or greater.

4. Sources of Gravitational Waves

Gravitational wave detectors will unveil dark secrets of the Universe by helping us to study sources in extreme physical environs: Strong non-linear gravity, relativistic motion, extremely high density, temperature and magnetic fields, to list a few. We shall focus our attention on *compact objects* (in binaries or isolation) and *stochastic backgrounds*.

4.1. *Compact binaries*

Compact binaries, consisting of a pair of black holes and/or neutron stars are an astronomer's standard candles[33]: a parameter called the *chirp mass* ($\mathcal{M} \equiv \eta^{2/3} M$) completely fixes the absolute luminosity of, and hence the luminosity-distance to, the source. If we can measure the chirp mass then we will be able to extract the luminosity-distance. The key idea is to use the chirping of a binary, that is its changing frequency, to measure the chirp mass. Hence, by observing GW from a binary we can measure the luminosity-distance to the source provided the source *chirps* by at least as much as $1/T$ during an observational period T. A chirping binary, therefore, helps to accurately measure cosmological parameters and their variation as a function of red-shift.

The dynamics of a compact binary consists of three phases: (1) the *early inspiral phase* in which the system spends 100's of millions of years and the power emitted in GW is low, (2) the *late inspiral,* and early *merger phase* wherein the two stars are orbiting each other at a third of the speed of light and experiencing strong gravitational fields with the potential $\varphi \sim M/R \sim 0.1$, and (3) the *late merger phase* when the two systems have merged to form either a single NS or a BH, settling down to a quiescent state by radiating the deformations inherited during the merger. We do have a good knowledge of the nature of the waveform emitted during the early inspiral and late merger phases, but the challenge is to model and understand the merger phase which involves strongly non-linear and highly relativistic gravity and, in the case of objects with large spins, the dynamics is complicated by precessing orbits due to strong spin-orbit and spin-spin interactions.

4.1.1. *Stellar mass compact binaries*

The span of ground-based interferometers to stellar mass binaries, shown in the right panel of Fig. 4, varies with the masses as $\eta^{1/2} M^{5/6}$, greater asymmetry (i.e. smaller value of $\eta = m_1 m_2/M^2$) in the masses reduces the span but larger total mass M increases the span. However, for $M \gtrsim 100 \, M_\odot$ the sensitivity worsens as the seismic and thermal noise in interferometers begin to dominate the noise spectrum at lower frequencies. The inspiral phase will be visible in a ground-based detector during the last few mins of its life before the binary merges.

NS-NS binaries: Double NS can be seen to a distance of 20 Mpc, 350 Mpc and $z = 0.25$, in initial LIGO, advanced LIGO and EGO, respectively, as shown in Fig. 2. Based on the observed small Galactic population of double NS binaries which merge within the Hubble time, Kalogera *et al.*[22] conclude that the Galactic coalescence rate is $\sim 1.8 \times 10^{-4}$ yr^{-1} which would imply an event rate of NS-NS coalescences of 0.25 and 1500 yr^{-1}, in initial and advanced LIGO, respectively, and 100 day^{-1} in EGO. As the spins of NS are very small, and because the two stars would merge well outside the LIGO's sensitivity band, the state-of-the-art theoretical waveforms,[10] computed using post-Newtonian expansion of Einstein's equations, can be used as search templates to dig these signals out of detector noise by matched filtering. However, detailed relativistic hydrodynamical simulations[23] would be needed to interpret the emitted radiation during the coalescence phase, wherein the two stars collide to form a bar-like structure prior to merger. The bar hangs up over a couple of dynamical time-scales to get rid of its deformity by emitting strong bursts of GW. Observing the radiation from this phase should help to deduce the equation-of-state (EOS) of NS bulk matter. Also, an event rate as large as in advanced LIGO and EGO will be a valuable catalogue to test astrophysical models of γ-ray bursts and see if they are associated with NS-NS or NS-BH mergers.

NS-BH binaries: Initial and advanced LIGO and EGO will be sensitive to NS-BH binaries at a distance of about 40 Mpc, 680 Mpc and $z = 0.45$, respectively. The rate of coalescence of such systems is not known empirically as there have been no astrophysical NS-BH binary identifications. However, the population synthesis models give[19] a Galactic coalescence rate in the range 3×10^{-7}–5×10^{-6} yr^{-1}, giving an optimistic detection rate of NS-BH of 0.05 and 300 per year in initial and advanced LIGO, respectively, and 20 per day in EGO.

NS-BH binaries are very interesting from an astrophysical point of view: the initial evolution of such systems can be treated analytically fairly well, however, the presence of a BH with large spin can cause the NS to be whirled around in precessing orbits due to the strong spin-orbit coupling. The relativistic evolution of such systems is really not very well understood, complicated also by the tidal disruption of the NS. It should be possible to accurately measure the onset of the merger phase and deduce the radius of the NS to $\sim 15\%$ and thereby infer the EOS of NS.[40]

BH-BH binaries: The span of interferometers to BH-BH binaries varies from 100 Mpc in initial LIGO, to a red-shift of $z \sim 0.3$ in advanced LIGO, and $z = 0.90$ in EGO (cf. Fig. 2 and 2). As in the case of NS-BH binaries, here too there is no empirical estimate of the event rate. Population synthesis models are highly uncertain about the Galactic rate of BH-BH coalescences and predict[19] a range of 3×10^{-8}–10^{-5} yr^{-1}, which is smaller than the predicted rate of NS-NS coalescences. However, owing to their greater masses, BH-BH event rate in our detectors is larger than NS-NS by a factor $M^{5/2}$ for $M \lesssim 100\,M_\odot$. The predicted event rate is a maximum of 1 yr^{-1} in initial LIGO, 20 day^{-1} in advanced LIGO and 400 day^{-1} in EGO.

Thus, BH mergers are the most promising candidate sources for a first direct detection of GW. These sources are the most interesting from the view point of general relativity, constituting a pair of black holes experiencing the strongest possible gravitational fields before they merge with each other to form a single BH, and serve as a platform to test general relativity in the strongly non-linear regime. For instance, one can detect the scattering of GW by the curved geometry of the binary,[11,12] and measure, or place upper limits on, the mass of the graviton to 2.5×10^{-22} eV and 2.5×10^{-26} eV in ground- and space-based detectors, respectively.[41] High SNR events (which could occur once every month in advanced LIGO) can be used to test the full non-linear gravity by comparing numerical simulations with observations and thereby gain a better understanding of the two-body problem in general relativity. As BH binaries can be seen to cosmological distances, a catalogue of such events compiled by LIGO/VIRGO/EGO can be used to measure Cosmological parameters (Hubble constant, expansion of the Universe, dark energy) and test models of Cosmology.[17]

4.1.2. *Massive black hole binaries*

It is now believed that the centre of every galaxy hosts a BH whose mass is in the range[28] 10^6–$10^9\,M_\odot$. These are termed as *massive black holes* (MBH). There is observational evidence that when galaxies collide, the MBH at their nuclei get close enough to be driven by gravitational radiation reaction and merge within the Hubble time.[24] For a binary with $M = 10^6 M_\odot$, the frequency of GW at the last stable orbit is $f_{\mathrm{LSO}} = 4$ mHz, followed by merger at 10 mHz, and quasi-normal mode ringing at 20 mHz (if the spin of the black holes is not close to 1). This is in the frequency range of LISA which has been designed to observe the MBH, their formation, merger and activity.

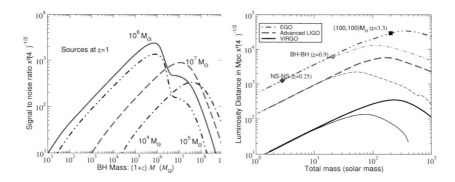

Fig. 4.   The span for binary coalescences of current and planned interferometers, VIRGO, Advanced LIGO and EGO are shown in the right panel, together with the signal-to-noise ratio in LISA for massive black hole binary mergers at a red-shift of $z = 1$ in the left panel. On the right panel thin lines correspond to the spans when including only the inspiral part of the signal in our search templates while thick lines show the span achieved when including both the inspiral and merger parts of the signal.

The SNR for MBH-MBH mergers in LISA is shown in the left panel of Fig. 4. These mergers will appear as the most spectacular events in LISA requiring no templates for signal identification, although good models would be needed to extract source parameters. MBH mergers can be seen to redshifts of $z = 5$ and higher. Therefore, one could study the merger-history of galaxies throughout the Universe and address astrophysical questions about their origin, growth and population. The recent discovery of a MBH binary[24] and the association of X-shaped radio lobes with the merger of MBH[27] has raised the optimism concerning MBH mergers and the predicted rate for MBH mergers is the same as the rate at which galaxies merge, about $1 \ \mathrm{yr}^{-1}$ out to a red-shift of[20] $z = 5$.

4.1.3. *Extreme mass ratio inspirals*

The MBH environment of our own galaxy is known to constitute a large number of compact objects and white dwarfs. Three body interaction will occasionally drive these compact objects, white dwarfs and other stars into a capture orbit of the central MBH. The compact object will be captured in an highly eccentric trajectory ($e > 0.99$) with the periastron close to the last stable orbit of the MBH. Due to relativistic frame dragging, for each passage of the apastron the compact object will experience several turns around the MBH in a near circular orbit. Therefore, long periods of low-frequency, small-amplitude radiation will be followed by several cycles of high-frequency, large-amplitude radiation. The apastron slowly shrinks, while the periastron remains more or less at the same location, until the final plunge of the compact object into the MBH. There is a lot of structure in the waveforms emitted by extreme mass ratio inspirals which arises as a result of a number of different physical effects: contribution from higher order multipoles, precession of the orbital plane that changes the polarisation of the waves observed by LISA, etc. This complicated structure smears the power in the signal in the time-frequency plane[31] as compared to a sharp chirp from a non-spinning BH binary and for this reason this *spin modulated chirp* is called a *smirch*.

As the compact object tumbles down the MBH it will sample the space-time geometry in which it is moving and the structure of that geometry will be imprint in the GW emitted in the process. By observing a smirch, LISA offers a unique opportunity to directly map the spacetime geometry around the central object and test whether or not this structure is in accordance with the expectations of general relativity.[29] Indeed, according to Einstein's theory the geometry of a rotating black hole is uniquely determined to be the Kerr metric involving just two parameters, the mass of the MBH and its spin. Thus, the various multipole-moments of the source are uniquely fixed once we have measured the mass and spin of the BH. With the observed smirch one can basically test (i) whether general relativity correctly describes the spacetime region around a generic BH and (ii) if the central object is indeed a BH or some other exotic matter.

The SNR from a smirch will be between 10–50 depending on the mass of the central object but it might be very difficult to match filter them due to their complicated shapes. The events rate is expected to be rather high. Indeed, a background population of smirches will cause confusion noise and only sources in the foreground will be visible in LISA. The event rate is as yet highly uncertain ranging from 1–10 yr^{-1} within 1 Gpc.[30]

4.2. *Neutron stars*

After BHs, NS are the most compact stars in the Universe. With a density of 2×10^{14} g cm^{-3}, and surface gravity $\varphi \equiv M/R \sim 0.1$, they are among the most exotic objects whose composition, equation-of-state and structure, are still largely unknown. Being highly compact they are potential sources of GW. The waves could be generated either from the various normal modes of the star, or because the star has a tiny deformation from spherical symmetry and is rotating about a non-axisymmetric axis, or because there are density inhomogeneities caused by an environment, or else due to certain relativistic instabilities. We will consider these in turn.

4.2.1. *Supernovae and birth of NS*

The birth of a NS is preceded by the gravitational collapse of a highly evolved star or the core collapse of an accreting white dwarf. Type II supernovae (SN) are believed to result in a compact remnant. In any case, if the collapse is non-spherical then GW could carry away some of the binding energy and angular momentum depending on the geometry of the collapse. It is estimated that in a typical SN, GW might extract about 10^{-7} of the total energy. The waves could come off in a burst whose frequency might lie in the range ~ 200–1000 Hz. Advanced LIGO will be able to see such events up to the Virgo supercluster with an event rate of about 30 per year.

4.2.2. *Equation of state and normal modes of NS*

In order to determine the equation of state (EOS) of a neutron star, and hence its internal structure, it is necessary to independently determine its mass and radius. Astronomical observations cannot measure the radius of a neutron star although radio and X-ray observations do place a bound on its mass. Therefore, it has not been possible to infer the EOS. Neutron stars will have their own distinct normal modes and GW observations of these modes should help resolve the matter since by measuring the frequency and damping times of the modes it would be possible to infer both the radius and mass of NS. The technique is not unlike helioseismology where observation of normal modes of the Sun has facilitated insights into its internal structure. In other words, GW observations of the normal modes of the NS will allow *gravitational asteroseismology*.[4]

Irrespective of the nature of the collapse a number of normal modes will be excited in a newly formed NS. The star will dissipate the energy in these modes in the form of GWs as a superposition of the various normal modes and soon the star settles down to a quiescence state. Normal modes could also be excited in old NS because of the release of energy from star quakes. The strongest of these modes, the ones that are important for GW observations, are the *p*- and *w*-modes for which the restoring forces are the fluid pressure and space-time curvature, respectively. Both of these modes will emit transient radiation which has a generic form of a damped sinusoid: $h(t; \nu, \tau) = h_0 \exp(-t/\tau) \sin(2\pi\nu t)$, where h_0 is the amplitude of the wave that depends on the external perturbation that excites the mode and ν and

τ are the frequency and damping time of the mode, respectively, and are determined by the mass and radius of the NS for a given EOS.

To make an order-of-magnitude estimate let us assume that the mass of the NS is $M_\star = 1.4\,M_\odot$ and that its radius is $R_\star = 10\,\mathrm{km}$. For the p-modes, which are basically fluid modes, the frequency of the fundamental mode, also called the f-mode, is simply the dynamical frequency of the fluid, namely $\nu_f \sim \sqrt{\rho}$, where ρ is the density of the fluid. For a NS of radius $R_\star$ and mass $M_\star$, detailed mode calculations for various EOS have been fitted to yield the following relations for f-modes[4]

$$\nu_f = \left[0.78 + 1.635\left(\frac{M_\star}{R_\star^3}\right)^{1/2}\right] \mathrm{kHz}, \qquad \tau_f^{-1} = \frac{M_\star^3}{R_\star^4}\left[22.85 - 14.65\frac{M_\star}{R_\star}\right] \mathrm{s},$$

(2)

and similarly for w-modes. The f- and w-mode frequencies nicely separate into two distinct groups even when considering more than a dozen different EOS: the f-modes are in the frequency range 1–4 kHz, w-modes are in the range 8–14 kHz, and therefore, detecting a signal at these frequencies places it in one or the other category. The frequency and damping time, together with the relations above, can then be used to fix the radius and mass of the star. Observing several systems should then yield a mass-radius curve which is distinct for each EOS and thereby helps to address the question of NS structure.

The amplitude of f- and w-modes corresponding to 12 different EOS from NS at 10 kpc to 15 Mpc is shown in Fig. 2 as two shaded regions. In a typical gravitational collapse the amount of energy expected to be deposited in f- or w-modes, $\sim 10^{-8}M_\odot$, makes it impossible to detect them in initial LIGO and barely in advanced LIGO instruments, even for a Galactic source. However, EGO should be able to detect these systems with a high SNR. The event rates for these systems would be at least as large as the supernova rate, i.e. about 0.1–0.01 yr^{-1} in our galaxy, increasing to 10–100 yr^{-1} within the Virgo supercluster.

4.2.3. *Relativistic instabilities in NS*

NS suffer *dynamical* and *secular* instabilities caused by *hydrodynamical* and *dissipative* forces, respectively. What is of interest to us is the secular instability driven by gravitational radiation. GW emission from a normal mode in a non-spinning NS would always lead to the decay of the mode. However, the situation might reverse under certain conditions: imagine a NS spinning so fast that a normal mode whose angular momentum (AM) in the star's *rest frame* is *opposite* to its spin, appears to an *inertial observer* to be *co-rotating* with the spin. In the inertial frame, GW extracts positive AM from the mode; therefore the mode's own AM should become more negative. In other words, the amplitude of the mode should grow as a result of GW emission, and hence the instability. The energy for the growth of the mode comes from the rotational energy of the star, which acts like a pump field. Consequently, the star would spin down and eventually halt the instability. It was expected that this instability, called the *CFS instability*,[14,18] might drive the f-modes in a NS unstable, but the star should spin at more than 2 kHz (the smallest f-mode frequency) for this to happen. Moreover, it has been shown that due to viscous damping in the NS fluid the instability would not grow sufficiently large, or sustain for long, to be observable.[4]

It was recently realized[3] that modes originating in current-multipoles, as opposed to mass-multipoles which lead to the f-mode, could be unstable at all rotational speeds of a NS. These modes, called the r-modes, have received a lot of interest because they could potentially explain why spin frequencies of NS in low-mass X-ray binaries are all clustered in a narrow range of 300–600 Hz or why no NS with spin periods smaller than 1.24 ms have been found. The role of r-modes in these circumstances is as yet inconclusive because the problem involves very complicated astrophysical processes (magnetic fields, differential rotation, superfluidity and superconductivity), micro-physics (the precise composition of NS – hyperons, quarks) and full non-linear physics of general relativity. It is strongly expected that r-modes will be emitted by newly formed NS during the first few months of their birth.[4,25] The frequency of these modes will be 4/3 of the spin frequency of the star and might be particularly important if the central object in a low-mass X-ray binary is a strange star.[5] The radiation might last for about 300 years and the signal would be detectable in initial LIGO with a few weeks of integration.

4.2.4. *NS environment*

A NS with an accretion disc would be spun up due to transfer of AM from the disc. Further, accretion could lead to density inhomogeneities on the NS that could lead to the emission of GW. The resulting radiation reaction torque could balance the accretion torque and halt the NS from spinning up. It has been argued[9] that GW emission could be the cause for spin frequencies of NS in low-mass X-ray binaries to be locked up in a narrow frequency range of 300–600 Hz. It is also possible that r-modes are responsible for the locking up of frequencies instead, in which case the waves would come off at a different frequency.[5] These predictions can be tested with advanced LIGO or EGO as Sco-X1, a nearby low-mass X-ray binary, would produce quite a high SNR (marked as $\star$ in Fig. 2).

4.2.5. *Spinning NS with asymmetries*

Our galaxy is expected to have a population of 10^8 NS and they normally spin at high rates (several to 500 Hz). Such a large spin must induce some equatorial bulge and flattening of the poles. The presence of a magnetic field may cause the star to spin about an axis that is different from the symmetry axis leading to a time-varying quadrupole moment.[15] Gravitational waves emitted by a typical NS a distance of $r = 10\,\text{kpc}$ from the Earth will have an amplitude[39] $h \sim 8 \times 10^{-26} f_{\text{kHz}}^2 \epsilon_{-6}$, where f_{kHz} is the frequency of GW in kHz and ϵ_{-6} is the ellipticity of the star in units of 10^{-6}. Fig. 2 plots the signal strength expected from a NS with $\epsilon = 10^{-6}$ at 10 kpc integrated over 4 months.

The ellipticity of neutron stars is not known but one can obtain an upper limit on it by attributing the observed spin-down rate of pulsars as entirely due to gravitational radiation back reaction, namely that the change in the rotational energy is equal to GW luminosity. The ellipticity of the Crab pulsar inferred in this way is $\epsilon \leq 7 \times 10^{-4}$. The GW amplitude corresponding to this ellipticity is $h \leq 10^{-24}$. Noting that Crab has a spin frequency of 30 Hz (or an expected GW frequency of 60 Hz), on integrating the signal for 10^7 s one obtains the source strength to be $h_c = h\sqrt{T_{\text{obs}}} = 3.3 \times 10^{-21}\,\text{Hz}^{-1/2}$, which is easily reachable by LIGO. It is unlikely that the ellipticity is so large and hence the GW amplitude is probably much less. However, seeing Crab at a hundredth of this ellipticity is quite good with advanced LIGO as indicated by a diamond in Fig. 2. (Note that Crab is at 2 kpc, so with an ellipticity of $\epsilon = 7 \times 10^{-6}$ the signal strength would be 35 times higher than the NS line.)

4.3. *Stochastic background*

A population of background sources[19] and quantum processes in the early Universe produce stochastic signals that fills the whole space. By detecting such a stochastic signal we can gain knowledge about the underlying populations and physical processes. A network of antennas can be used to discover stochastic signals buried under the instrumental noise backgrounds. It is expected that the instrumental backgrounds will not be common between two geographically well-separated antennas. Thus, by cross-correlating the data from two detectors we can eliminate the background and filter the interesting stochastic signal. However, when detectors are not co-located the SNR builds only over GW wavelengths longer than twice the distance between antenna, which, in the case of the two LIGO antennas, means over frequencies $\lesssim 40$ Hz.[1] The visibility of a stochastic signal integrated over a period T and bandwidth f only increases as $(fT)^{1/4}$ since cross-correlation uses a 'noisy' filter. But the noise in a bandwidth equal to frequency f is $\sqrt{fS_h(f)}$. Thus, the signal effectively builds up as $(T/f)^{1/4}$.

4.3.1. Astronomical backgrounds

There are thousands of white dwarf binaries in our galaxy with their period in the range from a few hours to 100 seconds. Each binary will emit radiation at a single frequency, but over an observation period T each frequency bin of width $\Delta f = 1/T$ will be populated by many sources. Thus, unless the source is nearby it will not be possible to detect it amongst the *confusion* background created by the underlying population. However, a small fraction of this background population will be detectable as strong foreground sources. The parameters of many white dwarfs are known so well that we can precisely predict their SNRs in LISA and use them to calibrate the antenna. In the inset panel of Fig 2 the curve labelled WDB is the expected confusion noise from Galactic white dwarf binaries.[19,21] NS and BH populations do not produce a large enough background to be observable. Note that the white dwarf background imposes a limitation on the sources we can observe in the frequency region from 0.3 mHz to about 1 mHz – the region where we expect smirches to occur.

4.3.2. Primordial background

A cosmological background should have been created in the very early Universe and later amplified, as a result of parametric amplification, by its coupling to the background gravitational field.[19] Imprint on such a background are the physical conditions that existed in the early Universe as also the nature of the physical processes that produced the background. Observing such a background is therefore of fundamental importance as this is the only way we can ever hope to directly witness the birth of the Universe. The cosmic microwave background, which is our firm proof of the hot early phase of the Universe, was strongly coupled to baryons for 300,000 years after the big bang and therefore the signature of the early Universe is erased from it. The GW background, on the other hand, is expected to de-couple from the rest of matter 10^{-24} s after the big bang, and would therefore carry uncorrupted information about the origin of the Universe.

The strength of stochastic GW background is measured in terms of the fraction Ω_{GW} of the energy density in GW as compared to the critical density needed to close the Universe and the amplitude of GW is given by[39]: $h = 8 \times 10^{-19} \Omega_{\mathrm{GW}}^{1/2}/f$, for $H_0 = 65$ km s^{-1} Mpc. By integrating for $T_{\mathrm{obs}} = 10^7$ s, over a bandwidth f, we can measure a background density at $\Omega_{\mathrm{GW}} \simeq 4 \times 10^{-5}$ in initial LIGO, 5×10^{-9} in advanced LIGO and 10^{-11} in LISA (cf. Fig. 2 and 3 dot-dashed curves marked Ω_{GW}). In the standard inflationary model of the early Universe, the energy density expected in GW is $\Omega_{\mathrm{GW}} \lesssim 10^{-15}$, and this will not be detected by future ground-based detectors or LISA. Future space missions, currently being planned (DECIGO/BBO) to exploit the astronomically quiet band of 0.01–1 Hz, might detect the primordial GW background and unveil the origin of the Universe.

Acknowledgments

It is a pleasure to thank Nils Andersson, Kostas Kokkotas, and Bernard Schutz for useful discussions and Leonid Grishchuk, Joseph Romano, Michael Thompson, for discussions and a careful reading of the draft manuscript. I am greatful to Michele Punturo for the EGO sensitivity curve in Fig. 2.

References

1. B. Allen and J. D. Romano, 1999, PRD 59 102001.
2. A. Abramovici *et al.*, 1992, Science 256, 325.
3. N. Andersson, 1998, ApJ 502, 708.
4. N. Andersson and K. Kokkotas, 2001, IJMPD 10, 381.
5. N. Andersson, D.I. Jones and K. Kokkotas, 2001, MNRAS 337, 1224.
6. BBO, 2003, `http://universe.gsfc.nasa.gov/program/bbo.html`
7. *European Gravitational-Waves advanced Observatory design*, proposal No. 011788 to the European Commission under the Call FP6-2003-Infrastructures-4.
8. P. Bender, *et al.* 1998, *LISA: Pre-Phase A Report*, MPQ 208 (Max-Planck-Institut für Quantenoptik, Garching, Germany, Second Edition, July 1998).
9. L. Bildsten, 1998, ApJ Letters 501, 89.
10. L. Blanchet, 2002, Living Rev. Rel. 5, 3.
11. L. Blanchet and B.S. Sathyaprakash, 1994, Class. Quantum Grav. 11, 2807.
12. L. Blanchet and B.S. Sathyaprakash, 1995, PRL 74, 1067.

13. B. Caron *et al.*, 1997, Class. Quantum Grav. 14, 1461.
14. S. Chandrasekhar, 1970, PRL 24, 611.
15. C. Cutler, 2002, PRD 66, 084025.
16. C. Cutler and K.S. Thorne, 2001, *An Overview of Gravitational Wave Sources,* in Proceedings of 16th International Conference on General Relativity and Gravitation (GR16), Durban, South Africa, 15-21 Jul 2001; e-Print Archive: gr-qc/0204090.
17. L.S. Finn, 1996, PRD 53, 2878.
18. J.L. Friedman and B.F. Schutz, 1978, ApJ 222, 281.
19. L.P. Grishchuk, V.M. Lipunov, K. Postnov, M.E. Prokhorov, B.S. Sathyaprakash, 2001, Phys. Usp. 44, 1 (2001).
20. M.G. Haehnelt, 1998, In *Laser Interferometer Space Antenna,* ed. W.M. Folkner, 456, 45 (AIP Conference Proceedings, AIP, Woodbury, NY).
21. D. Hils, P.L. Bender and R.F. Webbink, 1990, ApJ 360, 75.
22. V. Kalogera, *et al.*, 2004, ApJL, 601, 179, astro-ph/0312101.
23. M. Kawamura, K. Oohara and T. Nakamura, 2003, *General Relativistic Numerical Simulation on Coalescing Binary Neutron Stars and Gauge-Invariant Gravitational Wave Extraction,* astro-ph/0306481.
24. S. Komossa, *et al.*, 2003, ApJ 582 L15-L20.
25. L. Lindblom, B.J. Owen and S.M. Morsink 1998, PRL 80, 4843.
26. H. Lück *et al.*, 1997, Class. Quantum Grav. 14, 1471.
27. D. Merritt and R.D. Ekers, 2002, Science 297, 1310-1313.
28. M.J. Rees, 1997, Class. Quantum Grav. 14, 1411.
29. F.D. Ryan, 1995, PRD 52, 5707.
30. S. Phinney, *private communication.*
31. B.S. Sathyaprakash, 2003, *Problem of searching for spinning black hole binaries,* Proceedings of the XXXVIII Recontres de Moriond, March 24-29, 2003 (in press).
32. B.S. Sathyaprakash and B.F. Schutz, 2003, Class. Quantum Grav. 20, S209.
33. B.F. Schutz, 1986, Nature 323, 310.
34. B.F. Schutz, 1985, A First Course in General Relativity, (Cambridge University Press, Cambridge)
35. B.F. Schutz, 1999, Class. Quantum Grav. 16, A131.
36. J. H. Taylor and Weisberg, 1989, ApJ 345, 434.
37. R. Takahashi and T. Nakamura, 2003, ApJ 596, L231-L234.
38. K. Tsubono, 1995, in *First Edoardo Amaldi Conference on Gravitational Wave Experiments,* (World Scientific, Singapore) 112.
39. K.S. Thorne, 1987, *Gravitational Radiation,* In *300 Years of Gravitation,* Eds. S.W. Hawking and W. Isreal (Cambridge University Press, Cambridge) 330.
40. M. Vallisneri, 2000, PRL 84, 3519.
41. C.M. Will, 1998, PRD 57, 2061.

B.S. Sathyaprakash

Sathyaprakash was born in Bangalore, India in 1959 where he attended the National College. He did his M.Sc. in Physics at the Indian Institute of Technology, Madras and PhD in Theoretical Physics from the Indian Institute of Science, Bangalore. He was a post-doctoral fellow at the Inter-University Centre for Astronomy and Astrophysics (IUCAA), Pune and International Centre for Theoretical Physics, Trieste. After a brief appointment at IUCAA during 1993-1995 he moved to the Cardiff University in 1996 where he is now a Professor of Physics and a Leverhulme Fellow. Sathyaprakash's research interest is primariy in gravitational waves – their generation, propagation and detection – and large-scale structure of the Universe – its formation, evolution and geometry. His recent research includes the dynamics of black hole binaries during their inspiral and merger phases, extracting information from the radiation emitted during that phase, testing general relativity with the aid of such signals and models of structure formation in the Universe and statistical descriptors of the geometry and topology of the large-scale structure. He collaborates with the British-German GEO600, American LIGO and the European Gravitational Observatory and has been a Visiting Professor/Scientist at the *Albert Einstein Institute,* Golm, *California Institute of Technology,* Pasadena and *Institut des Hautes etudes Scientifiques,* Bures-sur-Yvette. Married with two children he enjoys Indian classical music, swimming, badminton and talking to the general public about his research and has an interest in biology (especially neuroscience). The photorgraph shows him delivering the plenary lecture on Gravitational Radiation at the International Cosmic Ray Conference 2003, Tsukuba, Japan.

CHAPTER 8

STRONG-FIELD TESTS OF RELATIVISTIC GRAVITY
WITH PULSARS

Michael Kramer

University of Manchester
Jodrell Bank Observatory
Jodrell Bank, Macclesfield, Cheshire SK11 9DL, UK
E-mail: mkramer@jb.man.ac.uk

A new era in fundamental physics began when pulsars were discovered
in 1967. They are useful tools for a wide variety of physical and astrophysical problems, in particular for the study of theories of relativistic
gravity. Being precise cosmic clocks, pulsars take us beyond the weak-
field limit of the solar-system. Their contribution is crucial as no test
can be considered to be complete without probing the strong-field realm
of gravitational physics by finding and timing pulsars. In this review, I
will explain some of the most important applications of millisecond pulsar clocks in the study of gravity. Recent discoveries such as the double
pulsar, and prospects of finding a pulsar-black hole system are discussed.

1. Introduction

In the 37 years since the discovery of pulsars,[1] these unique objects have
been proven to be invaluable in the study of a wide variety of physical and
astrophysical problems. Most notable are studies of gravitational physics,
the interior of neutron stars, the structure of the Milky Way and stellar
and binary evolution. A number of these studies utilise the pulsar emission properties and/or the interaction of the radiation with the ambient
medium. Most applications, however, are enabled by a technique known as
pulsar timing. Here, pulsar astronomers make use of pulsars as accurate
cosmic clocks where a number of fast-rotating pulsars, so called millisecond pulsars, show long-term stabilities that rival the best atomic clocks
on Earth. Being compact massive objects with the most extreme states of
matter in the present-day Universe, a number of pulsars are also moving
in the gravitational field of a companion star, hence providing ideal con-

ditions for tests of general relativity and alternative theories of gravity. In this review, I briefly describe pulsars and their use as clocks, before I summarise the results of such tests. We review classical tests, recent progress such as discovery of the first double pulsar and look ahead in the future. Firstly, however, we briefly discuss why a continuing challenging of Einstein's theory of gravitation, the theory of general relativity (GR), with new observational data is still necessary.

2. Was Einstein Right?

The theory of GR was an immediate success when it was confronted with experimental data. Whilst being a bold move away from Newton's theory of gravity, which had reigned supreme for nearly three centuries, GR could naturally explain the perihelion advance of Mercury and predicted the deflection of light by the Sun and other massive bodies. Its predictive power motivated physicists to come up with new and more precise tests. Eventually, tests could be performed in space, using, for instance, Lunar laser ranging[2] or radar reflection measurements of inner planets.[3] These tests, providing some of the most stringent limits on GR, are joined by recent modern space probes like Gravity Probe-B.[4] So far, GR has passed all these tests with flying colours. Nevertheless, Einstein may not have had the final word after all. GR is a classical relativistic theory that describes the gravitational interaction of bodies on large scales. One aim of today's physicists is to formulate a theory of quantum gravity which would fuse this classical world of gravitation with the strange world of quantum mechanics. Quantum mechanics (QM) describes nature on small, atomic scales. The success of both GR and QM suggests that we know physics on both extreme scales rather well, but it seems difficult or even impossible to extrapolate one theory into the other's regime. Since the aim of quantum gravity is to account for all of the known particles and interactions of the physical world, it is conceivable that the correct theory of quantum gravity will predict some deviations from GR, eventually occurring in some limit. If our efforts can determine to what extend the as yet accurate theory of GR describes the gravitational interaction of the macroscopic world correctly, the current approaches to use GR as the basis of quantum gravity would be justified. On the other hand, a flaw discovered in GR would imply that other alternative lines of investigations may have to be followed.

How can we find a flaw in GR, should it exist? Certainly, no test can be considered to be complete without probing the *strong-field* realm of

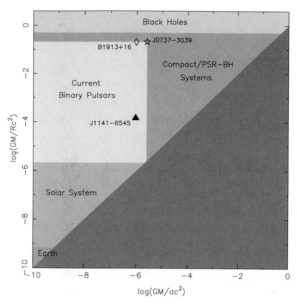

Fig. 1. Parameter space of gravitational physics to be probed with pulsars and black holes. Some theories of gravity predict effects that depend on the compactness of the gravitating body which is shown here (y-axis) as a function of orbital size and probed gravitational potential (x-axis). Note that the lower right half of the diagram is excluded as it implies an orbit smaller than the size of the body.

gravitational physics. In the *weak-field* limit of gravity we deal with small velocities and weak gravitational potentials. The effect of the latter can be estimated from the body's gravitational self-energy, ϵ. For a mass M with radius R, ϵ can expressed in a dimensionless way in units of its rest-mass energy, i.e. $\epsilon = E_{\mathrm{grav}}/Mc^2 \sim -GM/Rc^2$, where G is the gravitational constant and c is the speed of light. In the solar system we find $\epsilon \sim -10^{-6}$ for the Sun, $\epsilon \sim -10^{-10}$ for the Earth and $\epsilon \sim -10^{-11}$ for the Moon, so that we are performing solar system tests clearly in the weak-field limit. The problem is that we cannot rule out that the true theory of gravity "looks and feels" like GR in the weak field, but that it reveals deviations as soon as the strong-field limit is encountered where velocities and self-energies are large. For instance, for a neutron star we expect $\epsilon \sim -0.2$ whilst for a BH $\epsilon = -0.5$, i.e. five to nine orders of magnitude larger than encountered in the solar system. Indeed, Damour & Esposito-Farèse,[5] for instance, have shown that in the family of tensor-scalar theories, alternative solutions to GR exist that would pass all solar system tests but that would be violated as soon as the strong-field limit is reached. Even though in all metric theories like GR matter and non-gravitational fields respond only to a space-time metric, scalar-tensor theories assume that gravity is mediated by a tensor field and one or more scalar field. It is possible that deviations from GR manifest themselves in a "spontaneous scalarization"[5] if the strong field limit is approached. Therefore, solar system tests will not be able to replace the strong field tests provided by radio pulsars.[6]

Radio pulsars probe different regimes of the parameter space, for instance by allowing us to observe the effects of gravitational wave damping. We summarize the probed parameter space in Figure 1 where we concentrate on tests with binary pulsars. Whilst the recent discovery of the first double pulsar system, PSR J0737−3039 (see Sec. 6) pushes the probed limits into previously unexplored regimes, there is still a huge fraction of parameter space that needs studying. These areas can be explored by finding and studying radio pulsars in even more compact orbits, in particular in orbits about stellar and massive black holes. We discuss the exciting possibilities arising from such pulsar-black hole systems in some detail in Sec. 7.

3. Pulsars

Pulsars are highly magnetized, rotating neutron stars which emit a narrow radio beam along their magnetic dipole axis. As the magnetic axis is inclined to the rotation axis, the pulsar acts like a cosmic light-house emitting a radio pulse that can be detected once per rotation period when the beam is directed towards Earth.

3.1. *Neutron star properties*

Pulsars are born as neutron stars in supernova explosions of massive stars, created in the collapse of the progenitor's core. Neutron stars are the most compact objects next to black holes. From timing measurements of binary pulsars, we can determine the masses of pulsars to be within a narrow range around $1.35 M_\odot$,[7] where the smallest value is found for pulsar B in the double pulsar system with $M_B = 1.250 \pm 0.005 M_\odot$.[8] Modern calculations for different equations of state[9] produce results for the size of a neutron star quite similar to the very first calculations by Oppenheimer & Volkov,[10] i.e. about 20 km in diameter. Such sizes are consistent with independent estimates derived from modelling light-curves and luminosities of pulsars observed at X-rays.[11] Underneath a solid crust, the interior is super-conducting and super-liquid at the same time.[12]

As rotating magnets, pulsars lose rotational energy due to magnetic dipole radiation and a particle wind. This leads to an increase in rotation period, P, described by $\dot{P} > 0$. Equating the corresponding energy output of the dipole to the loss rate in rotational energy, we obtain an estimate for the magnetic field strength at the pulsar surface,

$$B = 3.2 \times 10^{19} \sqrt{P\dot{P}} \text{ G}, \tag{1}$$

with P measured in s and $\dot{P}$ in s s^{-1}. Typical values are of order 10^{12} G, although field strengths up to 10^{14} G have been observed.[13] Millisecond pulsars have lower field strengths of the order of 10^8 to 10^{10} G which appear to be a result of their evolutionary history (see Sec. 3.3). These magnetic fields are consistent with values derived from cyclotron absorption line features in X-ray spectra.[14]

3.2. *Radio emission*

The radio signal of a pulsar is of coherent nature but usually weak, both because the pulsar is distant and the size of the actual emission region is small. Despite intensive research, the responsible radiation processes are still unidentified. We believe that the neutron star's rotating magnetic field induces an electric quadrupole field which is strong enough to pull out charges from the stellar surface (the electrical force exceeds the gravitational force by a factor of $\sim 10^{12}$!). The magnetic field forces the resulting dense plasma to co-rotate with the pulsar. This *magnetosphere* can only extend up to a distance where the co-rotation velocity reaches the speed of light. This distance defines the so-called *light cylinder* which separates the magnetic field lines into two distinct groups, i.e. *open and closed field lines*. The plasma on the closed field lines is trapped and co-rotates with the pulsar forever. In contrast, plasma on the open field lines reaches highly relativistic velocities, produces the observed radio beam at some distance to the pulsar surface, and finally leaves the magnetosphere in a particle wind that contributes to the pulsar spin-down.[15]

3.3. *Evolution*

The observed rotational slow-down, $\dot{P}$, and the resulting evolution in pulsar period, P, can be used to describe the life of a pulsar. This is usually done in a (logarithmic) P-$\dot{P}$-diagram as shown in Figure 2 where we can draw lines of constant loss in rotational energy, magnetic field (Eq. (1)), and constant

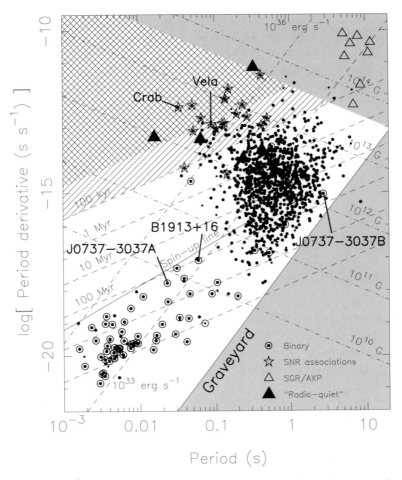

Fig. 2.   The $P - \dot{P}$–diagram for the known pulsar population. Lines of constant characteristic age, surface magnetic field and spin-down luminosity are shown. Binary pulsars
are marked by a circle. The lower solid line represents the pulsar "death line" enclosing
the "pulsar graveyard" where pulsars are expected to switch off radio emission. The gray
area in the top right corner indicates the region where the surface magnetic field appears
to exceed the quantum critical field of 4.4×10^{13} Gauss. For such values, some theories
expect the quenching of radio emission in order to explain the radio-quiet "magnetars"
(i.e. Soft-gamma ray repeaters, SGRs, and Anomalous X-ray pulsars, AXPs). The upper
solid line is the "spin-up" line which is derived for the recycling process as the period
limit for millisecond pulsars.

characteristic age, estimated from

$$\tau = \frac{P}{2\dot{P}}. \tag{2}$$

This quantity is a valid estimate for the true age under the assumption that the initial spin period is much smaller than the present period and that the spin-down is fully determined by magnetic dipole braking. Pulsars are born in the upper left area of Fig. 2 and move into the central part where they spend most of their lifetime.

Most known pulsars have spin periods between 0.1 and 1.0 s with period derivatives of typically $\dot{P} = 10^{-15}$ s s^{-1}. Selection effects are only partly responsible for the limited number of pulsars known with very long periods, the longest known period being 8.5 s.[16] The dominant effect is due to the "death" of pulsars when their slow-down has reached a critical state. This state seems to depend on a combination of P and $\dot{P}$ known as the *pulsar death-line*. The normal life of radio pulsars is limited to a few tens or hundreds million years or so.

The described evolution does not explain the about 100 pulsars in the lower left of the $P - \dot{P}$-diagram. These pulsars have simultaneously small periods (few milliseconds) and small period derivatives ($\dot{P} \lesssim 10^{-18}$ s s^{-1}). These *millisecond pulsars* (MSPs) are much older than ordinary pulsars with ages up to $\sim 10^{10}$ yr. It is believed that MSPs are born when mass and thereby angular momentum is transferred from an evolving binary companion while it overflows its Roche lobe.[17] In this model, MSPs are recycled from a dead binary pulsar via an accreting X-ray binary phase. This model implies a number of observational consequences: a) most normal pulsars do not develop into a MSP as they have long lost a possible companion during their violent birth event; b) for surviving binary systems, X-ray binary pulsars represent the progenitor systems for MSPs; c) the final spin period of recycled pulsars depends on the mass of the binary companion. A more massive companion evolves faster, limiting the duration of the accretion process; d) the ram-pressure of the magnetic field limits the accretion flow onto the neutron star surface. This results in the limiting *spin-up line* shown in Fig. 2; e) the majority of MSPs have low-mass white-dwarf companions as the remnant of the binary star. These systems evolve from low-mass X-ray binary systems; f) high-mass X-ray binary systems represent the progenitors for double neutron star systems (DNSs). DNSs are rare since these systems need to survive a second supernova explosion. The resulting MSP is only mildly recycled with a period of tens of millisecond. The properties of MSPs and X-ray binaries are consistent with the described picture. For instance, it is striking that $\sim 80\%$ of all MSPs are in a binary orbit while this is true for only less than 1% of the non-recycled population. For MSPs with a low-mass white dwarf companion the orbit is nearly circular.

In case of DNSs, the orbit is affected by the unpredictable nature of the kick imparted onto the newly born neutron star in the asymmetric supernova explosion of the companion. If the system survives, the result is typically an eccentric orbit with an orbital period of a few hours.

4. Pulsar Timing

By measuring the arrival time of the pulsar signals clock very precisely, we can study effects that determine the propagation of the pulses in four-dimensional space-time. Millisecond pulsars are the most useful objects for these investigations: their pulse arrival times can be measured much more accurately than for normal pulsars (the measurement precision scales essentially with spin frequency) and their rotation is much smoother, making them intrinsically better clocks. For both types we use pulsar timing to account for *every single* rotation of the neutron star between two different observations. Hence, we aim to determine the number of an observed pulse, counting from some reference epoch, t_0. We can write

$$N = N_0 + \nu_0 \times (t - t_0) + \frac{1}{2}\dot{\nu}_0 \times (t - t_0)^2 + ..., \tag{3}$$

where N_0 is the pulse number and ν_0 the spin frequency at the reference time, respectively. Whilst for most millisecond pulsars a second derivative, $\ddot{\nu}$, is usually too small to be measured, we expect ν and $\dot{\nu}$ to be related via the physics of the braking process. For magnetic dipole braking we find

$$\dot{\nu} = -\text{const.} \times \nu^n, \tag{4}$$

where the *braking index* takes the value $n = 3$. If ν and its derivatives are accurately known and if t_0 coincides with the arrival of a pulse, all following pulses should appear at integer values of N — when observed in an inertial reference frame. However, our observing frame is not inertial, as we are using telescopes that are located on a rotating Earth orbiting the Sun. Therefore, we need to transfer the pulse times-of-arrival (TOAs) measured with the observatory clock (*topocentric arrival times*) to the centre of mass of the solar system as the best approximation to an inertial frame available. The transformation of a topocentric TOA to such *barycentric arrival times*, t_{SSB}, is given by

$$t_{\text{SSB}} = \quad t_{\text{topo}} - t_0 + t_{\text{corr}} - D/f^2 \,, \tag{5}$$

$$+ \Delta_{\text{Roemer},\odot} + \Delta_{\text{Shapiro},\odot} + \Delta_{\text{Einstein},\odot} \,, \tag{6}$$

$$+ \Delta_{\text{Roemer,Bin}} + \Delta_{\text{Shapiro,Bin}} + \Delta_{\text{Einstein,Bin}}. \tag{7}$$

We have split the transformation into three lines. The first two lines apply to every pulsar whilst the third line is only applicable to binary pulsars.

4.1. *Clock and frequency corrections*

The observatory time is typically maintained by local Hydrogen-maser clocks monitored by GPS signals. In a process involving a number of steps, clock corrections, t_{corr}, are retroactively applied to the arrival times in order to transfer them to a uniform atomic time known as *Terrestrial Time*, TT (formerly known as Terrestrial Dynamical Time, TDT). The unit of TT is the SI second and may be regarded as the time that would be kept by an ideal atomic clock on the geoid. It is published retroactively by the *Bureau International des Poids et Mesures* (BIPM).

As the pulses are delayed due to dispersion in the interstellar medium, the arrival time depends on the observing frequency, f. The TOA is therefore corrected for a pulse arrival at an infinitely high frequency (last term in Eq. 5). The corresponding dispersion measure, DM, is determined during the discovery of a pulsar. For some pulsars, "interstellar weather" causes small changes in DM, which would cause time-varying drifts in the TOAs. In such cases, the above term is extended to include time-derivatives of DM which can be measured using multi-frequency observations.

4.2. *Barycentric corrections*

The *Roemer delay*, $\Delta_{\mathrm{Roemer},\odot}$, is the classical light-travel time between the phase centre of the telescope and the solar system barycentre (SSB). Given a unit vector, $\hat{s}$, pointing from the SSB to the position of the pulsar and the vector connecting the SSB to the observatory, $\vec{r}$, we find:

$$\Delta_{\mathrm{Roemer},\odot} = -\frac{1}{c}\, \vec{r} \cdot \hat{s} = -\frac{1}{c}\left(\vec{r}_{\mathrm{SSB}} + \vec{r}_{\mathrm{EO}}\right) \cdot \hat{s}. \tag{8}$$

Here c is the speed of light and we have split $\vec{r}$ into two parts. The vector $\vec{r}_{\mathrm{SSB}}$ points from the SSB to the centre of the Earth (geocentre). Computation of this vector requires accurate knowledge of the locations of all major bodies in the solar system and uses *solar system ephemerides*. The second vector $\vec{r}_{\mathrm{EO}}$, connects the geocentre with the phase centre of the telescope. In order to compute this vector accurately, the non-uniform rotation of the Earth has to be taken into account, so that the correct relative position of the observatory is derived.

The *Shapiro delay*, $\Delta_{\mathrm{Shapiro},\odot}$, is a relativistic correction that corrects for an extra delay due to the curvature of space-time in the solar system.[3]

It is largest for a signal passing the Sun's limb ($\sim 120\ \mu s$) while Jupiter can contribute as much as 200 ns. In principle one has to sum over all bodies in the solar system, but in practice only the Sun is usually taken into account.

The last term in Eq. (6), $\Delta_{\mathrm{Einstein},\odot}$, is called *Einstein delay* and it describes the combined effect of gravitational redshift and time dilation due to motion of the Earth and other bodies, taking into account the variation of an atomic clock on Earth in the varying gravitational potential as it follows its elliptical orbit around the Sun.[18]

4.3. *Relative motion & Shklovskii effect*

If the pulsar is moving relative to the SSB, the transverse component of the velocity, v_t, can be measured as the vector $\hat{s}$ in Eq. (8) changes with time. Present day timing precision is not sufficient to measure a radial motion although it is theoretically possible. This leaves Doppler corrections to observed periods, masses etc. undetermined. The situation changes if the pulsar has an optically detectable companion such as a white dwarf for which Doppler shifts can be measured from optical spectra.

Another effect arising from a transverse motion is the *Shklovskii effect*, also known in classical astronomy as *secular acceleration*. With the pulsar motion, the projected distance of the pulsar to the SSB is increasing, leading to an increase in any observed change of periodicity, such as pulsar spin-down or orbital decay. The observed pulse period derivative, for instance, is increased over the intrinsic value by

$$\left(\frac{\dot{P}}{P}\right)_{\mathrm{obs}} = \left(\frac{\dot{P}}{P}\right)_{\mathrm{int}} + \frac{1}{c}\frac{v_t^2}{d}. \tag{9}$$

For millisecond pulsars where $\dot{P}_{\mathrm{int}}$ is small, a significant fraction of the observed change in period can be due to the Shklovskii effect.

4.4. *Binary pulsars*

Equations (5) and (6) is used to transfer the measured TOAs to the SSB. If the pulsar has a binary companion, the light-travel time across the orbit and further relativistic effects need to be taken into account (see Eq. (7)). We can group the parameters that are derived from a least-squares fit to Eq. (3) into three categories: *astrometric parameters* (position, proper motion and parallax), *spin parameters* (ν and its derivatives), and *binary parameters*. Five Keplerian parameters need to be determined (see Fig. 3), i.e. orbital

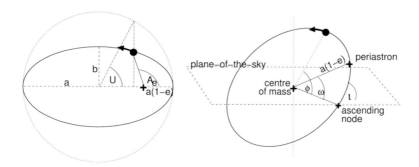

Fig. 3. Definition of the orbital elements in a Keplerian orbit and the angles relating both the orbit and the pulsar to the observer's coordinate system and line-of-sight. (left) The closest approach of the pulsar to the centre-of-mass of the binary system marks periastron, given by the longitude ω and a chosen epoch T_0 of its passage. The distance between centre of mass and periastron is given by $a(1-e)$ where a is the semi-major axis of the orbital ellipse and e its eccentricity. (right) Usually, only the projection on the plane of the sky, $a \sin i$, is measurable, where i is the orbital inclination defined as the angle between the orbital plane and the plane of the sky. The *true anomaly*, A_E, and *eccentric anomaly*, U are related to the *mean anomaly*. The orbital phase of the pulsar Φ is measured relative to the ascending node.

period, $P_{\rm b}$; the projected semi-major axis of the orbit, $x \equiv a \sin i$ where i is the (usually unknown) orbital inclination angle; the orbital eccentricity, e; the longitude of periastron, ω; and and the time of periastron passage, T_0. For a number of binary systems this Newtonian description of the orbit is not sufficient and relativistic corrections need to be applied, e.g. ω is replaced by $\omega + \dot{\omega}t$. The measurement of the *Post-Keplerian* (PK) *parameters* such as $\dot{\omega}$ allows a comparison with values expected in the framework of specific theories of gravity.

5. Tests of Relativistic Gravity

In the following we will summarize tests of relativistic gravity using pulsars as accurate cosmic clocks. Some of these tests involve studies of the parameters in the so-called *Parameterized Post-Newtonian* (PPN) formalism. Other tests make use of PK parameters.

5.1. *Tests of equivalence principles*

Metric theories of gravity assume *(i)* the existence of a symmetric metric, *(ii)* that all test bodies follow geodesics of the metric and *(iii)* that in local Lorentz frames the non-gravitational laws of physics are those of special relativity. Under these conditions we can study metric theories with the PPN formalism by describing deviations from simple Newtonian physics by a set of 10 real-valued PPN-Parameters.[19] Each of the parameters can be associated with a specific physical effect, like the violation of conservation of momentum or equivalence principles. Comparing measured PPN parameters to their theoretical values can tests theories in a purely experimental way. A complete description of the PPN formalism is presented by Will (2001),[19] whilst specific pulsar tests are detailed by Stairs (2003).[20]

Many tests of GR and alternative theories are related to tests of the *Strong Equivalence Principle* (SEP) which is completely embodied into GR. Alternative theories of gravity may predict a violation of some or all aspects of SEP. The SEP is, according to its name, stronger than both the *Weak Equivalence Principle* (WEP) and the *Einstein Equivalence Principle* (EEP). The WEP, included in all metric theories, states that all test bodies in an external gravitational field experience the same acceleration regardless of the mass and composition. The EEP also postulates Lorentz- and positional invariance, whilst the SEP includes both the WEP and the EEP, but postulates them also for gravitational experiments. As a consequence, both Lorentz- and positional invariance should be independent of the gravitational self-energy of the bodies in the experiment. Since all terrestrial bodies possess only a negligible fraction of gravitational self-energy, tests require the involvement of astronomical objects. A violation of SEP means that there is a difference between gravitational mass, M_g, and inertial mass, M_i. The difference can be written as

$$\frac{M_g}{M_i} \equiv 1 + \delta(\epsilon) = 1 + \eta\epsilon + \mathcal{O}(\epsilon^2), \qquad (10)$$

where the parameter η characterises the violation of SEP and ϵ is again the dimensionless self-energy. A possible difference can probed in the Earth-Moon system. Due to their different self-energies, Earth and Moon would fall differently in the external gravitational field of the Sun, leading to a polarization of the Earth-Moon orbit ("Nordvedt-effect"). Lunar-laser-ranging experiments can be used to put tight limits on η which is a linear combination of PPN parameters representing effects due to preferred locations, preferred frames and the violation of the conservation of momentum.[2] However, as discussed earlier, solar system experiments cannot make tests in the strong-field regime where deviations to higher order terms of ϵ could be present. We can probe this possibility using pulsar-white dwarf systems. since for neutron stars $\epsilon \sim -0.2$ is much larger than for white dwarfs,

$\epsilon \sim -10^{-4}$. A violation of the SEP would mean a acceleration that is different for the pulsar and the white dwarf in the external gravitational field of the Galaxy, and similar to the Nordvedt effect, this should lead to a polarisation of the orbit.[21] Since the orbit is also slowly precessing, a careful analysis of all relevant low-eccentricity pulsar-white dwarf systems in a statistical manner is needed but leads to limits that are comparable to those obtained in the weak-field.[22]

Some metric theories of gravity violate SEP specifically by predicting preferred-frame and preferred-location effects. A preferred universal rest frame, presumably equivalent with that of the Cosmic Microwave Background, may exist if gravity is mediated in part by a long-range vector field. A violation of local Lorentz-invariance would cause a binary system moving with respect to a preferred universal rest frame to suffer a long-term change in its orbital eccentricity. A statistical analysis of pulsar-white dwarf systems yields a limit of the PPN parameter $|\alpha_1| < 1.2 \times 10^{-4}$ (95% C.L.)[22] which is expected to be zero in GR.

Essentially any metric theory of gravity that embodies Lorentz-invariance in its field equations predicts gravitational radiation to be emitted by binary systems. In GR the *quadrupole* term is the lowest non-zero multipole component. In a theory that violates SEP gravitational *dipole* radiation can occur whereas the amplitude depends on the difference in gravitational binding energies, expressed by the difference in coupling constants to a scalar field, $(\hat{\alpha}_p - \hat{\alpha}_c)$, which are both zero in GR. For a white dwarf companion the coupling constant is much smaller than for a pulsar, $|\alpha_c| \ll |\alpha_p|$, so that the strongest emission should occur for short-orbital period pulsar-white dwarf systems. The currently best limit is provided by the binary pulsar PSR J1141−6545, consisting of a young pulsar and a heavy white dwarf companion ($M = 0.99 \pm 0.02 M_{\odot}$) in a compact eccentric 4.5-hr orbit.[23,24] The derived tight upper limit of $\alpha_p \lesssim 0.004$, is an order of magnitude better than the previous limits.[25]

5.2. *Tests of GR with double-neutron-stars*

If we study the compact orbits of double neutron stars, we enter a regime where we expect very much larger effects than previously discussed. In this case a PPN approximation is no longer valid. Instead, we can use an existing theory of gravity and check if the observations are consistently described by the measured Keplerian and PK parameters. In each theory, for negligible spin-contributions the PK parameters should only be functions of the a priori unknown pulsar and companion mass, M_p and M_c, and the easily measurable Keplerian parameters. With the two masses as the only free parameters during the test, an observation of two PK parameters will already determine the masses uniquely in the framework of the given theory. The measurement of a third or more PK parameters then provides a consistency check. In general relativity, the five most important PK parameters are given by the expressions in the box below.[26–29]

$$\dot{\omega} = 3T_\odot^{2/3} \left(\frac{P_b}{2\pi}\right)^{-5/3} \frac{1}{1-e^2} (M_p + M_c)^{2/3},$$

$$\gamma = T_\odot^{2/3} \left(\frac{P_b}{2\pi}\right)^{1/3} e \frac{M_c(M_p + 2M_c)}{(M_p + M_c)^{4/3}},$$

$$\dot{P_b} = -\frac{192\pi}{5}T_\odot^{5/3} \left(\frac{P_b}{2\pi}\right)^{-5/3} \frac{\left(1 + \frac{73}{24}e^2 + \frac{37}{96}e^4\right)}{(1-e^2)^{7/2}} \frac{M_p M_c}{(M_p + M_c)^{1/3}},$$

$$r = T_\odot M_c,$$

$$s = T_\odot^{-1/3} \left(\frac{P_b}{2\pi}\right)^{-2/3} x \frac{(M_p + M_c)^{2/3}}{M_c},$$

The five most important Post-Keplerian (PK) parameters as predicted by GR. All PK parameters can be expressed as a function of the Keplerian parameters and the pulsar and companion mass. P_b is the period and e the eccentricity of the binary orbit. The masses M_p and M_c of pulsar and companion, respectively, are expressed in solar masses $(M_\odot)$ where we define the constant $T_\odot = GM_\odot/c^3 = 4.925490947 \mu s$. G denotes the Newtonian constant of gravity and c the speed of light.

The PK parameter $\dot{\omega}$ is the easiest to measure and describes the relativistic advance of periastron. In GR, it provides an immediate measurement of the total mass of the system, $(M_p + M_c)$. The parameter γ denotes the amplitude of delays in arrival times caused by the varying effects of the gravitational redshift and time dilation (second order Doppler) as the pulsar moves in its elliptical orbit at varying distances from the companion and with varying speeds. The decay of the orbit due to gravitational wave damping is expressed by the change in orbital period, $\dot{P_b}$. The other two parameters, r and s, are related to the Shapiro delay caused by the curvature of space-time due to the gravitational field of the companion. They are measurable, depending on timing precision, if the orbit is seen nearly edge-on.

In the past, only two binary pulsars had more than two PK parameters determined, the 59-ms pulsar B1913+16 and the 38-ms B1534+12. For PSR B1913+16 with an eccentric ($e = 0.61$) 7.8-hr orbit, the PK parameters $\dot{\omega}$, γ and $\dot{P}_b$ are measured very precisely. Correcting the observed $\dot{P}_b$ value for the Shklovskii effect (see Sec. 4.3), the measured value is in excellent agreement with the prediction of GR for quadrupole emission (see Fig. 4). It demonstrates impressively that GR provides a self-consistent and accurate description of the system as orbiting point masses, i.e. the structure of the neutron stars does not influence their orbital motion as expected from SEP. The precision of this test is limited by our knowledge of the Galactic gravitational potential and the corresponding correction to $\dot{P}_b$. The timing results for PSR B1913+16 provide us with the currently most precise measurements of neutron star masses ever, i.e. $M_p = (1.4408 \pm 0.0003) M_\odot$ and $M_c = (1.3873 \pm 0.0003) M_\odot$.[30]

The 10-hr orbit of the second DNS B1534+12 ($e = 0.27$) is observed nearly edge-on. Thereby, in addition to the three PK parameters observed for PSR B1913+16, the Shapiro-delay parameters r and s can be measured, enabling non-radiative aspects of gravitational theories to be tested, as $\dot{P}_b$ is not needed. In fact, the observed value of $\dot{P}_b$ seems to be heavily influenced by Shlovskii-terms, i.e. it is larger than the value expected from GR by a term $\propto v_t^2/d$ (see Eq. (9)). Assuming that GR is the correct theory of gravitation, the measured proper motion and deviation from the predicted value can be used to derive the distance to the pulsar, $d = 1.02 \pm 0.05$ kpc.[31]

5.3. *Tests with pulsar structure data*

In addition to the use of pulsars as clocks, strong gravity effects can also be tested using pulse structure data, namely the effects of *geodetic precession* seen for PSRs B1913+16[32,34] and B1534+12.[35] In both cases, the pulsar spin axis appears to be misaligned with the orbital angular momentum vector, so that general relativity predicts a relativistic spin-orbit coupling. The pulsar spin precesses about the total angular momentum, changing the relative orientation of the pulsar towards Earth. As a result, the angle between the pulsar spin axis and our line-of-sight changes with time, so that different portions of the emission beam are observed[36] leading to changes in the measured pulse profile (Figure 4). In extreme cases, the precession may even move the beam out of our line-of-sight and the pulsar may disappear as predicted for PSR B1913+16 for the year 2025.[32]

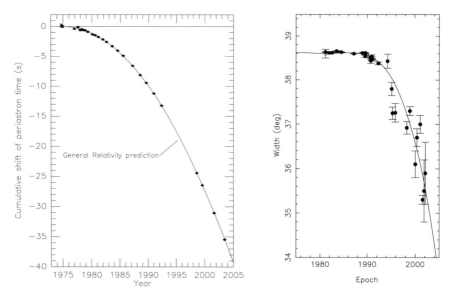

Fig. 4.   (left) Shift in the periastron passage of the DNS PSR B1913+16 plotted as a function of time, resulting from orbital energy loss due to the emission of gravitational radiation. The agreement between the data, now spanning almost 30 yr, and the predicted curve due to gravitational quadrupole wave emission is now better than 0.5%. Figure provided by Joel Weisberg and Joe Taylor. (right) Component separation as measured for PSR B1913+16 (Kramer 1998, Kramer et al. 2003). The component separation is shrinking as the line-of-sight changes due to geodetic precession. Extrapolating the model leads to the prediction that the pulsar should disappear in the year 2025, reappearing again around the year 2245.

6. The Double-Pulsar

The most remarkable discovery was made in 2003 with the DNS J0737−3039.[8,37] This system is not only the most relativistic one ever discovered, but it is also unique because both neutron stars are visible as radio pulsars. A recycled 22.8-ms pulsar, J0737-3039A, is in a 2.4-hr orbit with a young 2.8-s pulsar, J0737-3039B. Whilst this discovery clearly confirms the evolution scenario presented in Sec. 3.3, it also provides a truly unique testbed for relativistic gravity. The short orbital period and an eccentricity of $e = 0.09$ leads to a remarkably fast relativistic precession of the orbit with a periastron advance of $\dot{\omega} = 16.88 \pm 0.09$ deg yr^{-1}, which is four times larger than for PSR B1913+16! A short coalescing time of only 85 Myr also increases the estimated detection rates for a merger of two neutron stars with first-generation ground-based gravitational wave detectors by about an order of magnitude to one event every $2 - 3$ years.[37] Moreover, with a geodetic precession period of only 70 yr (compared to 300 yr for B1913+16), profile changes are expected soon. Meanwhile, the PK parameter γ and a Shapiro delay have also been measured. The latter provides a precise measurement of the orbital inclination of $\sin i = 0.9995^{+4}_{-32}$, i.e. as another strike of luck, we are observing the system almost completely edge-on which allows us to also probe pulsar magnetospheres for the very first time by a background beacon. The measurements of four PK parameters are summarized in a $M_A - M_B$ plot shown in Fig. 5.[a] The detection of B as a pulsar opens up opportunities that go well beyond what has been possible so far. With a measurement of the projected semi-major axes of the orbits of both A and B, we obtain a precise measurement of the mass ratio, $R(M_A, M_B) \equiv M_A/M_B = x_B/x_A$, providing a further constraint displayed in Fig. 5. The R-line is not only theory-independent,[38] but also independent of strong-field (self-field) effects which is not the case for PK-parameters. This provides a stringent and new constraint for tests of gravitational theories as any intersection of the PK-parameters *must* be located on the R-line.

The equations for the PK parameters given in Sec. 5.2 are all given to lowest Post-Newtonian order. However, higher-order terms may become important if relativistic effects are large and timing precision is sufficiently high. Whilst this has not been the case in the past, the double pulsar system may allow measurements of these effects in the future.[8] One such effect involves the prediction by GR that, in contrast to Newtonian physics, the neutron stars' spins affect the orbital motion via spin-orbit coupling. This effect would be visible clearest as a contribution to the observed $\dot{\omega}$[39,40] which for the J0737−3039 system should be about an order of magnitude larger than for PSR B1913+16. The exact value depends on the neutron star moment of inertia which could be determined for the first time if this effect is measured.[41]

7. Future Tests Using Pulsars and Black Holes

Whilst the currently possible tests of GR are exciting, they are only the prelude to what will be possible once the Square Kilometre Array (SKA) comes online. The SKA project is a global effort to built a radio telescope interferometer with a total collecting area of $10^6 \mathrm{m}^2$. It will be about 100 times more sensitive than the VLA, GBT or Effelsberg and about 200 times

---

[a]At the time of writing, the orbital decay due to gravitational wave emission is already visible in the data, but the uncertainties are yet too large to provide a useful constraint.

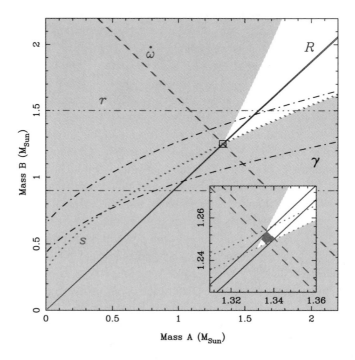

Fig. 5. "Mass-mass" diagram showing the observational constraints on the masses of the neutron stars in the double-pulsar system J0737–3039. The shaded regions are those which are excluded by the Keplerian mass functions of the two pulsars. Further constraints are shown as pairs of lines enclosing permitted regions as predicted by GR: (a) advance of periastron, $\dot{\omega}$; (b) mass ratio $R = m_A/m_B = 1.07$; (c) gravitational redshift/time dilation parameter γ; (d) Shapiro delay parameters r and s. Inset is an enlarged view of the small square encompassing the intersection of the three tightest constraints, representing the area allowed by general relativity and the present measurements. From these measurements the masses of the neutron stars are determined to be $M_A = (1.337 \pm 0.005)M_\odot$ and $M_B = (1.250 \pm 0.005)M_\odot$, respectively, making B the least-massive neutron star ever observed.

more sensitive than the Lovell telescope. Pulsar surveys with the SKA will essentially discover all active pulsars in the Galaxy that are beamed toward us. In addition to this complete Galactic Census, pulsars will be discovered in external galaxies as far away as the Virgo cluster. Most importantly for probing relativistic gravity is the prospect that the SKA will almost certainly discover the first pulsar orbiting a black hole (BH). Strong-field tests using such unprecedented probes of gravity have been identified as one of the key science projects for the SKA.[42] The SKA will enable us to measure both the BH spin and the quadrupole moment using the effects of classical and relativistic spin-orbit coupling – impossible with the timing precision affordable with present-day telescopes.[43] Having extracted the dimensionless spin and quadrupole parameters, χ and q,

$$\chi \equiv \frac{c}{G} \frac{S}{M^2} \quad \text{and} \quad q = \frac{c^4}{G^2} \frac{Q}{M^3}, \tag{11}$$

where S is the angular momentum and Q the quadrupole moment, we can use these measured properties of a BH to confront the predictions of GR[42,43] such as the "Cosmic Censorship Conjecture" and the "No-hair theorem".

In GR the curvature of space-time diverges at the centre of a BH, producing a singularity, which physical behaviour is unknown. The Cosmic Censorship Conjecture was invoked by Penrose in 1969[44] to resolve the fundamental concern that if singularities could be seen from the rest of space-time, the resulting physics may be unpredictable. The Cosmic Censorship Conjecture proposes that singularities are always hidden within the event horizons of BHs, so that they cannot be seen by a distant observer. Whether the Cosmic Censor Conjecture is correct remains an unresolved key issue in the theory of gravitational collapse. If correct, we would always expect $\chi \leq 1$, so that the complete gravitational collapse of a body always results in a BH rather than a naked singularity.[45] In contrast, a value of $\chi > 1$ would imply that the event horizon has vanished, exposing the singularity to the outside world. Here, the discovered object would not be a BH as described by GR but would represent an unacceptable naked singularity and hence a violation of the Cosmic Censorship Conjecture.[46]

One may expect a complicated relationship between the spin of the BH, χ, and its quadrupole moment, q. However, for a rotating Kerr BH in GR, both properties share a simple, fundamental relationship,[47,48]

$$q = -\chi^2. \tag{12}$$

This equation reflects the "no-hair" theorem of GR which implies that the external gravitational field of an astrophysical (uncharged) BH is fully determined by its mass and spin.[12] Therefore, by determining q and χ from timing measurements with the SKA, we can confront this fundamental prediction of GR for the very first time.

Finally, about 1000 millisecond pulsars to be discovered with the SKA can also be used to directly detect gravitational radiation in contrast to the indirect measurements from orbital decay in binaries. Pulsars discovered and timed with the SKA act effectively as the endpoints of arms of a huge, cosmic gravitational wave detector which can measure a stochastic background spectrum of gravitational waves predicted from energetic processes in the early Universe. This "device" with the SKA at its heart promises to detect such a background, at frequencies that are below the band accessible even to LISA.[42]

8. Conclusions

Pulsars provide some of the most stringent and in many cases the only constraints for theories of relativistic gravity in the strong-field limit. Being precise clocks, moving in deep gravitational potentials, they are a physicist's dream-come-true. With the discoveries of the first pulsar, the first binary pulsar, the first millisecond pulsar, and now recently also the first double pulsar, a wide range of parameter space can be probed. The SKA will provide yet another leap in our understanding of relativistic gravity and hence in the quest for quantum gravity.

Acknowledgements

I am grateful to Norbert Wex and Ingrid Stairs for many useful discussions.

References

1. A. Hewish, S. J. Bell, J. D. H. Pilkington, P. F. Scott, R. A. Collins, *Nature* **217** 709–713 (1968).
2. K. Nordtvedt, *Phys. Rev.* **170** 1186–1187 (1968).
3. I. I. Shapiro, *Physical Review Letters* **13** 789–791 (1964).
4. S. Buchman, C. W. F. Everitt, B. Parkinson, et al., *Advances in Space Research* **25** 1177–1180 (2000).
5. T. Damour, G. Esposito-Farèse, *Phys. Rev. D* **53** 5541–5578 (1996).
6. T. Damour, G. Esposito-Farèse, *Phys. Rev. D* **58** (**042001**) 1–12 (1998).
7. I. H. Stairs, *Science* **304** 547–552 (2004).
8. A. G. Lyne, M. Burgay, M. Kramer, et al., *Science* **303** 1153–1157 (2004).
9. J. M. Lattimer, M. Prakash, *ApJ* **550** 426–442 (2001).
10. J. R. Oppenheimer, G. Volkoff, *Phys. Rev.* **55** 374–381 (1939).
11. V. E. Zavlin, G. G. Pavlov, *A&A* **329** 583 (1998).
12. S. L. Shapiro, S. A. Teukolsky, Black Holes, White Dwarfs and Neutron Stars. The Physics of Compact Objects, Wiley–Interscience, New York, 1983.
13. M. A. McLaughlin, I. H. Stairs, V. M. Kaspi, et al., *ApJ* **591** L135–L138 (2003).
14. G. F. Bignami, P. A. Caraveo, A. D. Luca, S. Mereghetti, *Nature* **423** 725–727 (2003).

15. A. G. Lyne, F. G. Smith, Pulsar Astronomy, Cambridge University Press, Cambridge, 1990.
16. M. D. Young, R. N. Manchester, S. Johnston, *Nature* **400** 848–849 (1999).
17. M. A. Alpar, A. F. Cheng, M. A. Ruderman, J. Shaham, *Nature* **300** 728–730 (1982).
18. D. C. Backer, R. W. Hellings, *Ann. Rev. Astr. Ap.* **24** 537–575 (1986).
19. C. Will, *Living Reviews in Relativity* **4** 4 (2001).
20. I. H. Stairs, *Living Reviews in Relativity* **6** 5 (2003).
21. T. Damour, G. Schäfer, *Phys. Rev. Lett.* **66** 2549 (1991).
22. N. Wex, Small-eccentricity binary pulsars and relativistic gravity, in: M. Kramer, N. Wex, R. Wielebinski (Eds.), Pulsar Astronomy - 2000 and Beyond, IAU Colloquium 177, (Astronomical Society of the Pacific, San Francisco, 2000), pp. 113–116.
23. V. M. Kaspi, A. G. Lyne, R. N. Manchester, *et al.*, *ApJ* **543** 324–327 (2000).
24. M. Bailes, S. M. Ord, H. S. Knight, A. W. Hotan, *ApJ* **595** L49–L52 (2003).
25. G. Esposito-Farèse, Binary-pulsar tests of strong-field gravity and gravitational radiation damping, in: Proceedings of the 10th Marcel Grossmann meeting, (gr-qc/0402007, 2004).
26. H. P. Robertson, *Ann. Math.* **38** 101 (1938).
27. R. Blandford, S. A. Teukolsky, *ApJ* **205** 580–591 (1976).
28. T. Damour, N. Deruelle, *Ann. Inst. H. Poincaré (Physique Théorique)* **44** 263–292 (1986).
29. P. C. Peters, *Phys. Rev.* **136** 1224–1232 (1964).
30. J. M. Weisberg, J. H. Taylor, The Relativistic Binary Pulsar B1913+16, in: M. Bailes, D. J. Nice, S. Thorsett (Eds.), Radio Pulsars, (Astronomical Society of the Pacific, San Francisco, 2003), pp. 93–98.
31. I. H. Stairs, S. E. Thorsett, J. H. Taylor, A. Wolszczan, *ApJ* **581** 501–508 (2002).
32. M. Kramer, *ApJ* **509** 856–860 (1998).
33. M. Kramer, O. Löhmer, A. Karastergiou, Geodetic Precession in PSR B1913+16, in: M. Bailes, D. J. Nice, S. Thorsett (Eds.), Radio Pulsars, (Astronomical Society of the Pacific, San Francisco, 2003), pp. 93–98 . in: . S. T. M. Bailes, D. Nice (Ed.), Radio Pulsars (ASP Conf. Ser.), PASP, San Francisco, 2003, p. 99–105.
34. J. M. Weisberg, R. W. Romani, J. H. Taylor, *ApJ* **347** 1030–1033 (1989).
35. I. H. Stairs, S. E. Thorsett, J. H. Taylor, Z. Arzoumanian, Geodetic precession in PSR B1534+12, in: M. Kramer, N. Wex, R. Wielebinski (Eds.), Pulsar Astronomy - 2000 and Beyond, IAU Colloquium 177, (Astronomical Society of the Pacific, San Francisco, 2000), pp. 121–124.
36. T. Damour, R. Ruffini, *Academie des Sciences Paris Comptes Rendus Ser. Scie.Math.* **279** 971–973 (1974).
37. M. Burgay, N. D'Amico, A. Possenti, et al., *Nature* **426** 531–533 (2003).
38. T. Damour, J. H. Taylor, *Phys. Rev. D* **45** 1840–1868 (1992).
39. B. M. Barker, R. F. O'Connell, *ApJ* **199** L25–L26 (1975).
40. N. Wex, *Class. Quantum Grav.* **12** 983 (1995).
41. T. Damour, G. Schäfer, *Nuovo Cim.* **101** 127 (1988).

42. M. Kramer, D. C. Backer, J. M. Cordes, et al., Strong-field gravity tests using pulsars and black holes, in: C. Carilli, S. Rawlings (Eds.), Science with the Square-Kilometre-Array (SKA), Elsevier Science Publishers, 2004.

43. N. Wex, S. Kopeikin, *ApJ* **513** 388–401 (1999).

44. S. W. Hawking, R. Penrose, *Royal Society of London Proceedings Series A* **314** 529–548 (1970).

45. R. M. Wald, General relativity, Chicago: University of Chicago Press, 1984.

46. S. W. Hawkings, G. F. R. Ellis, The Large Scale Structure of Space-Time, Cambridge University Press, Cambridge, 1973.

47. K. S. Thorne, *Reviews of Modern Physics* **52** 299–340 (1980).

48. K. S. Thorne, R. H. Price, D. A. Macdonald, Black Holes: The Membrane Paradigm, New Haven: Yale Univ. Press, 1986.

Michael Kramer

Michael Kramer was born in Cologne, Germany, in 1967. He studied physics at the University of Cologne before he continued his Diploma studies at the University of Bonn, where he graduated with first class honours in physics in 1992. He obtained his PhD at the Max-Planck-Insititute for Radioastronomy, Bonn, in 1995. Continuing to work at the MPIfR as a staff astronomer, Michael was awarded the Otto-Hahn medal of the Max-Planck Society in 1997. He spent a year at the University of California at Berkeley in 1998. In 1999 Michael took up a lectureship at the University of Manchester, where he is currently appointed as a Reader. Recently, he wrote a book on pulsar astronomy with his colleague Duncan Lorimer. His scientific interests include pulsars and their applications over a wide range, including relativistic gravity and tests of general relativity. Michael is married to to his wife Busaba, also an astronomer, with whom he enjoys travelling, cooking and sports.

CHAPTER 9

GAMMA-RAY BURSTS AS COSMOLOGICAL PROBES

Nial R. Tanvir

Centre for Astrophysics Research, University of Hertfordshire
College Lane, Hatfield, AL10 9AB, UK
E-mail: nrt@star.herts.ac.uk

Gamma-ray bursts are immensely bright and therefore have great potential as probes of the universe. They reside in small, distant galaxies, and thus provide a route to investigating the properties of such galaxies, which are themselves too faint for detailed study by more direct means. The use of GRBs as cosmic beacons is still in its infancy, but will become much more important in the future as we find many more bursts at very great distances. Ultimately we hope GRBs will help us investigate the star formation rate, and the state of intergalactic matter through the history of the universe. They may even help us determine the large scale geometry of the universe and the microscopic structure of space-time.

1. GRBs – The Brightest Searchlights

Gamma-ray bursts are the ultimate cosmic lighthouses. When they occur, they are more luminous than any other astrophysical object. Indeed, the brightest can briefly rival the luminosity of the entire observable, extragalactic universe.

Discovered serendipitously by US military satellites in the late 1960s, for 30 years even the distance scale of GRBs remained unknown. Observations by the BATSE instrument on NASA's Compton Gamma-Ray Observatory satellite in the early 1990s showed that their distribution was highly isotropic on the sky. This provided the first strong indication that GRBs lie far away across the universe, since all other known classes of galactic object follow some kind of asymetric sky distribution.

1.1. *X-rays mark the spot*

The "smoking-gun" was finally provided in 1997 by the Italian-Dutch satellite, BeppoSAX, which utilized on-board x-ray telescopes to rapidly determine a fairly precise position for a burst, GRB 970228 (note, all GRBs are known by their date of discovery, in this case 28 February 1997). Quick reactions by a team of astronomers led by Jan van Paradijs of the University of Amsterdam triggered the William Herschel Telescope and obtained an optical CCD image within 24 hours of the burst going off. A week later, another image of the field was obtained at the Isaac Newton Telescope, and comparison the the two revealed a faint and fading optical "afterglow".[1] This proved the critical breakthrough since an optical detection gives a precise position and opens the door to distance measurements. The expansion of the universe means that the more distant an object is, the more rapidly it recedes from us. This profound fact is also a vital tool to the cosmologist – velocities are relatively easy to measure via spectroscopic redshifts, thus allowing distances to be inferred for objects for which direct distance measurements are extremely hard to obtain. Indeed, cosmologists almost invariably talk in terms of redshifts (given the symbol z) when discussing large scales in the universe, rather than translating to the corresponding distance measures.

Sure enough, redshifts have (to-date) been obtained for about three dozen bursts, and nearly all are at $z > 0.5$, corresponding to several billion light years in distance.

Now, when looking far across the universe, because light travels at a finite velocity (albeit very fast in human terms), we are also looking backward in time. This puts us in the remarkable position of being able to study the history of the universe by simply looking at GRBs and other bright objects far away. It also means that we have yet another way of describing distances, namely in terms of how far back in time we are looking. Figure 1 shows a scale of redshift, compared to the look-back time, along with the current records for most distant objects.[2-4]

1.2. *What are they?*

With the distance scale established, attention was focussed on the question of the progenitors of GRBs. What kind of event can produce such a dramatic explosion? A number of lines of evidence pointed to a connection with young stellar populations. For example, GRBs usually seem to lie in the brighter, bluer regions of their parent galaxies, where the newly-born

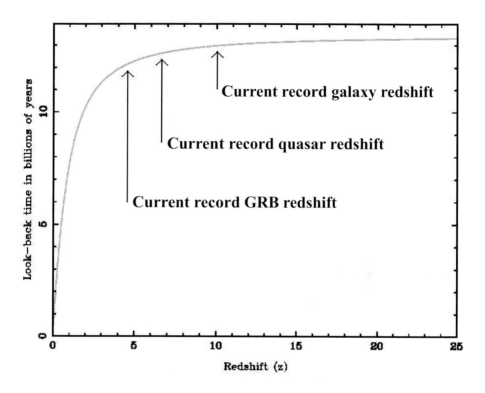

Fig. 1. Possibly the greatest discovery ever made in cosmology is that the entire universe is expanding. This is to say that the galaxies, on the average, are moving away from each other, and the velocities of recession increase with increasing distance. These velocities are measured by the positions of bright and dark lines in the spectra of the galaxies – recessional velocities being indicated by a shift of all the spectral lines to the red end of the spectrum. This is the famous "redshift" which, since it is closely related to distance, and is generally much easier to measure, is almost invariably used to describe the distance of remote objects. The figure shows how to translate redshift measurements into an equivalent "look-back time" – how far back from the present epoch we are seeing. The curve shown is for a particular "best-bet" set of cosmological parameters (specifically $h = 0.7$, $\Omega_m = 0.3$ and $\Omega_\Lambda = 0.7$). Notice that the last 90% of the history of the universe is at redshifts less than 5. The current records for the most distant, spectroscopically confirmed galaxy, quasar and GRB are also shown to set the scene. As discussed in the text, to-date GRBs have been studied in far smaller numbers than either galaxies or quasars. As larger samples are observed it is hoped that soon we shall find GRBs at redshifts of 15–20.

stars are concentrated. However, astronomers did not suspect these young stars of producing the GRBs directly, but rather that the GRBs were associated with the deaths of massive stars. The reasoning is that massive stars (say several tens of times more massive than the Sun) live for only a few million years. In that time they don't have chance to wander far from the star-forming gas cloud in which they were born. Thus, for all intents and purposes, the death of a massive star (often a dramatic and luminous event, as we shall see), is frequently taken as an indicator of star formation.

This picture was confirmed in 2003 when two groups[5,6] identified the presence of a powerful supernova underlying GRB 030329. It seems that, somehow, the collapse of a massive star produces an extremely high-energy jet of plasma moving at very close to the speed of light. Collisions of plasma shells within the jet itself, or when it impacts the surrounding medium, create shocks and hence radiation. Observers whose line-of-sight looks down the jet, see this radiation as a gamma-ray burst. Simultaneously the rest of the star explodes at a more leisurely 30000 km per second or so. This has the appearance of a violent supernova, although the light of the GRB afterglow is so intense that very careful observations have to be made to detect the SN light underlying it.

Further discussion of the astrophysics of GRBs is found elsewhere in this volume. In this article we are concerned with what GRBs can tell us about the universe at large.

2. Pointers to Distant Galaxies

As soon as the first GRB optical afterglows were identified astronomers began to search for galaxies in which they might be residing. Hubble Space Telescope pictures showed that in each case the GRB was located in a faint, distant galaxy. Usually the hosts were small, blue, irregular galaxies, with the GRBs located in regions which showed signs of recent star formation. An example is shown in Figure 2, although this host is in fact atypical in being the only instance found to-date of a GRB whose parent galaxy is a grand-design spiral.

This power of GRBs to point us to distant galaxies is tremendously exciting precisely because these galaxies are usually small and faint – and hence very hard to locate and study by other means. In fact, a fair proportion of the host galaxies are fainter than magnitude 25, which is about the limit for which redshifts can be obtained directly with even the largest telescopes. In other words, without their having hosted a GRB, we might have been able to detect them but wouldn't have been able to estimate a reliable distance.

The most extreme example so far is that of GRB 020124. This occured in a galaxy which wasn't visible at all in a very deep HST image of the region.

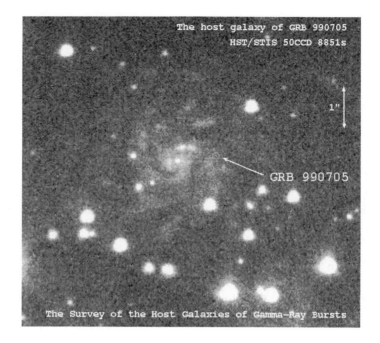

Fig. 2. The only GRB host galaxy (GRB 990705) so-far found to be a "grand design" spiral (URL http:/www.ifa.au.dk/~hst/grb_hosts/data/grb990705/index.html). The other hosts range from small, almost unresolved galaxies, to brighter galaxies which show evidence of disturbance or merging. This isn't surprising since galaxy interactions and mergers often go hand-in-hand with bursts of star formation, which in turn will increase the likelihood of GRB production.

Despite that, observations of the afterglow tell us the redshift (z=3.2) and also the density of gas[7] (see Figure 3).

The brightness of the afterglows means that their spectra also record the signature of other elements in the gas clouds within their parent galaxy (and indeed any other clouds along the line of sight). Thus we can obtain detailed information about the chemical and dynamical state of the gas in the immediate vicinity of the burst. Again, such information is essentially impossible to obtain from observations of such faint galaxies directly, and is important for understanding the formation and evolution of galaxy populations at high redshift. Although this approach has yet to be applied on a large scale, an excellent example[8] of what's possible is shown in Figure 4.

From optical spectra we can directly measure the amount of Hydrogen, and in some cases other chemical elements (usually referred to generically

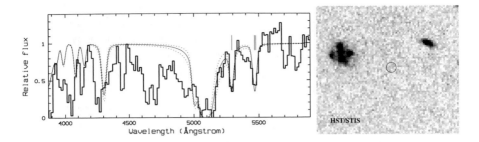

Fig. 3.   The most extreme example so far of a GRB (020124) occuring in a very faint
host galaxy. The right-hand panel is the deep Hubble Space Telescope image of the
region, showing no indication of optical light at the position of the afterglow, down to a
magnitude of $R = 29.5$. The left-hand panel shows the hydrogen absorption line which
provides the redshift ($z = 3.2$) and also confirms that the GRB lies within a galaxy
which contains a significant amount of gas.

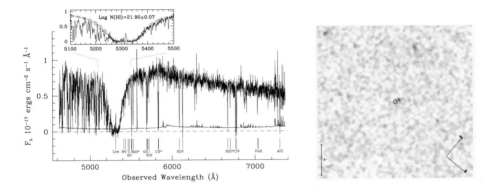

Fig. 4.   The host galaxy of GRB 030323 was also faint ($R = 27.5$) and small, quite
typical of many GRB hosts. Nonetheless, in this case the VLT spectrum of the afterglow
was sufficiently high signal-to-noise to reveal a host of absorption lines in addition to the
deep Hydrogen Lyman-α line.

as "metals", even though elements like Oxygen and Nitrogen are included).
The interest in the amount of metals is that they can only have been created
in the interiors of stars. Measuring the typical abundance of metals as a
function of redshift, therefore provides an indirect measure of the stellar
activity which has produced them.

   Occasionally the light from a distant quasar also passes through a
galaxy, providing a similar opportunity. The transience of GRBs makes

the observations more challenging in some respects. But it does also have the advantage that once the GRB afterglow has faded away, followup studies can more easily probe the host galaxy luminosity and morphology and identify faint galaxies on the line of sight to which the small gas clouds may belong. Furthermore, unlike quasars which most often just clip the edges of the intervening galaxies, GRBs by their nature lead us to the heart of their host galaxies. (To be sure, quasars too sit within host galaxies, but the bright quasar spectral emission lines, and the general effect the quasar has on its environment, mean that we generally learn little about the quasar hosts from their spectra.)

To date the most distant GRB host with a measured redshift is GRB 000131 at $z = 4.5$. However compared to galaxies and quasars, very few GRB redshifts have ever been determined, and in principle GRBs should be detected at much higher redshifts when it gets harder and harder to find galaxies by any other means.

2.1. *The history of cosmic star formation*

At first sight this seems all very well in that the properties of GRB hosts can be studied in detail, but this is not obviously of any wider importance. Afterall, any individual galaxy will probably only have a GRB every few tens or hundreds of thousands of years. Thus it would seem that the whole set of GRB hosts are simply a random ragbag selection of galaxies which just happen to have had GRBs observable by us in the past few years.

However, as we've already seen, each GRB signals the end point in the life of a massive star, and this, it turns, means that the GRB hosts provide a unique window onto the wider population of high redshift galaxies. To appreciate this point consider that amongst the GRB hosts we never see elliptical galaxies. This is easily understood since ellipticals typically exhibit very little ongoing star formation. The presence of the GRB ensures that there is some star formation happening in the galaxy in which it takes place. The same reasoning would lead us to expect that populations of galaxies with very high star formation rates should be found amongst the GRB hosts with enhanced frequency. In fact, if GRBs accompany the deaths of all (or a fixed proportion of) stars within a certain range of masses, and if the number of stars in that mass range is also a fixed proportion of all the stars which form, then statistically, we would expect the distribution of GRB host galaxy properties to reflect the populations of galaxies in the same proportions as they contribute to the global star formation in the universe. Thus GRBs allow us to select a special sample of high redshift galaxies, which are representative of the star formation which is taking place there.

So, for instance, since we find many of the GRB hosts to be small, faint galaxies which would not be included in most other studies of high redshift star formation, it may be telling us that a high proportion of star formation has been missed by these previous studies.

Of course, it is also quite possible that one or both of the above assumptions is incorrect. For example, there are various arguments to suggest that GRBs are both brighter and more likely to occur in regions of lower heavy element abundances (note, by heavy elements, astronomers mean more-or-less anything heavier than Helium). However, similar uncertainties afflict all other methods of estimating star formation rates at high redshift in one way or another. GRBs retain some important advantages over these other techniques, particularly that they can be detected (in γ-rays) even when the GRB is enshrouded within a dusty galaxy, and that the galaxies themselves don't even have to be detected to know that a star has died at some particular redshift.

The exact relationship between GRB rate and star-formation rate is not yet fully understood, and is the subject of ongoing studies.[9] However, whilst the stars which become GRBs are few and far between, the rate and location of GRBs at a given cosmic epoch is representative in a statistical sense of the amount of star formation ongoing at that epoch. Thus we hope that in the future it will be possible to estimate the global star formation history of the universe by counting at the number of GRBs as a function of redshift.

3. Moving Swiftly on – Where Do We Go Next?

In the seven years since the first optical afterglow was detected in 1997, astronomers have observed in the region of 50 afterglows. From this we have learnt a great deal, but much remains to be done. In particular, we'd like to gather larger samples, and also engage in more rapid and uniform followup campaigns. This isn't just to keep astronomers busy, rather the reason is that important science comes out from statistical studies. It is also because large samples and early observations make it much more likely that we'll manage to find and follow-up examples of particularly rare and exciting bursts.

To this end, a new GRB mission is planned for launch in autumn 2004.

The Swift satellite (which is named for its ability to slew rapidly round the sky), is a joint US/UK/Italian mission, although a large global network of astronomers is hoping to contribute to the followup of its discoveries. Swift is expected to provide accurate locations for many more GRBs than previous missions (prelaunch estimates are about 2 per week), and to considerably fainter flux limits. Moreover, Swift will relay those positions to ground stations in a matter of seconds, from where they will be immediately distributed over the internet. A combination of the high precision, rapid reaction and sensitive instrumentation is expected to lead to many breakthroughs, and in particular, to the identification of very high redshift GRBs. In fact, if high-redshift GRBs exist and have similar properties to those we've already studied at lower redshift, then they should be detected by Swift to $z = 20$ or more (Lamb and Reichart 2000).

Interestingly, when it comes to observing the afterglows of very high redshift GRBs, we benefit from a surprising additional advantage. This is because of a phenomenon known as "cosmic time-dilation". Partly a relativistic effect, this tells us that when we look at distant galaxies, in common astronomical parlance, "we see their clocks running slowly". Of course, we don't literally see any clocks, but physical processes, and events such as supernovae, are indeed observed to run slowly in the distant universe. For example, by redshifts of 10, time is stretched out by an order of magnitude. Thus, a GRB afterglow at this redshift, if observed 20 minutes after the burst will be seen as it was just 2 minutes after in the frame of the burst itself. Because GRBs are rapidly fading this time-dilation means that they will be a good deal brighter than one would otherwise expect at a given tiem post-burst. This is illustrated in Figure 5.

Furthermore, as we have seen, GRBs, are triggered by the collapse of massive stars. As it happens, there are theoretical reasons to believe that, at the earliest times, a greater proportion of all stars should be very massive, perhaps leading to a higher GRB rate. In addition, these stars would also have had a lower heavy element abundance, which it is thought may also be conducive to more numerous and brighter GRBs. It is easy to see why there is currently great excitement over the prospect of using GRBs to explore the distant universe.

4. Illuminating the Dark Ages

Being so bright, GRBs, we might expect, should provide other opportunities to map and probe the universe on very large scales. This is indeed the case,

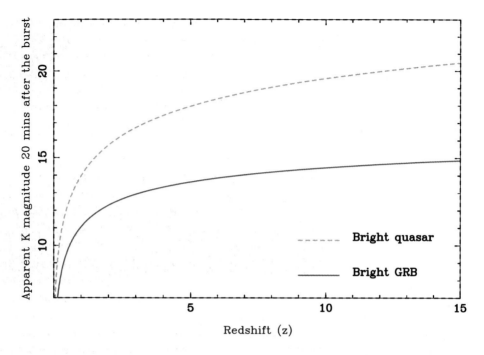

Fig. 5. An illustration of how the advantage of GRB afterglows over quasars, providing they are observed shortly after burst, actually increases with redshift due to cosmological time-dilation. This is because at a given observer time, we see a much earlier time in the frame of the GRB. Typical parameters are assumed for the GRB afterglow light curve (a pure power-law decline and no prompt component of the flux). Note that the infrared K-band is shown, since, as discussed in the text, optical light is almost completely absorbed from all objects beyond a redshift of about 6.

but before discussing these in detail, we shall digress briefly to consider our emerging understanding of the physics of the early universe, and how structures initially form from the smooth plasma of the Big Bang.

4.1. *A brief history of the first billion years of the universe*

It is widely accepted that our universe was born in an unimaginably extreme event, about 14 billion years ago. In the popular mind, this "Big Bang" is pictured as a super violent explosion from which all matter flew out across the universe. Unfortunately, such an image does not really correspond very closely to what the theory really says, or indeed to what we observe. In particular, cosmologists have found that the universe on large scales seems essentially the same everywhere as far as they look, without any indication of a "centre" or "edge". We therefore suppose that the "explosion" was of a curious kind, in which the matter by some means comes into existence at the same instant as space and time. Mathematically, the density is infinite at this moment, but so is the expansion rate, and this initial inertia blows the matter apart, giving rise to the (probably infinite) expanding universe we find ourselves in today.

That said, it is also widely suspected that our cosmological models, which are based on Einstein's General Theory of Relativity, will ultimately be shown to be wrong; particularly in regard to the behaviour at very early times when the universe was so incredibly dense. So infinite density might thus be avoided – although past experience suggests that whatever theory replaces it will seem no less weird!

Anyhow, these very earliest events are not our concern here. Rather, we are interested in the processes by which the primordial plasma cooled and condensed into the objects we see around us now. For approximately 400000 years after the Big Bang, the universe expanded and cooled, but remained essentially a sea of hot plasma. In this relatively dense plasma, electrons are separated from nuclei (chiefly simply free protons), and are very effective at scattering light photons. Hence the entire universe is essentially opaque – the ultimate pea-soup fog.

At 400000 years the gas has cooled sufficiently that free electrons combine to make neutral atoms, chiefly Hydrogen and Helium. This neutral gas is very transparent to low energy photons, so from this point onwards the photons from the Big Bang, now mostly of infrared wavelengths, stream on unimpeded through the universe (to be observed by us now as the cosmic microwave background radiation).

There were no sources of radiation during this early era, so the universe was essentially dark. The expansion of this tenuous gas follows simple physical laws, but eventually some regions collapse under their own gravity to form bound, condensed objects. Now the physics becomes much more complicated, involving an interplay between small scale phenomena, and the large scale force of gravity. As it stands, we have very little observational evidence to guide our theoretical explorations. However, whatever the first stars looked like, they would have certainly emitted the first light in the universe since the Big Bang. As transparent as the neutral gas is to low energy photons, so it is highly effective at absorbing energetic, ultra-violet photons. In particular, any photons with enough energy to excite Hydrogen atoms from their first to second energy level (the Lyman-α transition) are highly likely to be absorbed before they could travel very far through space. Thus if we were able to observe the spectra of these early objects, then we would expect to see a distinct step – a cutoff bluewards of which no photons would be detected. In fact, due to redshifting, the UV step in question should appear in the optical, or even the infra-red band.

Now, these same early sources are also expected to emit copious amounts of even higher energy UV radiation, which would begin to re-ionize the neutral gas again. We expect that around each source (star cluster or protogalaxy), an expanding bubble of ionized gas will form. Eventually the bubbles overlap, to leave the intergalactic medium completely ionized, as we find it today.

4.2. *Slicing through time*

A crucial question is, when did the reionization event occur? Observations of the microwave background by NASA's WMAP probe[11] suggest that the universe was first reionized around $z = 15$–20, corresponding to about 200 million years after the Big Bang. However, we can also probe this era from the lower redshift side, by studying individual objects at increasingly high redshifts. Indeed, one of the main motivations for observing high redshift quasars and GRBs is to look for the spectral signature of this reionization event. In the last two years, the first indications that we are beginning to come into contact with the neutral era have come from observations of quasars above a redshift of 6. However, the gap between $z = 6$ and $z = 20$ is currently very hard to explore, and, as illustrated in Figure 6 different scenarios may agree with the available evidence.

To appreciate the potential of GRBs as probes of the very distant universe it is worth considering the competition, namely quasars and galaxies. The former are brighter, but rarer than the latter. However, both these classes of object are likely to have been fainter and smaller at high redshifts, since they both take time to build up their much larger masses. Today's large galaxies appear to have been built up from smaller subunits, and the supermassive black-holes which power the brightest quasars, seem to reside in the most massive galaxies. GRBs by contrast have stellar progenitors and hence plausibly will have comparable brightness whenever they occur. In the future, then, GRBs may provide the searchlights by which we learn of the state of the universe when the very first objects began to condense.

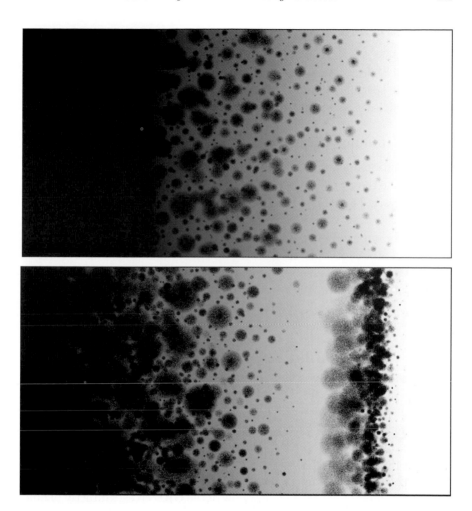

Fig. 6. This schematic figure illustrates how the universe may have evolved (time running right to left) from a largely neutral state (white in this representation) at early times to being completely ionized between the galaxies at the present day (coded black here). UV photons from the first stars and to some extent the first quasars, probably were responsible for the reionization, and therefore the process is thought to have proceeded through the gradual merging of growing regions of ionized gas surrounding nascent galaxies. However, the details are still murky, due to the extreme difficulty of observing these early objects. For instance, it has been suggested[12] that the process may in fact have taken place in two relatively rapid stages (lower panel) rather than a single, drawn-out phase (upper panel). Potentially GRBs may provide the search lights which allow us to see into this region of the universe for the first time.

5. The Structure of the Universe

We finish with a round up of several methods by which GRBs may help us probe some aspect of the structure of the universe. These techniques are all either more speculative or their realisation is a little further into the future than those discussed so far.

5.1. *Gravitational lensing*

The gravitational lensing of light by massive objects, an important consequence of the theory of General Relativity, has become a important industry in astrophysics in the last two decades. Since the amount of bending depends on the mass rather than the luminosity of the lensing object, it is particularly important for quantifying the presence of dark matter – the mysterious and exotic particles which are thought to make up 80-90% of the matter in the universe. Astronomers also make use of the magnifying effect to study galaxies which would be too faint to observe otherwise.[2]

The manifestations of gravitational lensing come in a number of different flavours. Most famously, so-called "strong lensing" occurs when the images of background sources are brocken up into several components, and is clearly seen in many HST images of rich clusters of galaxies. Here the tremendous mass of the cluster produces multiple images of many background galaxies. In the case of GRBs, calculations suggest they are too rare to provide many instances of such strong-lensing. However, these calculations depend on extrapolating the numbers of GRBs which may be found by future missions such as Swift. If these turn out to be underestimates, then prospects obviously improve.

In any case, there is also a possibility of GRBs exhibiting a "microlensing" effect. Here the expanding afterglow of the GRB crosses directly behind a star, which is most likely also within the GRB's host galaxy. The small gravitational bending effect of the star on the light of the GRB results in a small, but measurable, amplification of the afterglow light. The interest in this phenomenon is to learn something about the structure of the fireball itself.

5.2. *Cosmological parameters*

Historically one of the major goals of extragalactic astronomy has been the determination of the so-called "cosmological parameters" which govern the large scale evolution of the universe. A famous example is the Hubble constant, which describes the present day expansion rate. For much of the 20th century, astronomers struggled to estimate accurate distances to other nearby galaxies, from which the Hubble constant is inferred by comparing the distance to the redshift. Distances to more remote objects are even harder to estimate, but were also sought because they would provide a measure of the expected deceleration of the cosmic expansion. In 1998 two groups[13,14] announced a remarkable and surprising result that far from slowing down, the universe is actually accelerating in its expansion. Both groups based their conclusions on studies of high-redshift type Ia supernovae. The method relies on the fact that such supernovae have very predictable properties. They are, in astronomical jargon, "standard candles". Thus, from their observed luminosities we obtain a good estimate of their distance.

The obvious question is, might GRBs be used in the same way? If so it would be very exciting since they are so much brighter then supernovae. At first sight this doesn't look a promising line of attack as GRBs seem to have a much broader range of intrinsic properties. They are distinctly non-standard candles! However, a variety of ideas have been suggested which may improve this situation. For example, there it appears that GRBs with more fluctuations in their early γ-ray emission are intrinsically more powerful. It is still early days, but if we can develop a good understanding of the factors which effect GRB intrinsic luminosity, then potentially they may be as predictable as supernovae, but observable to much higher redshifts.

5.3. *Testing quantum gravity with GRBs*

Going from the structure of the universe on the very largest scales to its structure on the very smallest scales. In quantum theory, the vacuum of space is not as empty as it appears. In fact, it is described as a seething foam of so-called "virtual" particles, which flit in and out of existence too rapidly to be noticed by the laws of physics. As theories go, this may sound like the one about the crocodile under the bed which disappears by magic everytime you look! However this active vacuum is central to our understanding of the sub-atomic world and the description of the forces between particles.

The most ambitious attempts to unify the forces of nature in a single framework attempt to incorporate gravity with the other forces. This introduces new microscopic structure in the fabric of space-time. Unfortunately, to probe these tiny scales directly requires very high energies, which are far beyond the capabilities of any accelerator experiment yet conceived. However, the effects might be witnessed in a slightly less direct way due to the effect of the space-time foam on the propogation of photons. Specifically, high energy photons are predicted to travel somewhat more slowly through space than lower energy photons. The effect is very small, but would build up over the immense distances that light travels across the universe to us from a γ-ray burst.

Apart from the fact that GRBs do produce photons with sufficiently high energies, their other crucial characteristic is the very short duration (around a milli-second in some cases) of some of the pulses in the light curves of their γ-ray emission. By searching for very small systematic delays in the arrival time of pulses as a function of photon energy, predictions of quantum gravity models can be tested.

Applications of this technique to date have only produced fairly weak results, but a new satellite, GLAST (scheduled for launch in 2006) will provide much more significant constraints.

6. Conclusions

We have seen that γ-ray bursts are so bright that they should be detectable right back to the time when the first objects collapsed in the universe. Although GRBs are rare, and at the moment observed in far smaller numbers than high-redshift galaxies and quasars, they potentially open up a slew of new ways of exploring the universe. GRBs are already beginning to tell us about populations of small, distant galaxies which are too faint to study by other means. The imminent launch of the Swift satellite will provide exciting new possibilities to observe many more GRBs and their afterglows. In particular, GRBs may be the best way of probing the period when the intergalactic gas of the universe was reionized, which in many respects is the "last frontier" of galaxy evolution.

References

1. J. van Paradijs *et al.*, *Nature* **386**, 686 (1997).
2. R. Pello, D. Schaerer, J. Richard, J.-F. Le Borgne, J.-P. Kneib, *Astr. & Astrophys.* **416**, 416 (2004).
3. X. Fan *et al.*, *Astr. Journal* **122**, 2833 (2001).
4. M. I. Andersen *et al.*, *Astr. & Astrophys.* **364**, L54 (2000).
5. J. Hjorth *et al. Nature* **423**, 847 (2003).
6. K. Stanek *et al. Astrophys. Journal* **591**, L17 (2003).
7. J. Hjorth *et al. Astrophys. Journal* **597**, 699 (2003).
8. P.M. Vreeswijk *et al. Astr. & Astrophys.* **419**, 927 (2004).

9. N.R. Tanvir *et al. Mon. Not. Roy. Astr. Soc.* in press (2004).

10. D.Q. Lamb & D.E. Reichart *Astrophys. J.* **536**, L1 (2000).

11. A. Kogut *et al. Astrophys. J. Supp.* **148**, 161 (2003).

12. R. Cen *Astrophys. Journal* **591**, 12 (2003).

13. S. Perlmutter *et al. Astrophys. Journal* **517**, 565 (1999).

14. A. Riess *et al. Astr. Journal* **116**, 1009 (1998)

Nial R. Tanvir

Born and brought up in Derby, England, Nial Tanvir developed an interest in science, particularly astronomy, at an early age. He read Maths and Physics at Durham University, and stayed there as a post-graduate, obtaining his PhD in cosmology in 1992. The same year he moved to the Institute of Astronomy at the University of Cambridge as a post-doc in the UK Hubble Space Telescope Support Facility. In 1999 became a Lecturer at the University of Hertfordshire. Now a Reader in Astronomy, his primary research interests are the extragalactic distance scale, the structure and evolution of galaxies, and increasingly the study of gamma-ray bursts. He was joint winner of the Descartes Prize for Research in 2002 for his contributions to the discovery and understanding of GRB afterglows. When not at work, Nial spends time with his wife and three children, who he forces into going on long camping tours around Europe.

CHAPTER 10

NEW RADIO INTERFEROMETERS AND DATA ACCESS: INVESTIGATIONS OF STAR FORMATION

A. M. S. Richards[1] and R. A. Laing[2]

1. Jodrell Bank Observatory, University of Manchester,
Holmes Chapel Road, Macclesfield, Cheshire SK11 9DL, UK
2. European Southern Observatory, Karl-Schwarzschild-Straße 2
D-85748 Garching-bei-München, Germany
E-mail: amsr@jb.man.ac.uk

There are many natural sources of radio waves in the cosmos, from magnetised plasmas to molecular gas. These have two things in common; their emission can penetrate dust and clouds which block out optical light, and it can be studied using interferometry giving unparallelled angular resolution. This is invaluable for the study of star formation in our own and other galaxies and its place in stellar and cosmological evolution. Interferometric data comes at a price of complex data processing and very high data rates. The volume of radio astronomy data will increase over a thousandfold in the next few years as optical fibres are used to link ambitious new arrays. Fortunately the results will become more accessible thanks to the development of Virtual Observatories giving remote access not only to data but to data handling tools. Internationally agreed standards for describing images, spectra, catalogues and other data products allow astronomers to extract and compare results from the radio to the X-ray domain and beyond, without leaving their desks. These developments are illustrated by application to two problems; classification of young stellar objects in the Milky Way and the relationship between active galactic nuclei and starburst activity at high redshift.

1. Introduction

Even from the ground, radio astronomy produces diffraction-limited images. The largest fully steerable dish today, at Greenbank (USA), has a diameter only 33% larger than the 76-m Lovell telescope, built in 1957. In the same decade radio interferometry (first proposed by Ryle[1]) was developed at Jobrell Bank and Cambridge in the UK and Dover Heights and elsewhere in

Australia. This gives resolution λ/D where the effective telescope diameter D can be up to the diameter of the Earth (or more if an orbiting antella is used). Radio observations probe highly obscured or optically invisible regions from sub AU (astronomical unit) protostellar cores to some of the largest discrete objects in the universe, DRAGNs (k–Mpc-scale double radio jets originating from active galactic nuclei, AGN). They also give access to ubiquitous but elusive phenomena such as cool hydrogen (the 21-cm H I line) and the cosmic microwave background (CMB). Radio interferometry is particularly useful in the study of star formation in Galactic dark clouds and the phenomenon of starburst galaxies.

We introduce radio interferometry in Sec. 2. Section 3 describes some radio arrays currently in use and plans for the next generation of interferometers and Sec. 4 examines what they reveal about star formation and its place in cosmological evolution. We concentrate on sub-mm and longer wavelength interferometry. For general reviews see (including single dish work) *An Introduction to Radio Astronomy*[2] and *Internet Resources for Radio Astronomy*[3] (covering examples such as CMB arrays, omitted here for lack of space). The specialised techniques for pulsar astronomy are covered in Lorimer and Kramer[4] and for optical and infra-red interferometry see the recent review by Monnier.[5]

Section 5 describes how Virtual Observatories (VOs) tackle the challenges involved in remote access to large datasets which originate in a range of formats and require diverse processing routes. Section 6 gives some examples of science cases which take advantage of these advances in availability of multi-wavelength data.

2. Radio Interferometry Data Products

Cosmic radio emission flux density is measured in Jansky[a] (Jy) where 1 Jy = 10^{-26} W m^{-2} Hz. The Pioneer 10 spacecraft, now leaving the Solar System, produces signals of a few W (slightly more powerful than a mobile phone), detectable at a few mJy as a test 'alien' for the Search for Extra-Terrestrial Intelligence (SETI). In more conventional interferometry, an array of six 25-m radio telescopes with a 15 MHz bandwidth would detect about 10^4 $\lambda = 6$ cm photons per second from a supernova remnant (SNR) in a galaxy at a distance of 3–5 Mpc, e.g. M82, Fig. 1. This corresponds to a ~ 1 mJy signal at each telescope, which can be amplified and split six ways to allow

---

[a]In 1931 Karl Jansky detected the first extraterrestrial radio emission, from the centre of our Galaxy

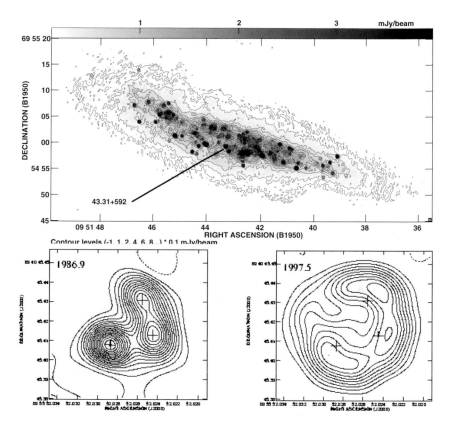

Fig. 1.   A MERLIN+VLA image of M82 at 18 cm (top, courtesy T.W.B. Muxlow) shows diffuse synchrotron emission with knots which can be resolved into individual SNR shells. Images of 43.31+592 made using the EVN (bottom left) and global VLBI (bottom right, angular resolution $\equiv 0.06$ pc)[6] show that the remnant is expanding at 9500 km s^{-1}.

correlation with the signal from each other telescope, giving 15 independent baselines. At 6 cm an array with baselines of up to $B \sim 200$ km has a resolution (synthesised beam width) of 50 mas, similar to that of the Hubble Space Telescope. Spectral line emission can be divided into very narrow channels, giving a resolution often $\sim 5 \times 10^6$ (~ 0.1 km s^{-1} in Doppler velocity) or even finer — SETI typically uses 2×10^9 1-kHz centred on 2 GHz (15 cm).

The primary radio interferometry data product is a set of complex visibilities which are calibrated to establish a flux density scale and remove atmospheric and instrumental artefacts (see e.g. *Synthesis Imaging in Radio Astronomy*,[7] the latest proceedings from the NRAO summer school[8] and

Getting results from radio interferometry[9]). Observations are often made using a compact quasar as a phase reference source. The positions of at least 212 such sources are known to better than 1 mas[10] and most modern radio astronomy data is aligned with the ICRF[b] to better than a synthesised beam width, with negligible rotation.

For the example shown in Fig. 2, left, the effective resolution is equivalent to a telescope 200 km in diameter but its aperture is poorly filled. A spectrum or a light curve can be measured in a few minutes or less for bright variable objects (e.g. a flaring X-ray binary) but several hours on-source are needed to image an extended object accurately. The observation needs to be completed in order to perform calibration and imaging.

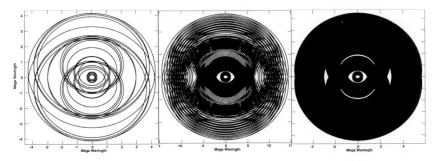

Fig. 2.   The left panel shows the present visibility plane coverage of MERLIN at 6 cm (16 MHz bandwidth centred at 5 GHz); the centre and right panels shows the coverage in a 2-GHz band divided between 4–8 GHz and continuous from 4–5 GHz, respectively.

Maps are made by 2-D Fourier transformation of the visibilities followed by incremental deconvolution to remove artefacts due to the incomplete baseline coverage, usually involving repeated Fourier transformation (CLEANing, Cornwell (1999)[11] and references therein). The fidelity of the image can be improved by several techniques including weighting the data, but increasing resolution decreases the sensitivity to extended flux — the user cannot optimise both simultaneously. Fig. 3 shows the ultra-luminous IR galaxy and megamaser source Markarian 231; low resolution (1) shows extended emission to the S, partly due to the dissipation of the jet. At high resolution (2) this is missing but the compact nature of the peak is more apparent, aligned with the dynamic centre of the maser disc (supporting

---

[b]International Celestial Reference Frame http://rorf.usno.navy.mil/ICRF/

its AGN nature); the arc is probably of starburst origin.[12] The shortest baselines present in the data (the 'pupils' in Fig. 2) limit the largest spatial scale to which the array is sensitive. This can be overcome by combining data from more than one array (as in Fig. 1), usually done in the visibility plane before imaging.

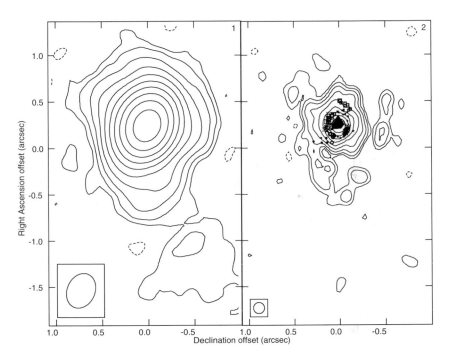

Fig. 3. Two MERLIN images of Markarian 231 at 18 cm made from the same visibility data. In both images contours are at (-1,1,2,4...)×0.25 mJy beam^{-1}. Image 1 uses natural weighting; image 2 uses partial uniform weighting (interpolating into the missing spacings of the visibility plane), giving restoring beams of 362×284 and 120×120 mas^2 respectively. The positions of OH masers are superimposed on image 2.

The response to sources offset from the pointing centre will be smeared by frequency- and time-averaging unless multiple narrow channels and short integration times are used. The ultimate field-of-view constraint is the primary beam of the individual antennas, giving ~ 8 arcmin for a 25 m dish at 5 GHz. For $\leq 10\%$ degradation, 0.5 GHz channels and 2 s integrations would be needed. It is impractical to make a single image of the whole primary beam as this would contain $\sim 5 \times 10^8$ pixels for this example. Moreover, a phase-shift is needed to prevent distortion at separations where the curvature of the sky is significant. Usually only limited regions are of interest, but in special cases a mosaic of manageable-sized images is built up as in Fig. 4. Full polarization data are usually available. Spectra covering small angular regions can be extracted from image cubes with frequency (or velocity) as the third axis. This leads to the conclusion that it is desirable to store partly processed data in a form where images and other products can be extracted easily as and when they are needed to meet the demands of the orginal investigator or archive users.

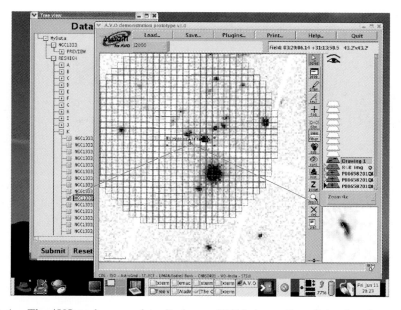

Fig. 4.   The AVO tool was used to display an XMM observation of star-forming region NGC 1333, ~ 8′ wide. The grid shows the position of a mosaic of MERLIN images (see Sec.3.1) generated dynamically by reading a directory containing the FITS files. Clicking on the main image allows the relevant tile of the mosaic to be loaded, see zoom at lower right. This shows a 200-mas (80-AU) microjet from YSO MM3.

3. Developments in Interferometry

Table 1 lists cm-wave interferometers. VLBI, very long baseline interferometry, can include almost any or all of these arrays and other single dishes and the data are usually tape-recorded for later correlation. Arrays operating at (sub)-mm wavelengths include the SMA (USA/Taiwan), CARMA (USA), IRAM (France/Germany/Spain), Nobeyama (Japan), and the CMA (Coordinated Millimetre (VLBI) Array).

Table 1. Some interferometry arrays open to external users at cm to m wavelengths. See home pages for each instrument for more details.

Name	Country	Antennas		Range	Baseline	Optical
		No.	D (m)	λ (m)	B_{max} (km)	fibres?
WSRT	Netherlands	14	25	0.36–0.92	2.7	
GMRT	India	30	45	0.003–0.2	25	yes
ATCA	Australia	6	25	0.003–0.2	6	yes
VLA	USA	27	25	0.006–4	$36^{(i)}$	future$^{(i)}$
MERLIN	UK	7	25–75^{ii}	0.012–0.23	217	future$^{(ii)}$
VLBA	USA	10	25	0.003–0.9	≤ 8600	
EVN	Europe+$^{(iii)}$	14	14–100	0.05–0.5	≤ 12000	adhoc$^{(iii)}$

(i) Can now include the Pie Town link giving $B = 150$ km. The next upgrade will be to connect all antennas with optical fibres to form the EVLA. The NMA will include 8 more antennas, $B \leq 300$ km.

(ii) Can include the 75-m Lovell at $\lambda \leq 0.025$ m. e-MERLIN will operate with all antennas linked with optical fibres by 2007 (Sec. 3.2).

(iii) The EVN also includes South African and Chinese antennas and the JIVE correlator. Not all antennas operate at all wavelengths. See Sec. 3.2.

3.1. *Improving resolution*

The most sensitive sub-arcsec-resolution radio images to date are 'deep fields' such as the 21-cm MERLIN+VLA observations of the Hubble Deep Field (North) (HDFN)[13,14] see Sec. 4.1, which took 18 days of MERLIN time and 42 hours of VLA time to reach a noise level of 3.5 μJy at 200 mas resolution. A similar observation of the star-forming region NGC 1333 at 6 cm (Fig. 4) using MERLIN alone produced 10 GB of target data after calibration. The 8-arcmin primary beam was mapped into a mosaic of 812 images each of 1024×1024 17.5-mas pixels, which took 2×1 month 2-GHz processor time on computers. Each image is only a few MB.

Millimetre VLBI gives beam sizes down to ~ 0.1 mas and by component fitting the positions of of H_2O and SiO masers at 13, 7 and 3.5 mm can be measured to within a few μas (e.g. H_2O masers in a sub-pc orbit showing a black hole lies in the centre of NGC 4258[15]), but this resolution is only achieved for non-thermal emission, $T_b > 10^{10}$ K. This technique can also resolve the jet collimation region in the nearest AGN, within 50 Schwarzschild radii of the black hole.

3.2. *Wide-band sensitivity: effects on data rates*

The radio telescope parameters used in Sec. 2 are similar to MERLIN. Long observations are needed for deep fields because its sensitivity is at present limited by the 16-MHz microwave links used to return data from the telescopes. Figure 2 shows the effect of the *e*-MERLIN 2-GHz bandwidth using optical fibres; this will increase sensitivity tenfold and also improve imaging fidelity as the visibility plane is filled better. At $3 < \lambda < 6$ cm the total improvements, including use of the resurfaced Lovell telescope, will produce a 30-fold increase in sensitivity. The VLA upgrades will see similar improvements. Micro-Jy sensitivity will be reached in a single day instead of the weeks used for the Deep Fields. The MERLIN+VLA observations[13,14] were complete to 27–40 μJy (7σ). Figure 5 shows that there is a statistically significant excess of $1 - 6\sigma$ radio flux at the positions of optical galaxies with $I < 24^{\mathrm{m}}$ which will be imaged clearly by the upgraded instruments.

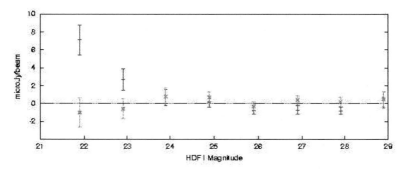

Fig. 5. The black symbols show the 1.4-GHz radio flux density at 1-arcsec resolution at the positions of HST galaxies in the HDF(N), plotted against their I magnitude, in 1-mag bins. The grey symbols are a control sample at a random 10 arcsec offset.[13]

For VLBI the first challenge is to correlate any bandwidth in real time. EVN stations have already developed disc-based recording, which allows test data to be transferred by ftp. On 28 April 2004 JIVE produced the first image (of the gravitational lens B0218+357) using data correlated in real time from 3 antennas at Onsala (Sweden), JBO and WSRT. This followed a month after the first real-time fringes from a transatlantic baseline, made by the Haystack (USA) correlator. The EVN network tests are using GÉANT, the European federation of National Research and Education Networks, to transport 0.5 Gb s^{-1}, ten times the normal traffic of $\sim$ 0.05 Gb s^{-1}. *e*-MERLIN and the EVLA, will require rates of 30–120 Gb s^{-1} per site, necessitating the use of dedicated fibres. In 2003 the *total* UK traffic was 10–20 Gb s^{-1} and at projected growth rates radio interferometry will continue to produce 20–50% of all IP traffic.

3.3. *Increased collecting area for total sensitivity: the SKA*

Increasing bandwidth to increase sensitivity is of most benefit to continuum observations at $\lambda < 8$ cm. In order to produce similar orders-of-magnitude improvements in spectral images an interferometer with greater collecting area is needed. The Square Kilometre Array is the most ambitious project, currently involving 34 institutes in 15 countries all around the world. The SKA was originally motivated by the requirements for imaging H I at high redshift; it will also provide a very rapid response for transient objects and highly efficient surveys. The prime science drivers are:

(i) Strong field tests of gravity using pulsars and black holes;
(ii) Probing the dark ages (re-ionisation, high-z star-formation and AGN);
(iii) The origin and evolution of cosmic magnetism;
(iv) The cradle of life (terrestrial planets, organic chemistry and SETI);
(v) The evolution of galaxies and large-scale structure.

A number of technological solutions are being investigated, minimising costs by using mass-produced components where possible, parallelising data processing and following an incremental approach. Low-interference, fine-weather sites spread over hundreds of km are needed. A choice will be made in 2006 between proposals from Australia, North America, China and Southern Africa. Construction will be started in 2012 and completed in 2020. Prototypes appropriate to various wavelength ranges include:

λ **1.5–30 m** LOFAR has recently received a funding commitment from the Dutch government. It will use almost a quarter of a million dipoles in up to 165 stations with a maximum baseline of 360 km. The development is in partnership with IBM providing supercomputing technology to cope with a data rate of 768 Gb s^{-1}.

λ **0.2–2 m** The European design for a 1:200-scale prototype uses commercial twin-polarized TV (yagi) antennae, combined to form individual ~ 30 m^2 phased arrays. This defines a primary beam of about 30 square degrees at 46 cm, within which up to 1000 beams of 5–10 arcmin will be synthesised simultaneously. Beam-forming and data processing will be done using a scaled-up version of the JBO Beowulf cluster COBRA.

λ **0.027–0.6 m** The Allen Telescope Array (ATA) is a project led by the SETI institute, under construction at Hat Creek, USA. It will consist of 350 6.1-m diameter satellite-type dishes giving a primary beam of ~ 2.5 degrees at 21 cm. It employs an innovative receiver design to observe anywhere from 0.5–11.2 GHz in 4 separate 1-GHz bands. Each can be split into four synthesised beams in different sky directions. The data acquisition control systems and data processing will be accessible over the internet using standard networking tools.

λ **0.01–0.15 m** The Faraday project is exploring the use of cryogenic multibeam receivers at 5–15 mm and non-cryogenic phased arrays at longer wavelengths. A dual-beam 1-cm receiver (built at JBO based on a Planck design) has already been installed on the 32-m telescope at Toruń in preparation for the 100-beam survey instrument OCRA.

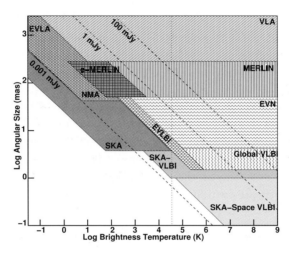

Fig. 6.   The sensitivity and resolution of arrays at 5 GHz in normal operation. Deep field-type observations would push the sensitivity of the SKA an order of magnitude deeper. Other factors are also significant; for example the added reliability of eVLBI and the wide field of view, synthesis speed and broad wavelength coverage of SKA. The first upgrades to be completed, such as *e*-MERLIN, will provide the first tests of the science drivers for supersensitive array technology.

Designs involving a smaller number of very large dishes (also including the Chinese KARST plan for Arecibo-like telescopes) are most suited to longer wavelengths. The many small dishes required at dm and shorter wavelengths could produce e.g. 2 Pb s^{-1} for 5000 antennas, 10^5 channels and 0.04 s integrations.[16] Using Moore's Law (see Sec. 5), by 2020 a correlator will barely be able to handle Tb s^{-1}. The data rate will probably have to be reduced at the expense of band-width or field of view, or by pre-correlation of groups of antennas, degrading the visibility plane coverage — correlator design seems at least as significant a challenge as antenna technology.[17] The growth of computing power implies that calibration and imaging of a single μJy wide-field (e.g. NGC 1333 described in Sec. 3.1) will take ~ 20 min processor time on an advanced computer. It will still be desirable to use the fastest parallelised computing available, especially for multi-beam or datacube experiments.

3.4. *ALMA – a new window on the mm domain*

The Atacama Large Millimetre Array, ALMA, is currently under construction, involving the US, Canada, European countries, Japan and its host country, Chile. When complete, it will cover λ 0.3–10 mm at resolutions ≥ 5 mas, able to image the CO $J(2-1)$ transition (for example) in a few hours with a spectral resolution of 1 km s^{-1} and a sensitivity of 1 mJy. This will be achieved by 64 12-m antennas plus a smaller compact array — up to 2206 baseline pairs. The brevity of atmospheric stability even at 5000 m above sea level requires integration times which may be as short as 0.1 s. The input data rate to the correlator is 96 Gb s^{-1} *per antenna*. Innovative digitisation and resampling techniques will enable the correlator to produce a spectral resolution $> 10^7$. The correlator output will be ~ 1 Gb s^{-1} data. It is due for completion around 2011, although data will start flowing in 2007.

4. Radio Emission as a Probe of Star and Planet Formation

4.1. *Star formation in active galaxies*

Radio spectral index ($S(\nu) \propto \nu^{\alpha}$, where S is flux density at frequency ν) measurements are a powerful means to determine whether continuum emission from distant galaxies is from AGN ($\alpha \sim 0$) or from the cumulative effects of SNe and winds from massive hot stars on the interstellar medium in starbursts ($\alpha < 0$). Interferometry is needed to resolve galaxies where these coexist e.g. Mrk 231 (Fig. 3).

In the HDF(N) (including flanking fields), almost all the radio sources are resolved over 1–2 arcsec,[13] suggesting emission of starburst origin. The brighter sources ($S > 100$ μJy) are more likely to contain compact, flat-spectrum radio AGN; the majority of faint sources do not. 60% of the radio sources are also Chandra detections[18] but there is no obvious correlation between X-ray and radio luminosity. The first VO science paper[19] used optical/X-ray comparisons to identify 44 new obscured AGN in the HDF(N). Radio emission is detected from 7 (15%) but, interestingly, in five of these it appears to be of starburst origin with possible SCUBA counterparts.[13,20] The other two have some signs of both starburst and AGN emission. There is separate evidence, from mm CO and dust emission, that optically luminous quasars at $1.5 < z < 6.4$ are also undegoing major starbursts.[21]

Sanders[22] suggested that, in violent galaxy mergers, gas and dust fall towards the nucleus, triggering starbursts and fuelling AGN. If this is also true at high z it would have implications for hierarchical models of galaxy formation. At $z \ll 1$ where the phenomena are well resolved there appear to be distinct stages; for example OH megamasers are associated with starburst galaxies which have compact nuclei but rarely show incontrovertible evidence for an AGN. In contrast, powerful extra-galactic 1.3-cm H_2O masers are almost always found associated with AGN or jets. Within the next 5 years we will know if this also pertains at $z > 1$, using e-MERLIN and also ALMA, which will be able to detect shorter wavelength H_2O masers which could sample conditions more similar to OH.[21]

The SKA will detect starburst emission out to the earliest galaxies, providing estimates of the massive star-formation rate (SFR)[23] which can be compared with mm–IR estimates to constrain the initial mass function. At present, CO and dust have been imaged in a few (probably lensed) galaxies at $z = 2 - 5$. Is their high metal content typical of the large population of dusty galaxies at high redshift known only as unresolved sub-mm sources? ALMA will resolve these (with the eventual help of the SKA at $z = 5 - 10$). The differences between silicate and carbon dust will distinguish the contributions of low- and high-mass stars. Spectral observations will also overcome confusion and provide redshift estimates. Multiple CO (and other) lines will provide a thermometer, reveal the kinematics and chemistry and test the fine structure constant.

Radio SNe are of types Ib/c and II. They are mostly found in regions of high optical extinction in nearby starburst galaxies. Present radio telescopes cannot detect type Ia optical SNe, used as standard candles. e-MERLIN, the EVLA and the SKA will increase the RSNe detection rate out to $z = 3$, providing distance-estimate links with SNR (Fig. 1) and with optical SNe.

4.2. *The birth of stars and planets*

The rate of dark cloud collapse is determined by the interplay between gravity, magnetic support and temperature; hence the chemistry (which determines the ionisation fraction and cooling rates) is crucial. Present interferometers can only image thermal spectral lines on arcsec scales. High resolution magnetic (Zeeman splitting), spatial and kinematic studies are

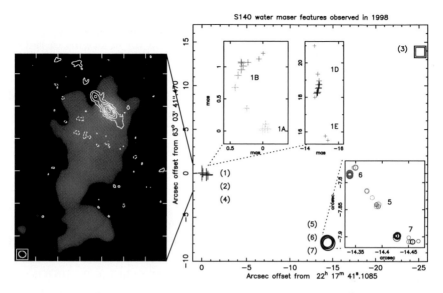

Fig. 7. The main right-hand image shows H_2O masers in the S140 star-forming region. Regions (5)-(6)-(7) trace a candidate protoplanetary disc, suggesting a YSO of a few $M_\odot$. Regions (1)-(2)-(4) are projected against the 6-cm radio continuum disc wind shown by white contours on the left. The greyscale shows a speckle interferometry image of NIR emission tracing a bipolar outflow[32], suggesting a massive YSO.

limited to masers. Methanol (CH_3OH) masers are uniquely associated with star formation, as CH_3OH forms in ices on interstellar grains and evapourates during cloud collapse. They were initially found by association with IR cores but recently dozens of CH_3OH masers have been found by blind surveys,[24–26] some at significant offsets. It has been suggested[27] that star formation is initiated along filaments of turbulent compression rather than in spherically symmetric collapse. Fractal analysis of H_2O maser distributions[28] in S128A finds some emission with distinctive velocity clustering which supports this. ALMA will reveal the composition as well as the kinematics of the dissipation scales. The Zeeman splitting factor for thermal lines is greater at cm wavelengths; the SKA (and possibly e-MERLIN) will be able to measure the magnetic field in pre-protostellar cores and hot cores using CCS and SO respectively.[29]

The best-understood maser emission traces the jets and discs of individual YSOs (young stellar objects). A few H_2O and CH_3OH maps already give tantalising suggestions of protoplanetary discs (e.g. Fig. 7). ALMA will map chemical species thought to be characteristic of various stages of star formation as well as giving detailed kinematic information to disentangle outflow, infall and orbital motions. The continuum spectral index will also help to settle arguments over dust-disc, disc-wind or jet origins of radio–IR emission[30] (such as alternative interpretations[31,32] of the S140 object in Fig. 7). Close to the star, dusty discs can be optically thick even at mm wavelengths.[33] Combinations of e-MERLIN, the EVLA, ALMA and eventually the SKA, will be necessary to image thermal emission. This will reveal the holes which protoplanets sweep in discs (as seen by SCUBA in ϵ Eri at 3.2 pc[34]). Thermal cm-wave emission also suggests the appearance of conglomerating pebbles of size comparable to the wavelength.[29]

4.2.1. Finding conditions for life

The familiar scenario for life needs planets (or moons) with the right sort of chemistry. How did the Solar System produce inner metallic or oxidised planets, at least one supporting oceans, and outer, reducing, hydrogen-rich giants? Protoplanetary discs posess radial and vertical temperature and density gradients; plausible chemical differentiation pathways have been suggested[35] but the processes of disc evaporation and condensation onto grains leave discs gas-depleted. The sensitivity of ALMA will be required to test the models by imaging the CH_4:CO and other ratios and the mm maser transitions of H_2O. This will be complemented by SKA observations of the lower maser transitions and NH_3. Detection of large interstellar or circumstellar organic molecules is favoured for the lowest energy states (cm transitions) since these are distributed over fewer molecular configurations.

Titan and Europa are the only extra-terrestrial locations with strong evidence for the current presence of oceans. Jupiter produces strong 10^6 Jy decameter bursts which are affected by the solar wind and by the passage of satellites through its magnetosphere. LOFAR may just detect extra-solar planetary emission at a distance of a few pc[29] and in any case lessons from its study of the Jovian system will help the SKA to identify giant planet moons. The SKA will also perform astrometry of radio stars with sufficient accuracy to show reflex wobbles due to jovian planets.

Of all SETI surveys to date, only Project Phoenix, searching in the direction of a few hundred stars with the Arecibo and Lovell antennas, would have (just) detected the Earth's most powerful transmitter (Arecibo planetary radar) at 300 pc. The SKA will go two orders of magnitude deeper, putting $> 10^5$ solar-type stars within reach in a few years; it would even be sensitive to TV-strength transmissions at the distance of the nearest stars.

5. Virtual Observatories and Radio Archives

Figure 8 shows the increase astronomical data from the days of hand-drawn images through photographic plates to the use of CCDs, radio interferometry and space-borne observatories. Al Sufi made the first recorded attempt at interoperability by cross-referencing star names in the unrelated Greek and Arabic constellations. Storage is keeping pace with data aquisition but Fig. 9 shows that processing power, although doubling every 18 months (Moore's Law), is lagging behind, and internet capacity is growing still more slowly. VOs[36] take advantage of exceptional high-speed links (e.g. between major archives and data processing centres) but allow astronomers anywhere to avail themselves of the results of data reduction without bulk movement of humans or raw bytes.

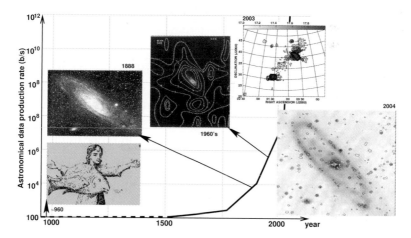

Fig. 8. The growth of astronomical data with time (estimates courtesy of P. Benvenuti). Insets show M31 as drawn by Al Sufi,[37] photographed by Roberts,[38] observed in the optical and radio using the Perkins and Big Ear telescopes[39] and mapped in H I by WSRT.[40] The most recent image is an X-ray/radio/IR composite showing contours from XMM (white) and the VLA NVSS (black) over an IRAS image, made using the AVO visualisation tool.

There is a multiplicity of VOs under construction, from national (e.g. AstroGrid[c] in the UK) to continental (e.g. AVO,[d] the European VO). All are committed to interoperability and the adoption of standards mediated

---

[c]http://www.astrogrid.org
[d]http://www.euro-vo.org/

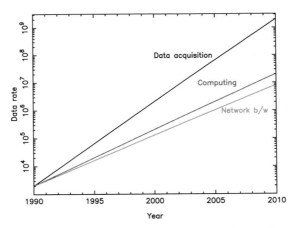

Fig. 9. Estimated growth-rate of electronic data handling (Mb s^{-1}) with time (based on ESO/AVO estimates kindly supplied by P. Quinn).

by the International VO Alliance (IVOA[e]). In principle any data can be published to a VO — so long as they are described accurately (including uncertainties), enabling the dynamic discovery of the contents of archives (illustrated by Fig. 4). One basic use is cross-identification of data across wavelengths and resolutions. This is being implemented for published radio catalogues,[41] using spectral index characteristics, in order to include radio identifications in the SIMBAD data base.

VO projects are building up a library of methods (including tools from external providers) for coordinate and flux scale conversion and more complex operations such as estimating photometric redshifts. These were used by Padovani et al.,[19] starting with multi-band images and building a workflow which included access to SExtractor and HyperZ. Heavy number-crunching or specialised tools are provided via interfaces to specialist data processing facilities.

VO facilities are ideal for radio interferometry. Sec. 2 explains why it is desirable to generate interferometric data products on demand. The next generation of sensitive instruments will find 30 sources of $S_{1.4} > 100$ μJy per 10-arcmin beam; it will be necessary to process the entire field of view and remove confusing sources in order to image any part of it with high fidelity. VOs will have an additional rôle in providing lists of known sources to facilitate this.

---

[e]http://www.ivoa.net

Remote access to data is a necessity for mountain-top instruments like ALMA, which already has an elaborate data model with VO compatibility designed in. The ATA is designed to be internet-accessible in operation as well as data retrieval. Currently operating arrays such as ATCA and MERLIN both have prototypes for extracting images on-demand from visibility data. An alternative approach adopted by the VLA is to make it possible to retrieve raw visibility data on-line according to a wide range of search parameters and to supply data reduction procedures. This could be adapted readily to allow users to request that a VO shipped the data to an interferometry processing centre. VLBI data reduction poses greater challenges as more complex judgements are needed (e.g. it is harder to distinguish between bad data and a rapid phase rate) but the EVN already offers pipelined calibration and diagnostic products on-line.

These examples were gleaned from a questionnaire circulated in the IVOA radio discussion forum[f] which showed that radio interferometry data providers are keen to use VOs to provide catalogues and data where possible. This will require authentication filters for proprietary data. Experienced radio astronomers may want to perform their own data reduction but many other users find it daunting to learn even one package, let alone the variety tailored to the needs of different observatories. Hence the emphasis was on standardising the form of data products (which should all come with a full history), adopting standard metadata and providing user-steered pipelines with jargon-free inputs. In Europe, RadioNet[g] offers a coordinating forum for radio-specific aspects of adopting (and influencing) VO practices as well as for radio engineering and software development.

6. VO Investigations of Star Formation

6.1. *Characterising galactic star formation*

The IMF is a crucial parameter for topics of major current interest such as interpretation of emission attributed to star formation in the cosmological and chemical evolution of galaxies or what fraction of Galactic stars might support terrestrial planets. Massive stars are thought to evolve too quickly to allow solar-type planetary systems to evolve and even the proximity of a young OB star can ablate protoplanetary discs around low mass stars (e.g. the effects of the Trapezium stars on the Orion proplyds[42]). How is the IMF

---

[f]http://www.ivoa.net/forum/radiovo/
[g]http://www.radionet.org

influenced by the trigger and the metallicity? Does the IMF evolve during a star-formation episode?

YSOs are characterised by evolutionary Classes 0–III, from warm IR cores, to the appearance of jets and the growth of accretion discs, their ablation and/or condensation into protoplanets and the emergence of young main sequence stars. There are ambiguities between evolutionary stage and mass; more massive YSO are initially more obscured and ultimately far brighter. It is not certain whether low-mass YSOs posess a large enough cool envelope to support molecular emission or whether its detection with current instruments indicates a high-mass star. In fact, no one wavelength regime can reveal every stage. The multiple scales involved are a further complication; a dynamically connected region can extend over many arcmin as can YSO jets, but individual discs and binary separations are on sub-arcsec scales.

VOs provide access to all available spectral and imaging information in order to compare the predictions for individual objects or unresolved multiples and develop a scoring system for YSO classification. This was started for the AVO 2004 science demonstration; part of the workflow is shown in Fig 4. In the near future data will be available from Spitzer, from ALMA pathfinders like APEX and the LMT and from *e*-MERLIN. This will lead to more reliable and flexible diagnostics to enable IMF models to be tested against any SFR, not just the most convient examples.

6.2. *Investigating the starburst-AGN link*

The bulk of cm-wave radio continuum emission from galaxies at $z > 1$ appears to be of starburst origin, as noted in Sec. 4.1. The majority of identified μJy radio galaxies have X-ray counterparts and in at least some cases the X-ray emission appears to be from AGN.[19] Methods similar to those employed by Padovani *et al.*[19] can be used to derive distance-independent radio/X-ray luminosity ratios and to look for correlations with spectral index or hardness ratio, redshift information and radio and optical morphology. The presence of faint background populations (of radio emission at X-ray positions and *vice-versa*) will be investigated using the method shown in Fig. 5. This will enable us to distinguish between (at least) three possibilities. A direct link between radio and X-ray emission would suggest that the latter may be due to X-ray binaries in starbursts, or that AGN associated with starbursts are very radio-weak. A non-correlation implies that any link between starburst strength and AGN activity takes longer to come to fruition than the duration of either individual stage. A complex correlation would suggest an evolutionary progression.

7. Conclusions

We have outlined the advances in radio interferometry planned for several decades to come but there will undoubtedly be unpredictable discoveries which motivate additional capabilities. VO tools are vital for extracting results from the huge data volumes predicted, as well as for multi-wavelength comparisons and tests against models. The examples in Sec. 5 describe investigations of star formation which exploit the VO capabilities and data available in 2004/5. In the longer term astronomers will simply use web-based tools to request images or spectra in compatible formats at the required wavelengths and resolutions and receive the results and analyses without separate consultation of the virtual and real observatories providing them. VOs will also provide interfaces to request intensive use of VO facilities and eventually to data provider proposal tools if the observations do not already exist (e.g. eSTAR[h]). This will in turn stimulate innovation in observational as well as data processing technology.

Acknowledgments

We thank the AVO and MERLIN teams for fruitful discussions. AMSR gratefully acknowledges the hospitality of CDS, Strasbourg where she was a 'professeur invité' during part of the writing of this article.

References

1. M. Ryle, *Vistas in Astronomy*, **1**, 532 (1955).
2. B. F. Burke and F. Graham-Smith, *An Introduction to Radio Astronomy*, (Cambridge Univ. Press, 2002),
3. H. Andernach, in *Astrophysics with Large Databases in the Internet Age*, eds. M. Kidger, I. Perez-Fournon and F. Sanchez, (Cambridge Univ. Press, 1999) p. 67.
4. D. R. Lorimer and M. Kramer, *Handbook of pulsar astronomy* (CUP, Cambridge, 2004).
5. J. D. Monnier, *Rep. Prog. Phys.* **66**, 789 (2003).
6. A. Pedlar, T. W. B. Muxlow and K. A. Wills, in *Winds, Bubbles & Explosions*, Eds. S.J. Arthur and W. J. Henney (RevMexAA (Serie de Conferencias) 15, 2003), p. 303.

---

[h]http://www.estar.org.uk/

7. A. R. Thompson, J. M. Moran and G. W. Swenson, Jr., *Interferometry and Synthesis in Radio Astronomy* (Wiley, New York, 2001).
8. G. B. Taylor, C. L. Carilli and R. A. Perley, Eds., *Synthesis Imaging in Radio Astronomy* (ASP Conference Series 180, San Francisco, 1999).
9. A. M. S. Richards and R. A. Laing *Getting results from radio interferometry* (C.U.P., Cambridge, in prep.)
10. C. Ma *et al.*, *Astron. J.* **116**, 516 (1998)
11. T. Cornwell, R. Braun and D. S. Briggs, in Ref.[8], p. 151.
12. J. S. Ulvestad, J. M. Wrobel and C. L. Carilli, *Astrophys. J.*, **516**, 127 (1999).
13. T. W. B. Muxlow *et al.*, *Mon. Not. R. Astron. Soc.*, **358**, 1159 (2005).
14. E. A. Richards, *Astrophys. J.*, **533**, 611 (2000)
15. J. R. Herrstein *et al.*, *Nature*, **40**, 539 (1999)
16. C. Lonsdale, *Data Rate and Processing Load Considerations for the LNSD SKA design*, (SKA Memo 32, 2003).
17. M. Wright, *SKA Imaging*, (SKA Memo 46, 2004).
18. D.M. Alexander *et al.*, *Astron. J.* **126**, 539 (2003)
19. P. Padovani, M. G. Allen, P. Rosati and N. A. Walton, *Astron. and Astrophys.*, **424**, 545 (2004).
20. C. Borys, S. Chapman, M. Halpern and D. Scott, *Mon. Not. R. Astron. Soc.*, **344**, 385 (2003).
21. P. Cox, in *Exploring the Cosmic Frontier*, eds. A. Zensus, P. J. Diamond and C. Cesarsky (Springer-Verlag, Heidelberg, 2004).
22. D. B. Sanders, B. T. Soifer, J. H. Elias, B. F. Madore, K. Matthews, G. Neugebauer and N. Z. Scoville, 1988, ApJ, 325, 74.
23. J. J. Condon, *Ann. Rev. Astron. and Astrophys.* **30**, 575 (1992).
24. M. Szymczak, A. J. Kus, G. Hrynek, A. Kepa and E. Pazderski, *Astron. and Astrophys.* **392**, 277 (2002).
25. M. Pestalozzi, V. Minier, R. Booth and J. Conway, in *Cosmic Masers: From Proto-Stars to Black Holes*, eds. V. Mignenes and M. Reid (IAU Symposium 206, ASP San Francisco, 2002) p. 139
26. S. P. Ellingsen, M. L. von Bibra, P. M. McCulloch, R. P. Norris, A. A. Deshpande and C. J. Phillips, *Mon. Not. R. Astron. Soc.*, **280**, 378 (1996).
27. T. Henning, in *Exploring the Cosmic Frontier*, eds. A. Zensus, P. J. Diamond and C. Cesarsky (Springer-Verlag, Heidelberg, 2004).
28. A. M. S. Richards, R. J. Cohen, M. Crocker, E. E. Lekht, E. Mendoza and V. A. Samodourov, in *Dense Molecular Gas around Protostars and in galactic nuclei*, ed. Y. Hagiwara, *Astrophys. and Space Sci.*, **295**, 19 (2004).
29. L. B. G. Knee in *Star Formation at High angular Resolution*, eds. M. Burton, R. Jayawardhana and T. Bourke, (IAU Symposium 221, ASP, San Francisco, 2004), p. 461.
30. D. J. Wilner, P. T. P. Ho and L. F. Rodgriguez, *Astrophys. J.*, **470**, L117 (1999).
31. G. Weigelt, *Astron. and Astrophys.*, **381**, 113 (2002).
32. M.G. Hoare, A. Glindemann and A. Richichi in *The Role of Dust in the Formation of Stars, Garching, 1995*, eds. H. U. Käufl, R, Siebenmorgen (Springer-Verlag, Berlin 1996) p. 35.

33. H. Chen, J.-H. Zhao and N. Ohashi, *Astrophys. J.*, **450**, L71 (1995).

34. J. S. Greaves *et al.*, *Astrophys. J.*, **506**, L133 (1998).

35. Y. Aikawa and E. Herbst, *Astron. and Astrophys.*, **351**, 233 (1999).

36. P. J. Quinn, in *Emerging and Preserving: Providing Astronomical Information in the Digital Age (LISA IV)*, eds. B. G. Corbin, E. P. Bryson, and M. Wolf, (U. S. Naval Observatory, Washington DC, 2003).

37. Abd-al-Rahman Al Sufi (903–986) reproduced in K. G. Jones, *J. B.A.A.*, **78**, 256 (1968).

38. I. Roberts *Photographs of Stars, Star Clusters and Nebulae Vol. II*, (The Universal Press, London, 1899). brynjones@beeb.net

39. Big Ear Radio Observatory and North American AstroPhysical Observatory (NAAPO) (1960's) webmaster@www.naapo.org

40. R. Braun and D. A. Thilker, *Astron. and Astrophs.*, **417**, 421 (2004).

41. B. Vollmer, E. Davoust, P. Dubois, F. Genova, F. Oschenbein and W. van Driel, *Astron. and Astrophys.*, **431**, 1177 (2005).

42. W. J. Henney, C. R. O'Dell, J. Meaburn, S. T. Garrington and J. A. Lopez, *Astrophs. J.*, **566**, 315 (2002).

Anita M. S. Richards

Anita was born in the Sudan in 1953. After studying biochemistry and working at a range of printing industry and public sector jobs, she completed an Open University degree and commenced a Radio Astronomy PhD at Jodrell Bank in 1993. Since then she has studied radio spectral line and continuum observations of stellar evolution and active galaxies. Her work has included supporting visiting astronomers working on MERLIN radio interferometry data and developing, documenting and teaching data reduction techniques. Anita helped to establish the MERLIN on-line data archive and now works for the Astrophysical Virtual Observatory. At present (2004) she is treasurer of the Astrophysical Chemistry Group and a member of RAS Council.

Robert A. Laing

Robert has worked in the field of cm-wavelength radio astronomy, primarily with the VLA, since 1978. He is currently the European Instrument Scientist for ALMA, which will be the premier instrument for mm and sub-mm interferometry. He has wide experience at other wavelengths, having been Project Scientist for the William Herschel (optical/IR) telescope on La Palma. He is a staff member at ESO, Garching and holds a visiting professorship at Oxford University. Robert's research interests include active galaxies, relativistic jets in extragalactic and galactic sources, synchrotron and inverse Compton emission, polarization and extragalactic magnetic fields, Faraday rotation and radio surveys.

Stars and Conditions for Life

CHAPTER 11

GAMMA-RAY BURSTS

Davide Lazzati

Institute of Astronomy, University of Cambridge
Madingley Road, CB3 0HA Cambridge, U.K.
E-mail: lazzti@ast.cam.ac.uk

Gamma-Ray Bursts have been puzzling astronomers and astrophysicists for more than 30 years. In the late nineties, the discovery of afterglow emission created a quantum jump in their understanding. We now know that they are extremely bright events taking place at cosmological distances, that they involve the fastest expansion velocities in the known universe and that their explosion is probably associated with the death of a massive rotating star. Yet, many key issues are still calling for an explanation, and new more powerful experiments are being designed and built to deepen our insight into these extraordinary phenomena. I will describe our understanding of Gamma-Ray Burst physics and the observational basis that led to the development of the present model, with particular emphasis on the last few years of research. I will then discuss the open issues and how future generation satellites, such as Swift, will help us to clarify them.

1. Introduction

If we build an experiment with a detector sensitive to photons in the keV to MeV band and fly it outside the Earth's atmosphere, we soon discover that, roughly three times per day, the whole γ-ray sky is suddenly illuminated by an intense source. This source becomes rapidly brighter than any other source in the γ-ray sky, even brighter than the sum of all the other sources in the whole universe. After few a milliseconds to several hundreds of seconds, as mysteriously as it appeared, the source disappears. The next source will appear in a completely different location in the sky and will be as elusive as all the others. These flashes of γ-rays are called Gamma-Ray Bursts (hereafter GRBs) by astronomers.

GRBs were discovered in the late sixties by the Vela satellites, launched by the US army to monitor possible nuclear experiments of the communist bloc in the sky or even on the dark side of the moon. Nuclear explosions can be revealed as bright flashes of light in the γ-ray domain. These satellites indeed discovered that there were bright flashes of γ-rays in the sky, but, after an initial stage of confusion, it was realised that these flashes were not coming from the Earth's atmosphere nor from the moon or any other known object in the solar system. They were coming from outside and were therefore of interest to the astrophysical community. Six years after the first GRB detection, the observations were therefore released to the public in a famous paper by Klebesadel, Strong & Olson.[1]

After that, several experiments have been flown on satellites to gather more information on the events and to understand the burst's origins. The main difficulty in the search for the origins of GRBs was the fact that GRB explosions last for a small interval of time, disappearing without leaving any detectable trace and without any repetition. To this we must add the intrinsic difficulty in the localisation of the direction from which γ-ray photons reach a detector.

A great contribution to the field was provided by the BATSE (Burst And Transient Source Experiment) on board the ComptonGRO satellite. BATSE did not have supreme localisation capabilities, but recorded about 3000 GRB light curves (Fig. 1) and spectra. It showed that GRBs are uniformly distributed on the sky and that their flux distribution is flattened at low fluxes. These two observations led to the first indication that GRBs would be of cosmological origin. At that time there was heavy debate whether GRBs were events of modest energy inside our Galaxy or rather extremely energetic events taking place at cosmological distances.

This issue, as well as many others, were addressed and solved by the discovery of afterglows made possible by the launch of the Italian-Dutch satellite *Beppo*SAX. This paper is dedicated to the discussion of how our comprehension of GRBs has evolved since the discovery of afterglows.

Before that, it is however worth remembering another important result obtained by BATSE. The distribution of the duration of the γ-ray phase of GRBs is shown in Fig. 2. The distribution of GRB durations is not uniform and is characterised by two peaks. The distinction of the GRB sample into two families is strengthened further by spectral considerations: the short-duration GRBs are harder (richer in MeV photons) than the long duration ones.[3] Due to trigger limitations on board the *Beppo*SAX satellite and to the intrinsic faintness of the prompt X-ray emission of short GRBs, it has

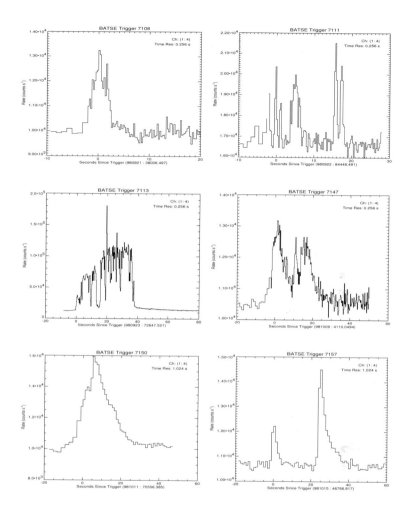

Fig. 1. An atlas of six GRB lightcurves from the BATSE archive. The selection shows the different character of lightcurves from different events. This diversity makes any classification impossible.

not been possible to date to detect any afterglow emission from short GRBs, and therefore measure their redshift and obtain all the information that led to the dramatic improvement of our understanding of the physics of GRBs. For this reason, the following of this paper is dedicated to the astrophysics of the family of long duration GRBs. Part of the last section is dedicated to a discussion on the possibilities of detecting afterglows from short GRBs in the near future.

D. Lazzati

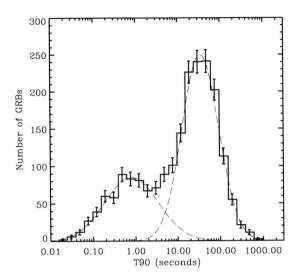

Fig. 2.   Distribution of the GRB γ-ray phase durations. The distribution is bimodal and defines two families of GRBs: the short GRBs (also richer in MeV photons) and the long GRBs (richer in keV photons). All the afterglow information we have is for the long GRBs. Data are from the final BATSE GRB catalogue.[2]

2. The General Picture

Figure 3 shows, in a single cartoon, our present understanding of the whole GRB phenomenon. The GRB evolution is divided into three phases: the energy release, the prompt emission and the afterglow.

The energy powering the GRB is released by a compact source, of the size of a Solar mass black hole over a time-scale comparable to the duration of the prompt γ-ray phase. The proper time-scale of the energy release has to be much smaller than that of the GRB, in order to produce the fast variability observed in GRB light curves. It is therefore inappropriate to define GRBs as "explosions" since the energy is released in a quasi steady-state process, in terms of the engine properties. This energy release produces an inhomogeneous relativistic wind made of a mixture of electromagnetic and baryonic components. This wind is also called a fireball.

The need for relativistic expansion is justified by the so-called compactness problem. Photons above an energy of ~ 511 keV can interact with each other to produce an electron-positron pair. The spectra of GRBs extend well above this threshold and contain many photons that could interact to produce pairs. If we compute the probability for such an interaction to

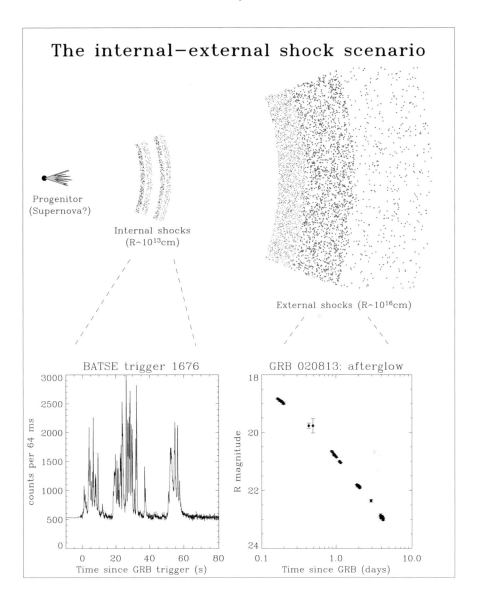

Fig. 3.   Cartoon of the canonical interpretation scheme for the GRB phenomenology.

take place in the GRB source, which must be compact in order to produce the observed γ-ray variability, we find that the overwhelming majority of photons with $h\nu > 511$ keV would be absorbed in this process. However we do not see any hint of absorption in the BATSE spectra. The only viable solution is that the source producing the photons is moving relativistically toward us, so that the photons we see at high energy are indeed, at the source, of lower energy and therefore unable to interact and produce pairs. In addition, a source moving towards the observer produces a faster variability, so that the photons are more dilute in the source. It turns out that, in order to solve the "compactness problem", the source of the γ-ray photons of GRBs must move with a Lorentz factor $\Gamma \geq 100$.

Inhomogeneities in the flow are averaged, by internal interactions, at a radius of $\sim 10^{13}$ cm from the source. Any of these interactions produces a shock wave within the flow that accelerates electrons and increase the magnetic component giving rise to synchrotron emission. Each of these interactions produces a spike in the γ-ray light curve.

Later, at a radius of $\sim 10^{16}$ cm, the interaction of the wind with the interstellar material drives a shock wave that produces the smoothly declining afterglow emission. In the following I will describe the three phases in reverse order, going from the afterglow phase, the better understood one, to the energy release by the inner engine, the most mysterious part. A recent observation has however shed new light on this phenomenon.

3. Afterglows

Until 1997 it had been possible to observe only the prompt phase of GRBs, which lasts between a few seconds and a few thousands of seconds. This was due to the fact that it is extremely difficult to measure with accurate precision the direction from which a γ-ray photon is coming. For this reason the only positional information that BATSE provided had an accuracy of several degrees in the sky, an area too large to be observed successfully with an optical telescope from the ground. Despite that, theoretical models to explain the burst emission predicted the possibility that a fainter but long living source should appear a few minutes after the burst and slowly decay with time.

In the first months of 1997, thanks to an appropriate combination of instruments and an adequate observation strategy, the X-ray telescopes on board the Italian-Dutch satellite *Beppo*SAX were able to discover the existence of such emission, called the afterglow. The GRB emission was detected by the γ-ray instrument which had no localisation capabilities but could trigger a search in the data of a relatively hard X-ray wide field camera (WFC). This camera provides localisation of the transient X-ray source with an accuracy of several arcminutes, an area small enough to be

GRB 971214

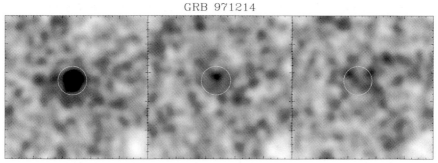

t$_{GRB}$+7 hours   t$_{GRB}$+17 hours   t$_{GRB}$+50 hours

Fig. 4.   Detection and follow-up observations of an X-ray afterglow (from GRB 971214) with the X-ray concentrators on board *Beppo*SAX. A bright new source is initially detected (7 h after the γ-ray event) within the error circle from which the GRB was detected. This source fades and eventually disappear under detectability after $\sim$ 2 days. Data from the *Beppo*SAXpublic archive www.asdc.asi.it/bepposax/.

observed with the X-ray telescopes on board the same satellite, which are much more sensitive than the WFC. Unfortunately the X-ray telescopes point in a direction perpendicular to the WFC, so that the whole satellite had to be rotated in order to image the location in the sky from which the GRB photons were coming. This rotation took about 6 hours. Despite this loss of time, observations revealed that a new bright X-ray source was coming from the same direction where the GRB was detected and that the source was fading in time.[4] This is shown in Fig. 4 where X-ray images of the burst location are compared at $\sim$ 7, $\sim$ 17 and $\sim$ 50 hours after the burst explosion.

The detection of an X-ray transient (or afterglow) allowed further refinement of the position from which the burst was coming, so that soon after an optical[5] and subsequently a radio[6] source were detected, coincident with the GRB location and fading in time.[a] After the optical source was dimmed enough, deep Hubble Space Telescope (HST) observations revealed also the presence of a faint galaxy.[7] The GRB did not explode in a random location of the universe, but inside a galaxy, an extremely important clue to understanding the nature of its progenitor.

Optical observations were of great importance in settling the debate on the distance of GRBs. The largest telescopes on Earth were used to de-

---

[a]The radio source started to fade in time only at a later time, as discussed below.

rive a spectrum of the optical transient,[8] and the spectrum revealed the presence of several systems of absorption features. Those features, if interpreted as absorption lines from metals such as iron and magnesium, implied a redshift close to unity. GRBs were therefore not taking place in our close neighbourhood but at incredibly large distances.

On the other hand, radio observations were decisive in confirming the presence of relativistic motions in the framework of the fireball model. At radio wavelengths, afterglows are observed to vary in a strong and unpredictable way at early times. As time goes on, the fluctuations are reduced in amplitude, until the radio lightcurve enters a phase in which the behaviour is smooth and resembles that at shorter wavelengths. The fluctuations of the early radio afterglow are not due to the source itself, but to the propagation of the radio waves into the interstellar medium of our Galaxy. If the photon source is small, diffraction of the radio waves in the interstellar clouds of the Galaxy causes a pattern of dark and bright areas, through which the Earth passes in its revolution around the Sun. It is a phenomenon similar to the twinkling of stars in the night sky: planets can be easily found not because we can see them as extended sources but since, being more extended, they do not twinkle. The analogous phenomenon cause the appearance of variability in the radio lightcurve. As the source expands, dark and bright areas begin to overlap until they eventually merge, suppressing the variability. In the case of GRB 970508, it was measured that, during the first month of life, the radio source had expanded to a radius of $> 10^{17}$ cm, implying an average Lorentz factor[6] $\langle \Gamma \rangle \sim 4$.

After these first results, afterglow observations have become more organised, and complete light curves in all wavebands are now routinely collected with telescopes and satellite instrumentation. Afterglow radiation is produced when the relativistically outflowing plasma released by the central object impacts the interstellar material[9,10] (hereafter ISM). A shock wave is driven in the ISM. At the shock front a huge amount of energy is converted from bulk outflow kinetic energy into internal (thermal) energy of the fluid. Electrons are accelerated to relativistic energies and a magnetic field component is built up by electromagnetic instabilities. Electrons gyrating around the magnetic field lines produce synchrotron photons with a spectrum that is made of several power-law branches connected by breaks.[10]

The model predictions are confirmed by observations. Collecting afterglow observations in different wavelengths (from radio to X-rays) it is possible to obtain the spectral shape and to follow its evolution with time.[11,12]

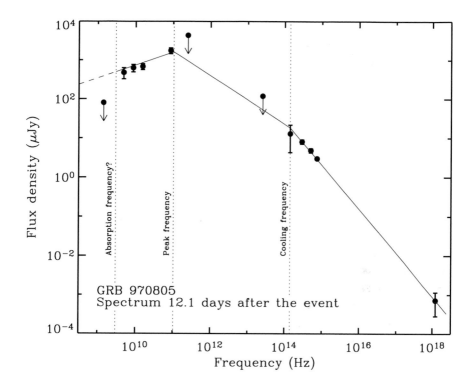

Fig. 5.   Broad band spectrum of the afterglow of GRB 970508 $\sim$ 2 weeks after the GRB explosion. From Galama *et al.*[11]

Figure 5 shows that the prediction of the synchrotron shock model (SSM) is supported by the data.

In addition to the spectral confirmation, the synchrotron origin of the afterglow emission has been confirmed by the observation of linear polarization of the optical emission.[13] If the magnetic field around which electrons gyrate to produce the afterglow photons has a sizable degree of alignment, the photon polarization will be aligned in an orthogonal direction. The detected degree of polarization, albeit small, allows us to confirm that the afterglow emission comes from a magnetised region with a magnetic field that, at least to some extent, has a definite structure.

Afterglow observations have been of great importance also in the study of the geometry of the fireball. If we measure the amount of photons that we receive from a bright GRB and correct them by the distance factor estimated through its redshift, we conclude that the energy involved in

D. Lazzati

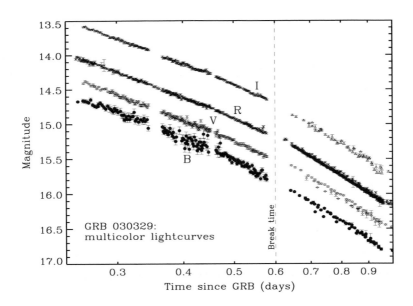

Fig. 6.   Multi-colour afterglow of GRB 030329, showing an achromatic break at ~ 0.6 d after the burst explosion. This behaviour is interpreted as the signature of a non spherical explosion. Data from Lipkin *et al.*[14]

a GRB explosion can be of the order of 10^{54} erg or, in other words, the entire mass of the Sun converted into pure energy within a time-span of few seconds. This conclusion holds under the assumption that the GRB explosion emits the same power in any direction. The surprisingly large energy inferred above led many scientists to consider the possibility that the GRB fireball is indeed beamed in a cone, that we see only if we happen to lie within its opening angle. It is not possible to tell a jet from a sphere during the prompt γ-ray phase. During the afterglow, on the other hand, the whole fireball becomes visible and the presence of a beam can be detected through the presence of a sudden decrease in the afterglow luminosity, in the form of a break in its temporal decay. This break is detected at the moment at which the observer perceives the edge of the jet, recognising that photons are not coming from all the possible directions but only from a limited region. Figure 6 shows one of the best examples of achromatic breaks detected so far, in the afterglow of GRB 030329.[14]

An important consequence of the identification of beaming and jets in the afterglow of several GRBs is the possibility to correct the energy output by the solid angle term. It turns out that GRB engines release a fairly standard amount of energy, but that this energy can be channelled in a narrow cone or in a wide one.[12,15] In the first case, a very bright GRB is detected, while in the second one a dim GRB is produced. The inferred energy release is of $E \sim 10^{51}$ erg, similar to the energy released in supernova explosions.

4. The Prompt Emission

Our understanding of the prompt emission of GRBs is far less accurate and complete than that of the afterglow phase. This is due to a combination of reasons. First of all, the prompt phase lasts for a very short span of time, making any complex observation difficult and making it impossible to define an observational strategy after the properties of the event have been understood. On top of this, relativistic effects play an important role in defining the spectral shape of the observed radiation, irrespective on the intrinsic spectrum produced by the emitting plasma. As a consequence it is very hard to tell which characteristics of the prompt emission are intrinsic and which are instead due to kinematic effects produced by fluid elements moving with a slightly different speed or direction.

The first issue one faces when addressing the properties of the prompt emission is how the energy of the flow is converted from kinetic to internal energy and eventually into photons. We know that, before the photons are radiated, the flow must be cold, in order to avoid the production of a powerful thermal component which is not observed in GRB spectra. The conversion of bulk kinetic energy into internal energy cannot be due to the interaction of the jet with the surrounding medium, since signatures of this effect (the progressive broadening of the GRB pulses) are not present in GRB light curves. In a baryon jet (dominated by the presence of baryon and neutrons) the easiest way to convert bulk kinetic energy into random internal energy is via shocks. In order for this process to be effective, the flow must present instabilities, in the form of faster and slower components. The fast components catch up with the slower ones and the collision releases a fraction of the energy in the form of heat. A sizable fraction of this energy is used to generate or amplify a magnetic field component and to accelerate electrons to relativistic energies, eventually producing synchrotron emission.[b] Problems of this scenario, which is the one on which a consensus is now being reached and is shown in Fig. 3, arise if one considers

---

[b]Synchrotron emission is due to relativistic electrons that rotate, and are therefore continuously accelerated, around magnetic field lines.

the efficiency of energy conversion. Shocks within the fluid can convert into internal energy only the energy that is associated with the inhomogeneities of the flow and not the energy of the flow itself. Observations seem however to indicate that most of the GRB energy is released in the prompt phase and not store for the afterglow.

Alternatively the energy may be carried by an electromagnetic component. In this case one would not need to have faster and slower parts in the jet. Magnetic energy, which is a form of "cold" internal energy, could be converted on the fly into hot internal energy through a number of magnetic and plasma instabilities, leading to an efficient energy dissipation without a sizable deceleration of the bulk motion.

5. The Progenitor

The progenitor of GRBs, i.e. the astronomical object that releases the enormous amount of energy powering the GRB emission, has been a mystery and a topic of deep speculation since the discovery of GRBs. The best candidate used to be the merging event that takes place when an old binary system made of two compact objects [two neutron stars (NS) or a NS with a black hole (BH)] shrinks to the point at which the two objects interact and disrupt each other.

In the early afterglow era much circumstantial evidence pointed to the association of GRB explosions with active galaxies, star forming regions and young stellar environments, contrary to what was expected for old binary systems. A new idea, linking the GRB explosion to the final evolutionary stages of massive stars, came into play, under various forms.[16,17] It is known that massive stars end their life in a spectacular explosion called a supernova (SN). A supernova is approximately as energetic as a GRB, but has a much slower evolution and is bright mainly at optical wavelengths.

The association of GRBs with SNe could be in one of two ways. In one form of the model, the SN explosion leaves a meta-stable object that, several months to years after the SN explosion, releases the energy for the GRB emission.[18] In the now more accepted scenario, the SN and GRB explosions are simultaneous (indeed they are a single explosion), but the explosion velocity is extremely high along the axis of rotation of the star and slower along the equator. In this case an observer along the axis would detect a GRB and a SN (outshined by the GRB and afterglow emission for most of the evolution) while an observer along the equator would see only a SN explosion.

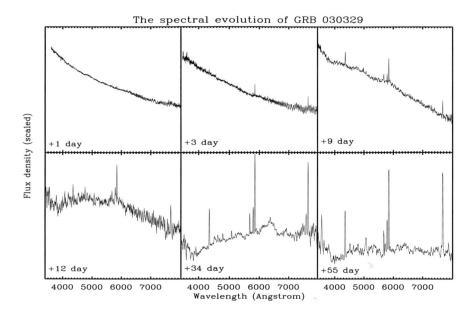

Fig. 7. Spectra of the optical transient associated to GRB 030329 at various epoch since the GRB onset. The spectrum progressively evolves from a featureless power-law (as expected in the afterglow model) to a curved thermal like spectrum. In the late phases broad deep absorption features appear, as expected if a SN was associated with the GRB event. Data from Matheson *et al.*[21]

Since the GRB and afterglow emission are brighter than the SN it is very difficult to detect the SN light in the overwhelming radiation of the burst. What makes the two components different is the spectrum of the emitted radiation. The afterglow photons are widely distributed in the electromagnetic spectrum, from radio waves to X-ray frequencies. SNe, on the other hand, have an almost thermal spectrum, concentrated at optical wavelengths. On top of this, the afterglow spectrum does not have any narrow feature, while the SN spectrum is rich with deep absorption troughs due to the presence of metals in the emitting material. For this reason astronomers had to wait for a very nearby GRB to perform an accurate analysis of the spectrum and detect signatures of a SN explosion simultaneous to the GRB. Figure 7 shows the result of this analysis. Spectra of the afterglow light were taken at different times after the GRB explosion, since the afterglow component decreases in time while the SN one increases. At early times (see the spectrum labelled +1 day) the spectrum is a featureless power-law, as predicted in the afterglow model and observed in many other GRB afterglows. As time progresses, the spectrum becomes curved and starts to develop deep and broad absorption features, typical of SN explosions (see the spectrum labelled +12 day).[19-21] The comparison of the late time spectra with SN spectra allows us to accurately time the two explosions (SN and GRB) which turn out to be simultaneous to within a few days. The most likely explanation of the observations of GRB 030329 is therefore that the GRB and the SN are a single event.[22]

A similar association, with better constrained timing, was claimed for GRB 980425 and SN1998bw.[23] In that case, however, the GRB properties were peculiar, and no afterglow emission was detected. In the case of GRB 030329 the burst and afterglow properties are much more similar to those of classical cosmological GRBs,[24] even though a deeper analysis shows that the event was under-energetic and the afterglow more complex than in other cases. Whether the association GRB/SNe can be extended to all long duration GRBs or not is therefore a matter of open debate.

6. Open Problems and the Future

Despite the wealth of new data gathered in recent years, there is still much to be understood and confirmed in the physics of GRBs. One of the open issues is of course the possibility to discover and study afterglows of short-duration GRBs. The last part of this section will be devoted to this.

Within the class of long duration GRBs there are several open issues awaiting new data and theoretical insight. It is quite natural to start this section by mentioning the problem of X-ray features. These are narrow absorption and/or emission features detected in the prompt X-ray emission or in the early X-ray afterglows of several GRBs.[25] They are problematic to explain in the framework of the fireball model since they are associated with ions at rest in the GRB host galaxy, while the afterglow emitting material is moving at relativistic speed toward the observer. One possibility is that the lines are produced by dense, metal rich material close to the burst site which intercepts part of the burst and afterglow emission and reprocesses it into line photons.[26] Such an explanation would be natural if the GRB and SN explosions were delayed by several months, a constraint which seems to be in contradiction with the observations of GRB 030329. Unfortunately, even though these features have been detected in different GRBs, a single incontrovertible detection is still lacking, and the reality of the features still awaits confirmation. Whether a distribution of delays between the GRB and SN explosions is therefore required to explain all the observations or not is still a matter of open debate.

Another aspect of GRB physics which is now under intense investigation is the structure and content of the relativistically expanding plasma (or fireball) that powers the GRB and its afterglow. In the past, GRB fireballs were supposed to be spherical. Robust evidence of conical beaming was gathered[27] in 2001 leading to a paradigm shift, and now it is customary to talk about GRB jets. But the nature of the main constituent of the jet is not yet clear. Possible components are baryonic material (protons and neutrons), electromagnetic energy (photons and electromagnetic field) or leptons (electrons and positrons). On top of that the energy distribution within the jet angle may be uniform, have a well defined non-uniform symmetric pattern or be characterised by a random occurrence of dark and bright uncorrelated spots. Combined photometric and spectral studies will help to understand the jet structure while polarimetric studies of the prompt and afterglow emission will give us important constraints on the contents of jets and their mutual interaction. Recently it has been claimed that the prompt γ-ray photons are characterised by a very large linear polarization, of the order of 80%.[28] This argues for the presence of a strong well-organised magnetic component in the jet, but the measurement has been under heavy debate and confirmation is required before any definitive conclusion can be drawn.

As anticipated above, a future challenge is that of detecting afterglows from the class of short GRBs. To date, it has been possible only to detect a hint of afterglow emission in the hard X-ray band. This detection was obtained by stacking a sample of 76 BATSE light curves of short GRBs. A tail of emission on a time scale of $\sim$ 100 s appears (see Fig. 8), with the softer spectrum typical of afterglow emission.[29] This detection tells us two important things: first that short GRBs have afterglows, and that it is therefore worth hunting for them. On the other hand, afterglows of short GRBs are on average a factor $\sim$ 10 fainter than those of long GRBs and consequently more difficult to observe. The weakness of the short GRB afterglows may be due to an intrinsic weakness of short GRB events, but also to the fact that the same energy as released in long GRBs is beamed in a wider cone. Answers to these questions and many others (such as the basic one on the distance scale of short GRB events) will have to await the prompt discovery of afterglow emission from an individual short GRB event. This will allow us to obtain an optical spectrum and therefore derive a redshift, and perform on the short GRBs all those studies that led to our deep understanding of the long GRB physics.

A great help in this will come with the launch of Swift, a dedicated

Channel 1+2

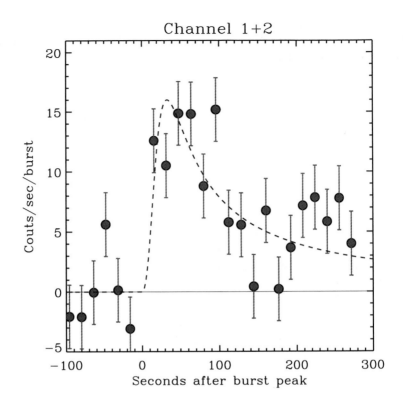

Fig. 8. Average hard X-ray afterglow of a sample of 76 short BATSE GRBs. The prompt GRB emission (concentrated at $0 < t < 1$ s) has been removed for clarity. From Lazzati et al.[29]

international mission that will be launched by NASA in the fall of 2004. Swift will have two unprecedented characteristics. On the one hand it will cover a broad band of the electromagnetic spectrum, from optical to soft γ-ray. The observations of the burst and early afterglow emission will be therefore pan-chromatic. Also, it will be able to slew to the position where the GRB is coming from in a time-scale of order ~ 60 s, to be compared to the ~ 6 h needed to re-point *Beppo*SAX. The extremely sensitive telescopes of Swift will therefore be on-target during the prompt emission and in the early bright phase of the afterglow, providing us with high quality data covering the critical phase of the early afterglow emission, which remains still poorly observed to date. Moreover, Swift will increase our database of GRBs with known redshift by one order of magnitude, allowing statistical studies on the GRB population to be performed for the first time and to use them in cosmological studies (to which a full chapter of this book, by N. Tanvir, is dedicated).

References

1. R. W. Klebesadel, I. B. Strong and R. A. Olson, *ApJ*, **182**, L85 (1973).
2. See http://www.batse.msfc.nasa.gov/batse/grb/catalog/current/
3. C. Kouveliotou *et al.*, *ApJ*, **413**, L101 (1993).
4. E. Costa *et al.*, *Nature*, **387**, 783 (1997).
5. J. van Paradijs *et al.*, *Nature*, **386**, 686 (1997).
6. D. A. Frail, S. R. Kulkarni, S. R. Nicastro, M. Feroci, G. B. Taylor, *Nature*, **389**, 261 (1997).
7. S. R. Kulkarni, *Nature*, **393**, 35 (1998).
8. M. R. Metzger *et al.*, *Nature*, **387**, 879 (1997).
9. P. Meszaros and M. J. Rees, *ApJ*, **476**, 232 (1997).
10. T. Piran, *Physics Reports*, **314**, 575 (1999).
11. T. J. Galama *et al.*, *ApJ*, **500**, L97 (1998).
12. A. Panaitescu and P. Kumar, *ApJ*, **571**, 779 (2002).
13. S. Covino *et al.*, *A&A*, **348**, L1 (1999).
14. Y. M. Lipkin *et al.*, *ApJ*, **606**, 381 (2003).
15. D. A. Frail *et al.*, *ApJ*, **562**, L55 (2001).
16. B. Paczynski, *ApJ*, **494**, L45 (1998).
17. S. E. Woosley, *ApJ*, **405**, 273 (1993).
18. M. Vietri and L. Stella, *ApJ*, **507**, L45 (1998).
19. K. Z. Stanek *et al.*, *ApJ*, **519**, L17 (2003).
20. J. Hjorth *et al.*, *Nature*, **423**, 847 (2003).
21. T. Matheson *et al.*, *ApJ*, **599**, 394 (2003).
22. A. I. MacFadyen and S. E. Woosley, *ApJ*, **524**, 262 (1999).
23. T. J. Galama *et al.*, *Nature*, **395**, 670 (1998).
24. D. Lazzati, proceedings of the 10th Marcel Grossmann meeting on General Relativity, Rio de Janeiro, Brazil, July 2003, astro-ph/0402058 (2004).
25. M. Böttcher, proceedings of the 10th Marcel Grossmann meeting on General Relativity, Rio de Janeiro, Brazil, July 2003, astro-ph/0312538 (2004).
26. D. Lazzati, S. Campana and G. Ghisellini, *MNRAS*, **304**, L31 (1999).
27. R. Sari, T. Piran and J. P. Halpern, *ApJ*, **619**, L17 (1999).
28. S. Coburn and Boggs, *Nature*, **423**, 415 (2003).
29. D. Lazzati, E. Ramirez-Ruiz and G. Ghisellini, *A&A*, **379**, L39 (2001).

Davide Lazzati

Davide Lazzati was born in Como, a small Italian town at the feet of the central Alps, on the shores of the homonymous lake. He graduated *cum laude* in Physics in 1994 and obtained his PhD in astronomy in 2001. His doctorate thesis, on Gamma-Ray Bursts, was awarded the "Livio Gratton" prize 2003, for the best PhD thesis on astronomical subjects presented in Italy in the years 2001/2002. He moved to Cambridge in the fall of 2000 to take up a research fellowship at the Institute of Astronomy of the University. He is still at the IoA, now with a PPARC postdoctoral fellowship, where he performs full time research on Gamma-Ray Burst theory and observations. His scientific interests include relativistic astrophysics and the high redshift universe; recreations include mountaineering, rock climbing, skiing and everything that can be performed on a mountain.

CHAPTER 12

ASTROPHYSICAL DUST

Malcolm Gray

Department of Physics, UMIST,
P.O. Box 88, Manchester, M60 1QD, United Kingdom
E-mail: Malcolm.Gray@umist.ac.uk

I consider the 'life-cycle' of astrophysical dust, from its formation and journey through the interstellar medium of the Galaxy, to its destruction. Likely structures for dust grains are examined, and the processes which control the composition of dust grains. I discuss the importance of astrophysical grain surfaces as sites of interstellar chemistry, and the effects of dust on the radiation which we observe from the Universe.

1. Introduction

Between the bright stars of the night sky (if you are lucky enough to live somewhere where the night is still dark enough to see any) lies a tenuous interstellar medium (ISM) made of gas and dust. The dust forms only about 1% of the mass of the ISM, but it has properties which make it much more important than its low abundance might imply.

ISM dust has profound effects on the light we observe from stars and galaxies; it also radiates strongly at much longer wavelengths, revealing the sites of current star and planet formation. Dust grains also contain a large proportion of the heavy elements of the ISM, and aggregations of dust produce the rocky material from which Earth-like planets must form. Grains often have active surfaces, which behave like miniature chemical factories, allowing many reactions to procede which would be vastly slower, even impossible, in the gas. This may include, perhaps, the production of simple biological molecules vital for the origins of life.

Astrophysical dust has a precarious existence, balanced between a set of processes which destroy grains, and other processes which allow them to grow. A typical lifetime can be estimated for grains, during which they may experience conditions as extreme as the shockwaves of supernovae, passage through several phases of the ISM to, perhaps, destruction in a new star, or incorporation into a rocky planet or asteroid.

Throughout the extensive odyssey of a grain through the galaxy, it is constantly being reprocessed by its environment. The surface is particularly vulnerable to change, and it is the surface of a grain which is most open to analysis: by observing features in the spectra of dust grains, we gain information about the composition of grains - their minerology. The study of dust grains has recently become very exciting because new technology has enabled us to observe at the infrared wavelengths where dust emits most of its radiation.

2. Dust and Light

Unlike atoms and molecules, which have discrete line and band spectra based on their internal energy levels, dust grains can absorb, emit and scatter light in a continuum, extending over many decades of wavelength or frequency. Dust is therefore very efficient at stopping and re-directing light, and the greatest efficiency is found at wavelengths comparable to the sizes of grains. Wavelength-dependent extinction (the combined effect of absorption and scattering) is the property that originally led to the discovery of dust in the ISM. Light from distant stars experiences extinction by dust, and they consequently appear dimmer than they ought to. Without correction for extinction, we would conclude that such a star is more remote than its true distance. The wavelength dependence of extinction can be deduced by looking at the relative effects on stars of different intrinsic colour (as deduced from their spectral types). It is found that bluer (hotter) stars are extincted more than redder (cooler) types. A typical interstellar extinction curve[1] is plotted in Fig. 1. Starlight is therefore not only extincted by dust, but reddened as well. The fact that ultraviolet and blue light are affected more by dust than red and infrared light gives us an idea of the size distribution of the interstellar dust grains: they are mostly smaller than near-infrared wavelengths (about 1 micrometre) and become more common (more grains per unit size range) down to sizes at least as small as a few nanometres.

When a dust grain absorbs a photon of light, the energy of the photon has to be redistributed within the grain material. Almost all of this energy appears as a slight increase in the temperature of the grain. A grain eventually adopts a temperature which corresponds to a balance between its

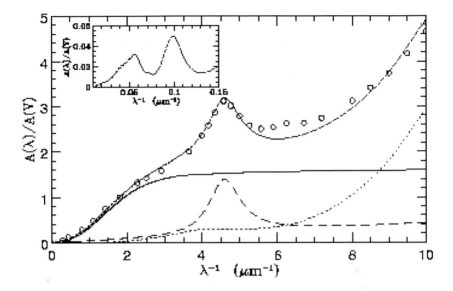

Fig. 1. The interstellar extinction curve: Extinction, relative to the value at visual wavelengths, is plotted as a function of reciprocal wavelength. The open circles show measured values, and the associated solid line is a theoretical fit. The inset shows the extinction in the infra-red, where it appears close to zero in the main graph. The three lower lines show the contributions of the three dust components in the model: large composite grains based on silicate cores (solid line); small carbonaceous grains (dashed line) and PAH macromolecules (dotted line).

heating processes and cooling processes. Absorbed light is often an important source of heating, but the grain will also radiate in accordance with the temperature of the material. The grains are not black-bodies, but still emit a continuous spectrum, which can be considered to be a convolution of a black-body at the temperature of the grain material, and an emissivity function, characteristic of the material. Because the equilibrium temperatures of grains are typically very low, rarely more than 100 Kelvin, the wavelength of maxium grain emission typically lies in the far-infrared (FIR) or sub-millimetre ranges of the electromagnetic spectrum. The ability to observe grain emission therefore relies on recent technology: sensitive bolometers, such as SCUBA on the James Clerk Maxwell Telescope in Hawaii, and satellite missions, like ISO (the infra-red space observatory). By contrast, observations of dust in absorption are typically made in the visible and ultraviolet regions of the spectrum, and have a considerably longer history.

2.1. *Spectral features*

As the paragraph above suggests, the atoms and molecules in grains still maintain some characteristic optical properties, and are not entirely subsumed under the continuum spectrum of the bulk material. Some molecular characteristics are obviously lost: molecules in a grain are not free to rotate like those in a gas, so we lose the detailed rotational band spectra of the molecules, but vibration is still possible to some extent. Vibrational band spectra, characteristic of particular chemical bonds, can therefore still be seen as features on top of the continuum spectrum. The vibrations, however, are distorted by their interaction with the grain lattice, and have frequencies which are generally offset from those known for gas-phase molecules. One feature is evident in the inset part of Fig. 1. This is the strong silicate feature near $0.1\,(\mu m)^{-1}$, which results from the stretching vibration of Si-O bonds. Other important features include the ice vibrations at 3 and $6\,\mu$m, and the C-O stretch at $4.7\,\mu$m.

2.2. *Polarization*

Dust grains produce polarization of background starlight as well as causing extinction. Although many simple models represent dust grains as spheres, real dust particles are elongated, and many are likely to be electrically charged. Charged grains that are set spinning - by a collision for example - will orient themselves in the interstellar magnetic field. The orientation which has the lowest energy, and is therefore the most likely, is to have the long axis of the grain perpendicular to the field. For light travelling along the field lines, there is little selective effect, but light travelling perpendicular to the field will have its electric field at an angle of between zero and 90 degrees to the long-axes of the grains. For light at visible wavelengths, light with the electric field parallel to the long-axes is preferetially extincted, and the transmitted light becomes preferentially polarized parallel to the short-axes, that is parallel to the magnetic field. Interstellar polarization by grains was first identified as such by its strong correlation with reddening and extinction. The planes of interstellar polarization along different lines of sight have been used to map the Galactic magnetic field.

3. Dust and Chemistry

The ISM has a rich chemistry, particularly in cold molecular clouds. However, it is very difficult to make some molecules in gas-phase reactions. The prime example is the most abundant interstellar molecule of all: molecular hydrogen, H_2. The apparently obvious formation reaction, involving the combination of two hydrogen atoms, is forbidden because the two atoms are the same, and a third body is required to conserve angular momentum. Three-body collisions are much rarer than two-body collisions in such a diffuse gas as the ISM, so an additional channel for H_2 formation is needed. Dust grain surfaces can provide this: two hydrogen atoms which separately land on the grain have some mobility over the surface and, if they meet, they can form an H_2 molecule. The angular momentum problem is solved by the presence of the lattice of grain material, which acts as the necessary third body. Formation of an H_2 molecule in this way releases energy, so formation of H_2 - and most other molecules - acts as a heating process for the grain. Depending on the tempearture of the grain, the newly formed molecule may be effectively trapped on the grain surface, or, if the grain becomes warmer, so that the thermal energy of the molecule becomes comparable to its binding energy to the grain surface, it may easily be able to escape into the gas. All particles stuck to a grain surface are bound in some potential well of energy-depth, D, in which they can vibrate with some characteristic frequency, ν. The dwell-time, t, of a particle at this site is given by $t = (1/\nu) \exp[D/(kT_g)]$, where T_g is the temperature of the grain surface. Under most interstellar conditions, hydrogen atoms and molecules are highly mobile, but atoms of carbon, and heavier species, have low mobility.

There are a variety of possible grain surfaces, and these are dependent on the grain material, and the amount of processing which the grain has experienced. Mildly processed surfaces may have quite regular, crystalline, structures, with the surface atoms arranged in predictable patterns. Processing of grains usually results in the development of an amorphous surface, full of dislocations, implanted impurity atoms, and vacancies, where atoms have been ejected by, for example, cosmic ray impacts. From the point of view of chemistry, the important property of a surface is the number of free chemical bonds per unit area: incoming atoms striking a surface with few free bonds are likely to be physisorbed, that is, they are weakly bound to the surface, and have high mobility (binding energy typically 10 meV). Conversely, atoms striking a surface with many free bonds are likely to be chemisorbed: attached strongly to the surface by a covalent bond (energy typically several eV), and almost becoming part of the grain. Chemisorbed atoms have low mobility. One likely grain material, graphite, can have both types of surface: physisorbing planes, but chemisorbing surfaces perpendicular to the planes, where many carbon atoms will have free bonds. Physisorbing surfaces are good for producing small molecules quickly: atoms that stick to the surface are likely to sample that surface quickly, meet each other and, if suitable, react. However, the product is also likely to return quickly to the gas-phase, and is unlikely to involve atoms of the grain material.

The chemisorbing surfaces have a much richer chemistry,[5] which often involves atoms from the grain material. In the case of graphite, the chemistry is driven by a number of possible reactive end groups. These include vacant carbon bonds, aromatic C-H, phenolic O-H, and the toluene type C-CH_3. Return of a product to the gas phase usually requires a slow loosening of an end group over a sequence of reactions, otherwise the new molecules are unlikely to be returned to the gas phase unless the grain becomes very hot, gets sputtered, or is shattered in a collision (see Sec. 5). An example reaction is shown in Fig. 2, in which successive hydrogen atoms latch onto an active site, with the eventual release of a molecule of methane.

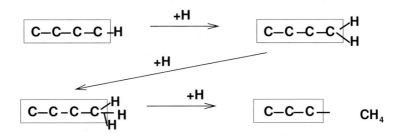

Fig. 2.   A grain surface reaction generating methane: Carbon atoms in the grain (represented by the box) have an end site occupied by a hydrogen atom. Following the arrows, successive hydrogen atoms are added, developing the end-group. When the final atom is added, a molecule of methane is released, removing one carbon atom from the grain. Addition of a further hydrogen atom to the vacant carbon bond will return the model to the beginning, but with one fewer carbon atoms in the grain.

Overall, grain surface chemistry is competitive, in speed, with many gas-phase channels for the production of several important interstellar molecules. However, the grain chemistry is exceptionally important for those species where gas-phase channels are weak or absent: homonuclear molecules like H_2, O_2 and N_2, and for making saturated hydrides such as water, ammonia and methane.

3.1. *Small grains*

When a small grain absorbs an ultraviolet photon, the energy absorbed per grain atom can be significant, and the sparse phonon spectrum of a small grain means that this energy will often not be efficiently shared, but remain localised, producing a large temperature spike. These high temperatures may enable endothermic reactions to operate on small grains.

3.2. *Cold cloud effects*

As clouds becomes cold, dwell times for some species become very long, and they effectively become trapped on the grain surface. This process is known as freeze-out, and is apparent from the appearance of spectral features of ices on the grains, and from the depletion of some molecular species from the gas phase. Freeze-out is progressive, with more and more volatile species becoming depleted from the gas as the cloud gets older and colder. The last, and most volatile, compound to freeze out is molecular nitrogen, which becomes an ice only in the coldest regions of clouds at 10 Kelvin or less. See Table 1 for the freeze-out temperatures of some common molecules. Reactions on a CO surface release H_2CO, CH_3OH and CO_2 which can act as feedstock for gas-phase reactions that go on to make complex organic molecules.

Table 1. Freeze-out or sublimation temperatures for some astrophysical molecules found in cold clouds.

Species	T_{sub} (K)	Type
H_2O	90	polar
CO_2	45	non-polar
CH_4	20	non-polar
CO	20	non-polar
N_2	10	non-polar

When star-formation begins, the grains are warmed, and begin to lose their icy mantles. The progress of warming is quite predictable, and reappearance of sulphur-bearing species in hot molecular cores can be used to estimate the age of the core. This sequence begins with the formation and release of H_2S, followed by S, SO, and then a more complex chemistry releasing SO_2, CS and H_2CS.

4. The Origins of Dust

Until recently, the main sources of interstellar dust have been assumed
to be long-period variable (LPV) and red supergiant stars.[4] These stellar
types have cool atmospheres and rapid mass-loss. The outflowing material
is dense enough that chemistry is fast compared to the outflow timescales,
giving dust a good chance of growth following any initial nucleation. Radiation pressure on the dust, once it has formed, helps to sustain the mass
outflow from the host star. LPV stars with oxygen-rich atmospheres have
all their carbon budget in strongly-bound CO molecules. The CO remains
gaseous, and the dust formed in the envelopes of these stars is of silicate
type. As these stars age, convection dredges up nuclear-processed material
from their interiors to their atmospheres, and they become carbon-rich. For
these stars, the CO binds all the oxygen, and carbon dust forms. Stars in
an intermediate state can form dust based on silicon carbide (SiC). Most
work on the actual nucleation of the dust uses a thermodynamic argument:
when the temperature of the outflowing gas falls below the condensation
temperature (related to the Gibbs free energy of the solid), then grains will
condense. Table 2 lists the condensation temperatures of some likely grain-
core materials. Much work remains to be done on the detailed kinetics of
grain formation.

Table 2. Condensation temperatures for
likely grain materials.

Material	T_{cond} (K)	Source
H_2O	150	ISM
C (graphite)	1850	C-rich star
SiC	1500	intermediate star
Fe_3O_4	400	O-rich star
Fe_3C	1100	intermediate star
Al_2O_3	1760	O-rich star
$MgSiO_3$	1350	O-rich star
Mg_2SiO_4	1440	O-rich star
Fe	1470	supernova

It has long been known that some grains have peculiar isotopic compositions. In particular, grains often contain rare isotopes that can only
be produced in supernova explosions. Astronomers therefore accepted that
some grains had to be formed in the aftermath of supernovae, but usually
believed that, though they produce some dust, supernova explosions destroyed far more; that is, they are net destroyers of dust. This seems logical

because of the violence of the shockwaves in supernova remnants, and the very high temperatures in the post-shock gas. However, recent observations using SCUBA on the JCMT[2] have shown that, at least in one remnant, a supernova has been a net producer of dust.

5. To Grow, Shrink or Shatter

Once a grain has condensed, it will be modified by its environment. Perhaps the most obvious interaction is with atoms and molecules of the surrounding gas. If these strike at an energy low compared to the binding energy of the atom, or molecule, to the surface of the grain material (typically a few electron volts for chemisorption, but only 0.001-0.01 eV for physisorption) then the gas particles will probably stick to the grain, increasing its size and mass. For the most part, the impacting atoms and molecules will not be like the grain material, but the common atoms and molecules in the local gas. Another obvious interaction is with photons of starlight. Energetic photons can eject photoelectrons from grains, leaving them more positively charged, and probably also leaving a chemically active bonding site on the grain surface. Both particle and photon interactions heat grains.

In the warm and hot phases of the interstellar medium, impacting gas particles have energies that are typically well above the binding energies of passive (physisorbing) surfaces, and we expect that grains will not grow due to gas bombardment. In addition, these ISM phases are normally open to energetic ultraviolet radiation, photons of which have sufficient energy to eject loosely bound atoms and molecules from grain surfaces. However, in dense, cold molecular gas, temperatures are typically less than 100 Kelvin, and there is considerable shielding from ultraviolet radiation. In these cold clouds, gas particles can often stick to grains. Polar molecules, for example water, bind more strongly to surfaces than non-polar species (like molecular hydrogen), and can stick to grains at higher temperatures. Grains in molecular clouds therefore accumulate mantles of ices, drawn from the gas. As a grain moves from a warm to a cooler region, it can build up a mantle of several layers of ice, usually starting with water ice, and then including less polar species as the temperature falls: methanol, carbon monoxide, carbon dioxide and, in the coldest zones, molecular oxygen and nitrogen. These ice mantles change the optical properties of the dust, and the ices present can be identifed from spectral features: see also Sec. 3.

If a grain escapes to the warmer ISM, the ice mantles become vulnerable. Total loss, however, may not always occur. Ultraviolet irradiation and higher temperatures drive chemistry in the mantle as well as ejecting atoms, and, at least for some grains, the icy mantles may be partially converted into a strongly-bound organic shell, which can protect the old mineral grain core beneath. Repeated passages through cold and warm gas can result in grains accumulating further icy mantles, followed by having the mantles lost or processed, leading eventually to a gobstopper-like accumulation of organic layers around the old core.

When the impact energies of gas particles rise above the surface binding energy of the core material then, providing the core is exposed, the impacts can damage the grain core by ejecting atoms from the mineral. This process is known as sputtering. At even higher energies, the impacting particles become able to eject grain atoms from the interior of the grain, and to do a lot more damage per impact. Sputtering can be thermal, where the impacting gas particles have energies commensurate with the local gas temperature, or non-thermal, where the bombarding particles have energies dictated by some other means: often the controlling parameter is a bulk velocity. Efficient sputtering requires a mean energy amongst the bombarding particles of at least 100 eV (there is a dependence on the grain material). For thermal sputtering, only the hot ISM phase qualifies, but this phase is so rarified, that impacts are too infrequent to erode a typical grain over the lifetime of the Galaxy. Only in the hot postshock gas of supernova remnants do we find gas with both the temperature and density to make thermal sputtering a quick mechanism for grain destruction. Non-thermal sputtering usually occurs in shockwaves. These vary from the mild (shocks bordering outflows from low-mass protostars, typically moving at $30 \, \mathrm{km \, s^{-1}}$) to the very severe (a young supernova blast-wave expanding at $10000 \, \mathrm{km \, s^{-1}}$). When a shockwave passes a grain, it is struck by a collection of particles which all have the same velocity, rather than the same energy (which is the case in thermal sputtering). The particle energy is therefore proportional to the mass, and heavy particles (though rare) become disproportionately effective sputterers. The sputtering potential of shocks is often further improved by the presence of a drift velocity between charged and neutral fluids in the gas behind the shock.

The details of sputtering depend on the grain material, the species, energy and impact angle of the incoming particle and the composition of the gas from which the bombarding particles are drawn. A typical angle and energy dependence is shown in Fig. 3. A useful point to note is that, even for the most optimistic angle and energy, no sputtering yield in Fig. 3 approaches anywhere near the 100% which is required to erode the grain absolutely. The sputtering process requires that the projectile particles, once deposited in the grain, can easily escape back into the gas, whilst the rare mineral atoms in the solid are irreplaceable. The enrichment of these heavy elements in the gas of supernova ejecta may explain how grains can survive these violent events: the rate of their implantation may exceed their rate of loss by sputtering. The process of sputtering should be seen as one of modification of the grain composition, rather than of simple erosion.

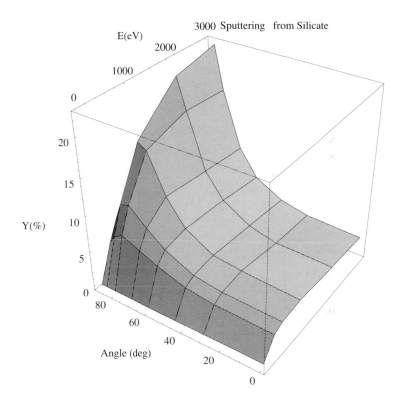

Fig. 3. The sputtering yield, as a percentage, for hydrogen ions impacting on a silicate grain, as a function of energy and impact angle.

So far, we have considered dust grains interacting with gas particles and light; grains can also collide with each other, with results that depend strongly on the collision energy. As with particle impacts, there is a hierarchy of energy ranges which give different typical consequences to a collision. If the relative velocity of the collision is lower than the sound speed in the grain material, then the collision is almost certainly either accumulative or neutral in its effect. In the latter case, we describe the collision as elastic: the grains bounce off each other, changing direction but unaltered in structure. An elastic collision has no effect on the size-spectrum of grains. The other alternative is for the colliding grains to stick together and form one larger grain. The likelihood of this outcome is increased if the grain structures are highly irregular, so that parts of the grains can act as hooks that prevent separation after impact. Collisions like this reduce the total number of grains, and increase their mean mass. Almost all grain-grain collisions in molecular clouds are neutral or accumulative.

Finally, for impact energies which correspond to supersonic motion in the grain material, a shockwave develops inside the grain following impact, and grain material can be melted, or even vapourised. The initial result of the impact is to generate a very high pressure in the shocked region. This compression is followed by isentropic expansion: if the isentrope runs through a pressure lower than the critical point of the material, a solid-liquid phase transition will result, but a solid-vapour transition occurs if the pressure is always above the critical point. Partial vapourisation requires a collision energy per atom of about twice the lattice binding energy; three times the binding energy is necessary for total vapourisation of the impact region. The energy coupled to the target region is highest for dense grain material, and we expect shattering effects to be dominated by high-velocity impacts of small, dense grains. Depending on the grain material, fragmentation sets in at relative velocities on impact between 1.5 and $20\,\mathrm{km\,s^{-1}}$. Collisions of this energy are likely to shatter the colliding grains, producing several fragments. High energy collisions, then, tend to increase the total number of grains, but to decrease their mean mass. Various additional processes can modify the effectiveness of grain-grain collisions. For example, a charged grain can be accelerated in a magnetic field (the betatron process) and aquire a much larger impact velocity than would be expected from the local kinetic temperature. At even higher relative velocities (between 75 and $1000\,\mathrm{km\,s^{-1}}$, depending on the grain material) catastrophic destruction of the colliding grains is likely, with complete vapourisation into a cloud of atoms and molecules instead of fragmentation into smaller grains.

6. Dust: A Space Odyssey

A typical survival time for a grain is a few hundred million years. Even travelling at a modest $1\,\mathrm{km\,s^{-1}}$, the grain will cover over $500\,\mathrm{pc}$ in this time.

When grains leave the site of their formation, they face an uncertain future. Their fate largely depends on which phases of the interstellar medium they pass through on their sometimes vast trajectories across the Galaxy. Much of the ISM, in terms of volume, is a hot, tenuous gas at about a million Kelvin. This sounds dangerous for the dust, but although this gas is very hot, leading to high-energy, sputtering impacts from gas particles, it is so tenuous that the impacts are insufficiently frequent to erode a grain over the lifetime of the Galaxy. Dust is fairly safe in the hot gas unless it is swept by shockwaves. However, since much of the hot gas lies well above and below the Galactic plane, dust in the hot gas is often remote from most shock events.

Much of the rest of the ISM (in terms of volume) comprises a 'warm' gas, with a temperature of a few thousand Kelvin. This is probably the most dangerous part of the ISM for dust grains. Although the temperatures are too low for thermal sputtering, this gas is regularly swept by shockwaves, which cause the destruction of grains by a combination of non-thermal sputtering and energetic grain-grain collisions. This gas is also sufficiently diffuse to be open to ultraviolet radiation that can eject atoms from the grain surface. Like shockwave-producing events, starlight is concentrated in the Galactic plane, making the warm gas a signifcantly more dangerous host than the hot gas in this respect. Even grains which avoid destruction, or significant erosion of their core material, are likely to lose any icy mantle material they have gained in cooler regions. Any surviving mantle material will be heavily processed (see Sec. 3 above).

The cooler phases of the ISM represent only a small volume, but still have significant mass. Rough pressure equilibrium with the hotter phases gives the cool gas high density. The warmer regions of this cold gas are largely atomic hydrogen, whilst the cooler parts are based on molecular hydrogen. Dust in these cooler gases is more protected than in the warm ISM: high density tends to damp the effect of shockwaves, and in the molecular gas, the environment is well shielded from UV photons which are energetic enough to dissociate H_2. Especially in the molecular gas, grains can accumulate icy mantles, possibly in several layers, depending upon the temperature of the gas through which the grain passes. The molecular gas is, however, where new stars are formed. The gas surrounding a protostar experiences a rapid rise in temperature, a large increase in radiation flux and, if the new star is massive, photoionization. Some dust disappears into the body of the new star. Other grains, especially near the poles of the star, usually become part of an outflow, partially accelerated by radiation pressure on the grains. Many grains get destroyed by the shockwave that forms as this outflow drives into the surrounding ISM. Yet other grains, which initially had high angular momentum, fall into a disc of material which slowly accretes onto the young star. This disc is thin, and can become very dense. Under the right conditions, usually for stars of modest mass, the grains in the disc can become very large, becoming planetesimals, and, perhaps, the asteroids and planets of a new planetary system.

7. Missing Metals

Dust is also important in helping to explain the heavy element (anything heavier than helium to an astrophysicist) budget of the Galaxy. From detailed spectra of the Sun, and other stars, we have a good idea of the relative abundances of the elements in the solar neighbourhood. These can be compared with abundances in other parts of the Galaxy by examining the visible and ultraviolet absorption spectra produced by interstellar clouds along the lines-of-sight to more distant stars, for example 23 Orionis[3] and references therein. The absorption spectra give us elemental abundances in the gas, and we find that in many places in the Galaxy, particular heavy elements are under-abundant in the gas phase, compared with what we expect from the solar neighbourhood. The most depleted metals in the gas are calcium (Ca) and titanium (Ti), which can be more than a thousand times less abundant in the gas (relative to hydrogen) than in the solar neighbourhood. Other strongly depleted elements include Al, Ni and Fe.

If we assume that the gas over most of the Galactic plane is well-mixed, then the relative abundances of the elements ought to be much the same whichever line-of-sight we choose in the Galaxy. As we observe gas-phase depletions of some metals, we conclude that they are present in some form, but not in the gas. Presumably, then, dust grains act as reservoirs of many metallic elements that are defficient in the gas. This view is supported by the observation that the 'missing' metals are less depleted in warm gas, particularly where there is evidence of high-velocity motion: these are exactly the conditions that we expect to erode and destroy grains, releasing some of the heavy elements back into the gas.

The observed metal depletions remain somewhat mysterious, however: on the basis of popular assumptions about the composition of grain cores, we would expect that the most depleted elements would be carbon and those that form part of the most likely silicate minerals - silicon, iron, magnesium and oxygen, but these, though depleted, are not the most defficient. We still do not know why elements like Ca and Ti are the most heavily depleted from the gas phase. One indicator is a correlation between depletion and the condensation temperature of the elements. We have to assume that either the highly depleted elements are extremely difficult to extract from grains, or that their atoms have a very high probability of becoming embedded in a grain if they collide with one. The exact mechanism, however, remains unknown.

8. Global Grain Models

Any successful model of grains has to be able to reproduce the observed extinction and polarization properties of the dust over a wide frequency range, from the radio through to the far ultraviolet region of the spectrum. The model also has to contain a sensible size-spectrum for the dust, and suggest a reasonable scenario for the 'life-cycle' of grains. Probably the best global model is by Greenberg.[1] Greenberg adopts three components for the dust, which are all inter-related, two of them strongly. The large-grain component is based on silicate cores. By repeated passage through cold and warm ISM phases, these cores develop icy mantles, and some of the mantle material survives as a processed organic shell, or shells, protecting the original core. These large, composite grains provide the extinction curve from the radio through to the near ultraviolet, and, providing they are not spherical, the same grain population can fit the necessary polarization properties. The model is layed out schematically in Fig. 4.

The Greenberg model further proposes that damage to the coatings of the large composite grains can release fragments of the organic shell into the ISM as a population of small carbonaceous grains. The presence of these small grains explains the 'ultraviolet bump' in the extinction curve near 0.22 micrometres wavelength (see Fig. 1). Degradation of the small carbon-grains, and condensation in the atmospheres of carbon-rich LPV stars is taken to be the source of a third dust component: polycyclic aromatic hydro-carbon (PAH) molcules. The PAHs have properties intermediate between those of a very small dust grain and a large molecule. Their main contri-bution to the model is to produce the strong extinction observed in the far ultraviolet region of the spectrum, shortward of $0.16\,\mu$m. Several species of PAH are known observationally from features in their spectra.

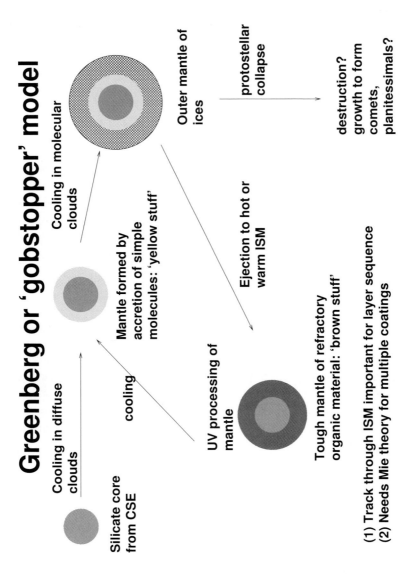

Fig. 4. A pictorial view of the Greenberg dust model.

References

1. A. Li and J. M. Greenberg, *Astron. Astrophys.* **323**, 566 (1997).
2. L. Dunne, S. Eales, R. Ivison, H. Morgan and M. G. Edmunds, *Nature* **424**, 285 (2003).
3. D. E. Welty, L. M. Hobbs, J. T. Lauroesch, D. C. Morton, L. Spitzer and D. G. York, *Astrophys. J. Supp.* **124**, 465 (1999).
4. D. C. B. Whittet, *Dust in the Galactic Environment* (IOP Publishing, Bristol, 1992).
5. T. J. Millar and D. A. Williams (Eds.) *Dust and Chemistry in Astronomy* (IOP Publishing, Bristol, 1993).

Malcolm Gray

Malcolm Gray was born in Barnstaple Devon. He read Physics with Astrophysics at Birmingham University, graduating with first class honours in 1983, and then moved on to Sussex University to do a DPhil on the origins of 'starburst' activity in galaxies. After working as a postdoctoral research assistant with David Field at Bristol University, and a brief period of employment as a physicist at the British Antarctic Survey, Malcolm was awarded a Royal Society university research fellowship in 1991. He held this fellowship in Bristol and Cardiff, before being appointed as a lecturer at UMIST in 2000. Malcolm is currently looking forward to being a senior lecturer at the newly merged University of Manchester, which will replace both UMIST and the Victoria University of Manchester in October 2004. Apart from his academic interests, Malcolm likes skiing, flying light aircraft, hill walking, and can occasionally be glimpsed in the chorus of the Manchester Universities' Gilbert and Sullivan Society. He obviously has little experience at driving the skidoo.

CHAPTER 13

THE ASTROPHYSICS OF CROWDED PLACES

Melvyn Davies

Lund Observatory, Box 43, 221–00 Lund, Sweden
E-mail: mbd@astro.lu.se

Today the sun is in a relatively uncrowded place. The distance between it and the nearest other star is relatively large (about 200,000 times the Earth–sun distance!). This is beneficial to life on Earth; a close encounter with another star is *extremely* unlikely. Such encounters would either remove the Earth from its orbit around the sun, or leave it on an eccentric orbit similar to a comet's. But the sun was not formed in isolation. It was born within a more–crowded cluster of perhaps a few hundred stars. As the surrounding gas evaporated away, the cluster itself evaporated too, dispersing its stars into the Galaxy. Virtually all stars in the Galaxy share this history, and in this review I will describe the role of "clusterness" in a star's life. Stars are often formed in larger stellar clusters (known as open and globular clusters) some of which are still around today. I will focus on stars in globlular clusters and describe how the interactions between stars in these clusters may explain the zoo of stellar exotica which have recently been observed with instruments such as the Hubble Space Telescope, and the X–ray Telescopes XMM–Newton and Chandra. In recent years, a myriad of planets orbiting stars other than the sun — the so–called extrasolar planets — have been discovered. I will describe how a crowded environment will affect such planetary systems and may infact explain some of their mysterious properties.

1. Introduction

Stellar clusters are ubiquitous. Globular clusters contain some of the oldest stars, whilst the youngest stars are found in clusters associated with recent star formation. Such crowded places are hostile environments: a large fraction of stars collide or undergo close encounters. Wide binaries are likely to be broken up, whilst tighter ones suffer major pertubations and possibly collisions from passing stars. Hydrodynamical computer simulations of

---

such encounters play a vital role in understanding how collisions affect the evolution of stellar clusters and produce the myriad of stellar exotica seen in stellar clusters. The cluster of stars at the centre of a galaxy may provide the material to form a massive black hole and fuel it as what is known as a quasar, where the central regions of a galaxy emit roughly as much energy as the rest of the galaxy. Encounters in very young clusters influence the fraction of stars contained in binaries and their nature. Such encounters affect the stellar population of the entire galaxy as all stars are formed in clusters.

In this review, we focus on the importance of stellar encounters within globular clusters, where collisions between two main–sequence stars similar to our sun may produce more–massive main–sequence stars known as blue stragglers. Encounters involving neutron stars (the remnants of the most–massive stars) and non–compact stars may help explain the large population of rapidly–rotating neutron stars known as millisecond pulsars. Stellar clusters are hostile environments for planets orbiting stars. Close encounters with other stars may perturb or breakup such planetary systems.

2. Stellar Encounters

Encounters between two stars are extremely rare in the low–density environment of the solar neighbourhood. However, in the cores of stellar clusters, and galactic nuclei, number densities are sufficiently high ($\sim 10^5$ stars/pc^3 in some systems [where 1pc $\equiv$ 3.25 light-years], as shown in Figure 1) that encounter timescales can be comparable to, or even less than, the age of the universe. In other words, a large fraction of the stars in these systems have suffered from at least one close encounter or collision in their lifetime.

We now consider how large a target (otherwise known as the cross section) one star is to another within a stellar cluster. The cross section for two stars, having a relative velocity at infinity of V_∞, to pass within a distance $R_{\min}$ is given by

$$\sigma = \pi R_{\min}^2 \left(1 + \frac{V^2}{V_\infty^2}\right)$$

where V is the relative velocity of the two stars at closest approach in an encounter where both stars had negligible relative velocity when they were far apart. For two solar–like stars passing within a few radii of each other, $V \simeq 100$km/s. The second term is due to the attractive gravitational force, and is referred to as gravitational focussing. The mutual gravity of the two stars is pulling them much closer together such that the minimum

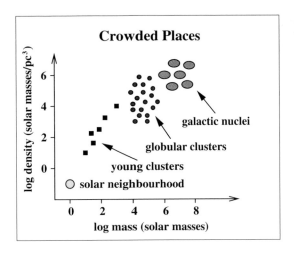

Fig. 1. A schematic diagram of crowded places.

distance between two stars is much smaller than the impact parameter for a particular encounter. If $V \gg V_\infty$, as is the case in most stellar clusters, the cross section for stellar collisions is enhanced enormously.

One may estimate the timescale for a given star to undergo an encounter with another star, $\tau_{\text{coll}} = 1/n\sigma V_\infty$. For clusters with low velocity dispersions, we therefore have

$$\tau_{\text{coll}} = 7 \times 10^{10}\text{yr} \left(\frac{10^5 \text{ pc}^{-3}}{n}\right)\left(\frac{V_\infty}{10 \text{ km/s}}\right)\left(\frac{R_\odot}{R_{\text{min}}}\right)\left(\frac{M_\odot}{M}\right) \text{ for } V \gg V_\infty$$

where n is the number density of single stars of mass M, and $M_\odot$ and $R_\odot$ are the mass and radius of the sun respectively. For an encounter between two single stars to be hydrodynamically interesting, we typically require $R_{\text{min}} \sim 3R_\star$ (see for example, Davies, Benz & Hills 1991). We thus see that for typical globular clusters, where $n \simeq 10^5$, and $V_\infty \simeq 10\text{km/s}$, $\tau_{\text{coll}} \simeq 10^{11}$ years. This is about ten times the age of a globular cluster, and indeed about ten times the age of the universe. What this means is that some 10% of the stars in the cluster cores will undergo a collision at some point during the lifetime of the cluster.

We may estimate the timescale for an encounter between a binary and a third, single star, in a similar manner where now $R_{\text{min}} \simeq d$, where d is the separation of two stars within the binary. The encounter timescale for a binary may therefore be relatively short as the semi–major axis can greatly exceed the size of the stars it contains. For example, a binary with $d \sim 1\text{AU}$

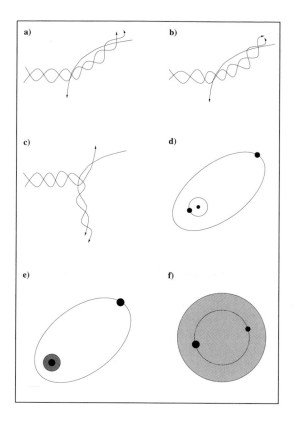

Fig. 2. The possible outcomes of encounters between binaries and single stars: a) a fly–by, b) a scattering–induced merger where a fly–by leads to a merger of the two stars in the binary, c) an exchange, d) a triple system, e) a merged binary where two stars have merged and remains bound to the third star, and f) a common envelope system where the envelope of the merged star engulfs the third star.

(1AU ≡ the average Earth–sun distance, ie 216 $R_\odot$), will have an encounter timescale $\tau_{\mathrm{enc}} \ll 10^{10}$ years in the core of a dense globular cluster. Thus encounters between binaries and single stars may be important in stellar clusters even if the binary fraction is small.

Encounters between single stars and extremely wide binaries lead to the break up of the binaries as the kinetic energy of the incoming star exceeds the binding energy of the binary. Such binaries are often referred to as being *soft*. Conversely, *hard* binaries are resilient to break up. Encounters between single stars and hard binaries have three main outcomes as shown in Figure 2. A fly–by may occur where the incoming third star leaves the original binary intact. However such encounters harden (ie shrink) the binary, and alter its eccentricity. Alternatively, an exchange may occur where the incoming star replaces one of the original components of the binary. During the encounter, two of the stars may pass so close to each other that they merge or form a very tight binary (as they raise tides in each other). The third star may remain bound to the other two as indicated in outcomes d), e) and f) in Figure 2.

The most likely outcome of an encounter between a single star and a wide (though hard) binary is either a fly–by or an exchange. Such an event hardens the binary by about 20%. After a number of these encounters the binary is therefore very much smaller and consequently the relative probability of some variety of merger occurring during later encounters is enhanced. For example, binaries containing solar–like stars which are just resilient to breakup in a typical globular cluster have initial separations $d \sim 1000 R_\odot$ but have separations $\sim 50 - 100 R_\odot$ today (see for example Davies & Benz 1995).

In encounters between two binaries, we again require the two systems to pass within $\sim d$ of each other. Hence binary/binary encounters will dominate over binary/single encounters only if the binary fraction is $\geq 30\%$. Unfortunately, the binary fraction in many stellar clusters is not well known.

3. The Stellar Zoo

Thus far we have seen that collisions and tidal encounters between two single stars occur in the cores of globular clusters and that encounters involving binaries occur both in globular and open clusters. These encounters are important for a number of reasons. They may produce the various *stellar exotica* that have been seen in clusters such as blue stragglers and millisecond pulsars. Stellar encounters also have a role in the dynamical evolution of clusters. Stellar collisions in clusters may also lead to the production of massive black holes. Once produced, these black holes may be fed by the gas ejected in subsequent stellar collisions. Stars begin their lives as main–sequence stars, powered by hydrogen fusion within their cores. Stars then evolve into red giants, ultimately forming a white dwarf (for low–mass stars like the sun), or a neutron star via a supernova explosion in the case of the most–massive stars. Given that each star involved in a collision may be either a main–sequence star, a red giant, a white dwarf, or a neutron star, there are ten distinct combinations of collision pairs. In this review we consider encounters involving two main–sequence stars, and encounters involving a red giant or main–sequence star and a compact star (*i.e.* either a white dwarf or a neutron star). Encounters between two main–sequence stars may be responsible for the observed population of blue stragglers within globular and open clusters, as will be discussed in a later section. Those involving compact objects and red giants or main–sequence stars may produce interacting binaries where material is transferred from the larger donor onto the compact object. Examples of interacting binaries include low–mass X–ray binaries (known as LMXBs, where material is transferred from a low–mass main–sequence star onto a neutron star) and cataclysmic variables (know as CVs, where material is transferred from a low–mass donor onto a white dwarf). Both classes of objects will be considered in subsequent sections.

Merging neutron stars have been suggested as the source of gamma–ray bursts (GRBs). A direct collision between two neutron stars is extremely unlikely. However, tight neutron–star binaries might be produced through the evolution of binaries, including those that have been involved in encounters with other stars. The neutron stars in such tight binaries may then ultimately merge as they spiral together as angular momentum and energy is emitted in the form of gravitational radiation. Any material ejected from the merger will be extremely dense and neutron–rich. Calculations suggest that this ejecta is a suitable site for the production of a subset of heavy elements, including platinum and gold, produced via nuclear reactions. Indeed, neutron–star mergers may be *the* source of such elements (Freiburghaus *et al.* 1999).

3.1. *Blue stragglers*

The globular clusters in the Milky Way are all old and we believe all star formation within them occurred early in their lives. As more–massive stars live relatively short lives compared to lower mass stars, due to much higher nuclear reaction rates at their centres, we would not expect to see any main–sequence stars more massive than ~ 0.8 $M_\odot$ in globular clusters today. However, observations reveal a small population of stars known as blue stragglers, which appear to be main–sequence stars with masses $\sim 1.0 - 1.4$ $M_\odot$. Blue stragglers have been observed in many globular clusters, often by making use of the high–resolution capabilities of the Hubble Space Telescope (HST): examples include ω Cen (Kaluzny *et al.* 1997), and M80 (Ferraro *et al.* 1999). These stars may have formed from the merger of two lower–mass main–sequence stars either in an encounter between two single stars or in encounters involving binaries when two main–sequences collide and merge as part of the encounter. The idea is that during the collision the interiors of the stars become mixed and the core of the newly–formed merged star is provided with more hydrogen fuel to fuse to helium. An important element of any computational investigation of these stellar collisions is to see how much mixing of material actually occurs during a collision.

Simulations of collisions between two main–sequence stars have been performed by many groups, most using the method known as Smoothed Particle Hydrodynamics (SPH). SPH is a lagrangian method where the fluid is modelled as an ensemble of particles which follow the flow of the fluid. Computational resources are not wasted in following the evolution of the voids, such as the gaps between two colliding stars, and the resolution can vary in a natural way; more particles being found in the places of most interest. Because SPH has no specific need for a computational box, we are able to follow the flow of gas completely. Thus we do not experience the Columbus Effect, where material is lost off the edge of a computational domain.

For a given value of the relative velocity, V_∞, between the two stars, one may compute how close the stars have to pass in order for a capture to occur. Even closer encounters will produce a single, merged object. Simulations yield values for the capture radius, $R_{capt} \simeq 3R_{ms}$, and provide a lower limit for the merger radius of $R_{merg} \simeq 2R_{ms}$. The mass lost from the system on the initial impact is small, typically $M_{lost} \leq 0.01 M_{ms}$. Early work suggested that the merged stars would be well mixed (Benz & Hills 1987, 1992). Subsequent simulations, using a more centrally concentrated model for the main–sequence stars, seem to suggest that the material in the cores will not (at least initially) mix with the envelope gas (Lombardi, Rasio & Shapiro 1995). More recently, the subsequent evolution of the merged objects has been considered (Sills & Bailyn 1999). This study showed that the distribution of blue stragglers in M3 were difficult to reproduce using a single set of assumptions, however if three particular bright blue stragglers were neglected, the remaining population could have been produced in mergers of stars occurring in encounters between binaries and single stars. Extremely–high resolution simulations of main–sequence star collisions are ongoing to study in more detail the internal structure of the merger products and a snapshot of one such simulation is shown below in Figure 3 (see also Sills *et al.* 2002). Looking to the future, these high–resolution simulations combined with stellar evolution programs will be able to model more

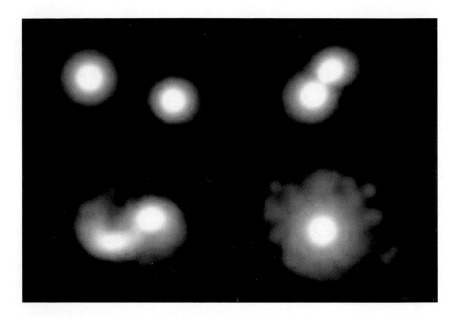

Fig. 3. A figure showing four snapshots of a collision between two main–sequence stars (top–left to bottom–right). The density of the gas in the plane of the collision is shown here (white is for the highest densities, black for the lowest densities). For this collision, the original stellar masses are 0.8 $M_\odot$ and 0.6 $M_\odot$ and the cores of the two stars passed within one quarter of the sum of the stellar radii. As a result of the collision, the two stars merge to form a single object.

accurately the subsequent evolution of collision products and compare them to state–of–the–art observations such as those obtained with HST.

3.2. *Low–mass X–ray binaries and millisecond pulsars*

Both millisecond pulsars (MSPs) and low–mass X–ray binaries (LMXBs) have been observed in relative abundance in globular clusters clearly indicating that their origin is related to stellar encounters. Under the standard model, MSPs are produced in LMXBs where the neutron star is spun–up as material is transferred from their companion. However observations suggest that there are far more MSPs than LMXBs which, given their comparable expected lifetimes, poses a problem for the standard model. One would therefore wish to investigate whether encounters may lead to other potential channels for MSP production which will not pass through a prolonged phase of X–ray emission.

Early work focussed on encounters between single neutron stars and either red giants or main–sequence stars. Fabian, Pringle & Rees (1975) suggested that such encounters would produce the observed X–ray binary population. Calculations of encounter rates suggest that encounters involving main–sequence stars are more frequent than those involving red giants. Numerical hydrodynamic simulations of encounters between neutron stars and red giants or main–sequence stars revealed that the $R_{\rm capt} \simeq 3.5R_\star$ for main–sequence stars, and $R_{\rm capt} \simeq 2.5R_\star$ for red giants (Rasio & Shapiro 1991; Davies, Benz & Hills 1992, 1993). Consideration of the subsequent evolution of the two stars suggested a lower limit of $R_{\rm merg} \geq 1.8R_\star$ in both cases. As all encounters in globular clusters are highly gravitationally focussed, the cross section for two stars to pass within a distance $R_{\rm min}$, $\sigma_{\rm rmin} \propto R_{\rm min}$. Hence approximately *half* of the bound systems form binaries and the rest form single merged objects. In the latter case encounters involving main–sequence stars produce a thick–disc system with the shredded remains of the main–sequence star engulfing the neutron star. The equivalent encounters involving red giants produce common envelope systems, where the red–giant envelope engulfs both the neutron star and the red–giant core.

Even if all the merged systems produced MSPs without passing through a prolonged X–ray phase, the expected MSP production rate is only a factor of $\simeq 2 - 3$ times larger than that for the LMXBs. The solution to the MSP enigma seems unlikely to lie with encounters involving single stars.

As was mentioned earlier, the binary fraction within globular clusters is highly uncertain. However, because of the larger cross section for encounters involving binaries, only a small fraction of binaries are required for binary–single encounters to be as important as encounters between two single stars. Calculation of the cross sections for fly–bys, exchanges, and mergers, lead to the predicted production rates for LMXBs and smothered neutron stars from encounters involving primordial binaries (Davies, Benz & Hills 1993; Davies & Benz 1995; Davies 1995). Although encounters today produce smothered neutron stars, they also produce LMXBs in roughly equal numbers; the MSP production rate problem remains.

The solution may lie in considering the *past*. The idea (developed by Davies & Hansen [1998]) is shown in Figure 4. Imagine a box – a very large box – containing a crowded population of stars. Imagine also that some fraction of these stars are in binaries. If the stars in the box resemble the population we see in our galaxy, then most stars will be relatively low mass, like the sun. This could also be true for the stars in binaries, *initially*. But

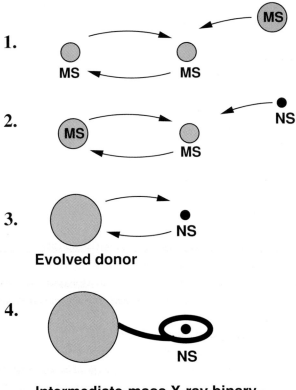

Intermediate-mass X-ray binary

Fig. 4. The evolutionary pathway to produce intermediate–mass X–ray binaries (IMXBs) in globular clusters (Davies & Hansen 1998). A more–massive main–sequence star exchanges into a binary containing two main–sequence stars (phase 1), a neutron star exchanges into the binary replacing the lower–mass main–sequence star (phase 2). The intermediate–mass star evolves of the main–sequence and fills what is known as its Roche Lobe (phase 3). The system has a relatively short phase as an IMXB (phase 4), possibly producing a millisecond pulsar.

imagine an encounter between a binary containing two low–mass stars and a single, more–massive, star. The most massive star is extremely likely to exchange into the binary replacing one of its less–massive brethren. This process will only occur in crowded places where the encounter timescale is sufficiently short, leaving the binaries containing at least one relatively massive star. When neutron stars exchange into these binaries, the less massive of the two main–sequence stars will virtually always be ejected.

The remaining main–sequence star will typically have a mass of $\sim 1.5 - 3$ $M_\odot$. The binary will evolve into contact once the donor star evolves up the giant branch.

The subsequent evolution of such a system will depend on the mass of the donor star and the separation of the two stars when the donor fills what is known as its Roche lobe, at which point material begins to flow from the donor towards the second star. For example, it has been suggested that the system may enter a common envelope phase (e.g. Rasio, Pfahl & Rappaport 2000). Alternatively, the system may produce an *intermediate–mass X–ray binary (IMXB)*. In such a system the neutron star may accrete sufficient material (and with it, angular momentum), for it to acquire a rapid rotation (ie millisecond periods). Because the donors are all more massive than the present turn–off mass in globular clusters, all IMXBs will have undergone their mass transfer *in the past*. If these systems evolve into MSPs, then we obtain, quite naturally, what is observed today, namely a large MSP population and a relatively small X–ray binary population.

Observations and modelling of the X–ray binary Cygnus X–2, provide important clues in helping determine the subsequent evolution of intermediate–mass systems. This binary is unusual in that its donor has the brightnesss and surface temperature of a slightly–evolved 3–5 $M_\odot$ star, yet its measured mass is much lower ($\simeq 0.5\ M_\odot$). The evolutionary history of Cygnus X–2 has been considered (King & Ritter 1999, Podsiadlowski & Rappaport 2000, and Kolb *et al.* 2000). The unusual evolutionary state of the secondary today appears to indicate that the system has passed through a period of high–mass transfer from an initially relatively–massive star ($\simeq 3.6 M_\odot$) which had just evolved off the main sequence. The neutron star has somehow managed to eject most of the $\simeq 3 M_\odot$ of gas transferred from the donor during this phase. This evolutionary history may also apply to the IMXBs formed dynamically in globular clusters. Vindication of this model also comes from studying the dynamical evolution of the binary within the Galactic potential (Kolb *et al.* 2000). A suitable progenitor binary originating in the Galactic Disc has sufficient time, and could have received a sufficient kick when the primary exploded to produce a neutron star, to reach the current position of Cygnus X–2.

Thus in order to explain the overabundance of millisecond pulsars in globular clusters, and the strange appearance of one star in a particular binary, we seem to have discovered an entirely new class of binary. Studies of the evolution of these *intermediate–mass systems* continues apace and will undoubtedly shape how we view binary evolution.

3.3. *Cataclysmic variables*

A cataclysmic variable (CV) comprises of a low–mass star in a binary with
a white dwarf. The low–mass star is filling its Roche lobe and is transferring material onto the white dwarf via an accretion disc. CVs are thus the
white–dwarf analogues of LMXBs. One might therefore imagine that the
CV population within a globular cluster may be boosted in a similar fashion
to the LMXB population. However detecting CVs in globular clusters has
proved to be difficult, although a few clusters are now known to contain
spectroscopically–confirmed CVs (for a review see Grindlay 1999). Ongoing
surveys of a number of globular clusters using Chandra and XMM are also
likely to boost the known population (see for example Grindlay *et al.* 2001).

CVs might be produced via tidal encounters between white dwarfs and
main–sequence stars (Verbunt & Meylan 1988). Calculations of encounters
between binaries and single white dwarfs demonstrate that white dwarfs
will exchange into primordial binaries producing CVs (Davies 1995; Davies
& Benz 1995). CVs are also produced in primordial binaries without the
outside interference of a passing single star. Indeed some 1% of all binaries
will produce CVs (de Kool 1992) by passing through the following stages.
Beginning with two main–sequence stars in the binary, the primary evolves
off the main–sequence, expands up the giant branch and fills its Roche
lobe. The subsequent mass transfer may be unstable and lead to the formation of a common envelope of gas around the red–giant core (which is
essentially a white dwarf) and the secondary (which is still on the main
sequence). This enshrouding envelope of gas is ejected as the white dwarf
and main–sequence star spiral together. If the initial separation of the two
stars is too small, the main–sequence star and white dwarf coalesce before
all the envelope is removed, however under favourable initial separations,
the entire common envelope can be removed leaving the white dwarf and
main–sequence star in a tight binary (with a separation of a few $R_\odot$). Such
a binary then comes into contact if angular momentum loss mechanisms
can work on sufficiently short timescales or when the main–sequence star
evolves into a red giant. A CV is then produced if the subsequent mass
transfer is stable.

The formation route for *primordial CVs (PCVs)* described above is *in-
hibited* in dense clusters if encounters with single stars or other binaries
disturb the PCV binary before the onset of the common–envelope stage
(Davies 1997), for example an intruding third star may break up a wide
binary. An interesting consequence is that PCVs are unlikely to be found

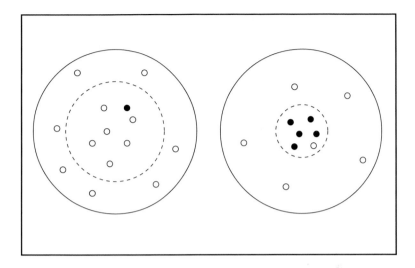

Fig. 5. A schematic illustration of the CV population within a low–density (left) and high–density (right) globular cluster. The dashed circles denote the cores of the clusters. Filled circles are CVs formed dynamically, whilst open circles are primordial CVs (PCVs) which have formed from primordial binaries.

within the cores of dense clusters, but will be found in their halos and throughout lower–density clusters. Conversely CVs formed via encounters between binaries and single stars are likely to be found exclusively in the cores (where the encounters occur) and are produced in greater numbers within higher–density clusters (Davies 1997) as is illustrated schematically in Figure 5. Today, relatively little is known about the actual distribution of CVs in globular clusters because such systems are relatively faint and earlier X–ray telescope technology lacked the resolving power to differentiate individual sources within a crowded core of a stellar cluster. Ongoing work with the XMM and Chandra telescopes will, in the near future, tell us much more about the actual CV population which combined with our theoretical work will tell us more about the population of binaries present in clusters.

4. Planets in Crowded Places

The last few years has seen a rapid growth in our knowledge of extra-solar planets (see for example Perryman 2000). Enigmatically, these systems contain Jupiter–mass objects at Earth–like distances (or closer) from their stars. Many planets are also observed to be on extremely eccentric orbits. As all stars are formed in clusters, one possible explanation lies in encounters between planetary systems and other stars in very young stellar clusters Such encounters may perturb the planets significantly, or break up the planetary system completely. The latter may explain why a recent survey of the core of the globular cluster 47 Tucanae has failed to reveal any tight planetary systems when about 20 were expected (Davies & Sigurdsson 2001). In this case an encounter between a star, with an orbiting planet, and a solitary star is the extreme limit of encounters between binaries and single stars described earlier. Planets on relatively wide orbits will be unbound as they receive some fraction of the kinetic energy from the incoming star. On the rare occasions when the binding energy of the planet exceeds the kinetic energy of the incoming star, the planetary system is still destroyed as the more–massive incoming star will virtually always replace the much less–massive planet in a bound orbit around the other star.

Such strong encounters are unlikely to happen in the less-crowded clusters more typical of young, star–forming regions, especially as these clusters will disperse in a relatively short time. However more distant encounters will happen between planetary systems and passing stars within the clusters. Such encounters may perturb the planetary orbits by small amounts. Small perturbations will also sometimes have an important effect as changing the orbital parameters of planets only very slightly can lead to enormous changes in the planetary orbits at a later time. To illustrate this, I considered the four gas giants of our solar system (Jupiter, Saturn, Uranus, and Neptune) and perturbed their orbits by relatively modest amounts. I then evolved their orbits using a computer program and in Figure 6, I plot the distance from the sun for each of the four planets as a function of time. As can be seen from the figure, the orbits evolve radically, in fact Uranus is ejected from the system entirely after 300,000 years, the energy required coming from a change in the orbital properties of the remaining planets. Saturn and Neptune are left on *extremely* eccentric orbits. Hence one might imagine that such dynamical interactions might explain the large eccentricities of many of the observed extra–solar planets, where some planets have been ejected whilst those remaining have been left on very tight, eccentric orbits. Although computer simulations of the dynamical interaction between planets often show the outcome described above, there are indications that the distribution of planetary orbits does not match that seen in observations (Ford, Havlickova & Rasio 2001). We have some idea what's happening within stellar clusters to affect planetary orbits however there

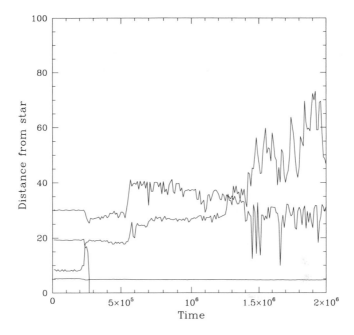

Fig. 6. A figure showing the chaotic evolution of the four gas–giant planets of our own solar system (ie Jupiter, Saturn, Uranus and Neptune) placed initially on perturbed, eccentric orbits. The time is given in years, while the distances from the star are given in AU, where 1 AU is the average Earth–sun separation.

are clearly still some missing pieces in the puzzle. Future work will include more detailed studies of chaos in planetary dynamics and further research into the evolution of young low–mass clusters in which the majority of stars seem to form.

5. Summary

We have seen that all stars are formed within clusters, and that some stars are still in clusters today. Clusters are sufficiently crowded that stars will collide with other stars, forming a variety of unusual objects. The high–resolution vision of the Hubble Space Telescope (HST) is able to peer into the cores of these clusters to catalogue the collision products, observing the so-called blue stragglers, which are main–sequence stars rather like the sun but a little more massive, formed when two lower–mass stars collide. LMXBs and CVs are binaries visible in the X–ray part of the spectrum which are currently being studied by XXM–Newton and Chandra – the

X-ray equivalents of HST. In order to explain the abundance of rapidly–rotating neutron stars within globular clusters, we discovered intermediate–mass binaries, which had previously been largely neglected. The recent discovery of extrasolar planets has lead to a flurry of computational research in an effort to explain the surprising distribution of planetary orbits which have been observed. Dynamical interactions between stars with planetary systems and other stars may leave the planets on suitably eccentric orbits. The study of the evolution of planetary systems around stars in stellar clusters will be very important in the coming years. As all stars were once in clusters, this work will have implications for our understanding of the frequency of habitable worlds beyond our own, and thus on the likelyhood of life elsewhere in the galaxy.

Acknowledgements

I gratefully acknowledge the support of the Royal Society via a University Research Fellowship. The work described here has also been supported by the UK Particle Physics and Astronomy Research Council. This chapter is based on a previous article written for the Philosophical Transactions of the Royal Society of London (Davies 2002).

References

1. Benz W., Hills J. G., 1987, *Astrophys. J.*, **323**, 614
2. Benz W., Hills J. G., 1992, *Astrophys. J.*, **389**, 546
3. Davies M.B., 1995, *Mon. Not. R. Astr. Soc.*, **276**, 887
4. Davies, M. B., 1997, *Mon. Not. R. Astr. Soc.*, **288**, 117
5. Davies, M. B., 2002, *Roy. Soc. of London Phil. Tr. A*, **360**, 2773
6. Davies M.B., Benz W., 1995, *Mon. Not. R. Astr. Soc.*, **276**, 876
7. Davies M. B., Benz W., Hills J. G., 1991, *Astrophys. J.*, **381**, 449
8. Davies M.B., Benz W., Hills J.G., 1992, *Astrophys. J.*, **401**, 246
9. Davies M.B., Benz W., Hills J.G., 1993, *Astrophys. J.*, **411**, 285
10. Davies, M. B., Hansen, B. M., 1998, *Mon. Not. R. Astr. Soc.*, **301**, 15
11. de Kool M., 1992, *Astron. Astrophys.*, **261**, 188
12. Fabian, A. C., Pringle, J. E., Rees, M. J., 1975, *Mon. Not. R. Astr. Soc.*, **172**, P15
13. Ferraro, F. R., Paltrinieri, B., Rood, R. T., Dorman, B., 1999, *Astrophys. J.*, **522**, 983
14. Ford, E.B., Havlickova, M., Rasio, F.A., 2001, *Icarus*, **150**, 303
15. Freiburghaus, C., Rosswog, S., Thielemann, F.-K., 1999, *Astrophys. J.*, **525**, L121

16. Grindlay, J. E., 1999, in ASP. Conf. Ser. 157, Annapolis Workshop on Magnetic Cataclysmic Variables, ed, C. Hellier & K. Mukai (San Francisco: ASP), 377

17. Grindlay, J. E., Heinke, C.O., Edmonds, P.D., Murray, S.S., Cool, A.M., 2001, *Astrophys. J.*, **563**, L53

18. Hut, P., Inagaki, S., 1985. *Astrophys. J.*, **298**, 502

19. Kaluzny, J., Kubiak, M, Szymanski, M., Udalski, A., Krzeminski, W., Mateo, M., Stanek, K., 1997, *Astron. Astrophys. Suppl.*, **122**, 471

20. King, A.R., Ritter, H., 1999, *Mon. Not. R. Astr. Soc.*, **309**, 253

21. Kolb, U., Davies, M. B., King, A., Ritter, H., 2000, *Mon. Not. R. Astr. Soc.*, **317**, 438

22. Lombardi, J. C., Rasio, F. A., Shapiro, S. L., 1995, *Astrophys. J.*, **445**, 117

23. Perryman, M.A.C., 2000, *Rep. Prog. Phys.*, **63**, 1209

24. Podsiadlowski, P., Rappaport, S., 2000, *Astrophys. J.*, **529**, 946

25. Rasio, F. A., Pfahl, E. D., Rappaport, S., 2000, *Astrophys. J.*, **532**, L47

26. Rasio, F. A., Shapiro, S. L., 1991, *Astrophys. J.*, **377**, 559

27. Sills, A., Bailyn, C. D., 1999, *Astrophys. J.*, **513**, 428

28. Sills, A., Adams, T., Davies, M.B., Bate, M., 2002, *Mon. Not. R. Astr. Soc.*, **332**, 49

29. Verbunt F., Meylan G., 1988, *Astron. Astrophys.*, **203**, 297

Melvyn B. Davies

Born in Leamington Spa, Warwickshire in 1967, Melvyn Davies studied at the University of Oxford, where he graduated with first–class honours in physics in 1988. He then moved to the US, obtaining his PhD in 1992 from Harvard University and becoming a Tolman Research Fellow at the California Institute of Technology. In 1995, he returned to Britain, taking up a Royal Society University Research Fellowship at the Institute of Astronomy, Cambridge. In 1998, he moved this fellowship to the University of Leicester, also becoming an Honorary Lecturer there. He was promoted to a Readership in April, 2003. In 2004, he moved to Lund, Sweden, taking up a Royal Swedish Academy of Sciences Research Fellowship.

CHAPTER 14

ASTROCHEMISTRY: FROM MOLECULAR CLOUDS TO PLANETARY SYSTEMS

Andrew Markwick

*Space Science Division, NASA Ames Research Center,
Moffett Field, CA 94035, U.S.A.
E-mail: ajm@ajmarkwick.com*

In this chapter I will describe aspects of the chemistry occurring in the transition of interstellar material from molecular clouds to planetary systems, and what we can hope to learn about this transition by studying chemistry.

1. Introduction

There are many different regions in the Cosmos - covering all regimes of physical conditions. What they all have in common, however, is that they are probed by observing molecules, and therefore, chemistry. Figure 1 shows schematically the life of cosmic material, which is recycled from the interstellar medium through stars and back again. Material processed during a star's life is subsequently ejected back into the interstellar medium when the star dies. All the atoms of heavy elements (e.g. carbon, oxygen etc.) in the Universe today, of which we too are made, were formed in stars. Studies of the various astrophysical regions depicted are almost separate disciplines today, but in reality they are all connected - by time.

2. The Interstellar Medium and Molecular Clouds

Many molecular species have been observed in the interstellar medium via ground-based and space-borne telescopes. The first molecule to be detected was CH, a simple diatomic, way back in 1937. The most abundant molecule in interstellar space after molecular hydrogen (H_2) is another diatomic, carbon monoxide (CO). At the other end of the molecular size scale, organic molecules such as ethanol (CH_3CH_2OH) and dimethyl ether (CH_3OCH_3)

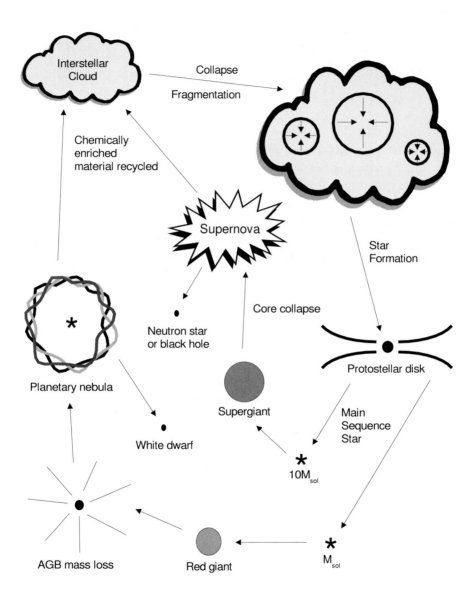

Fig. 1. Cosmic recycling. Material in the Universe is recycled through the interstellar medium, stars, and back again. M_{sol} is a solar mass, i.e. the mass of the Sun.

have been detected. Recently, astronomers have also identified a simple sugar (glycolaldehyde; CH_2OHCHO), and glycine (NH_2CH_2COOH) - the simplest amino acid. Although molecules (and hence chemical processes) are observed in many astronomical sources, there are three main regions of differing physical condition in which interstellar molecules are prevalent. These are *cold dark clouds*, which are so-called because they are at a temperature of around 10 K (hence *cold*) and are dense enough that their interiors are shielded from UV radiation by H_2 and dust (hence *dark*). This is important because it means that many molecules can exist in them which would otherwise be *photodissociated*, that is, destroyed by radiation. Secondly, *hot molecular cores*, which range in temperature from 50-300 K and have density in excess of 10^6 hydrogen molecules cm^{-3}, are rich sources of molecular line emission. These objects are associated with very young stars, and many of the complex organic molecules detected have only been observed in them. Finally, many molecules are detected in the *circumstellar envelopes* of old (so-called *late-type*) stars. Several metal-bearing molecules such as magnesium cyanide and isocyanide (MgCN; MgNC) have only been detected in these regions so far.

2.1. *Observations of interstellar molecules*

Table 1 shows an up-to-date list of molecules detected in interstellar space and comets, taken from `astrochemistry.net`. It is immediately apparent that interstellar space contains many *organic* (loosely, carbon-containing) molecules. There are difficulties associated with identifying molecules from their astronomical spectra, and therefore the organic molecules observed are not complex by terrestrial standards. Certain molecules are easier to detect than others, and so in observational astronomy there are *selection effects*. The energy level structure of linear carbon chain molecules, for example, is simple enough that for a given energy, there are not many ways for it to be distributed (the *partition function* is small), and so individual spectral lines can be quite strong. That is why the largest molecule unambiguously detected in space is $HC_{11}N$, a cyanopolyyne species which was detected astronomically before it could be produced and identified in the laboratory. When we move to non-linear molecules, the energy level structure becomes vastly more complicated, so much so that an 11-atom non-linear molecule would need to be about 10 000 times more abundant than an equivalent 11-atom linear molecule to produce lines of similar strength. The astronomical detection of molecules is also hampered by the atmosphere, which is why astronomers had to wait for space-based observatories like The Infrared Space Observatory (ISO) to detect carbon dioxide (CO_2). Table 1 is biased therefore towards diatomics, linear carbon-chains and simple non-linear organics. Many other molecules await detection - indeed new molecules are detected regularly - but some will remain undetected because their lines are too weak. Similarly, there are molecules which we know must be present, but which are impossible to detect because of their structure. The abundances of such molecules are inferred from observations of species that we think are chemically very closely related. Also detected are many unidentified lines which have so far defied assignment, perhaps because their carrier's spectrum has not yet been measured or calculated. Table 1 lists only gas-phase detections, but some molecules have been observed in the solid-state - that is, as the ice component of interstellar dust grains. They were detected through their infrared absorption features and they include CO_2, CH_4, H_2O, H_2CO, NH_3, CH_3OH and HCOOH.

2.2. *Gas-phase astrochemistry*

As more and more molecules are observed, theoretical astrochemistry has to evolve and adapt to explain how these molecules form and how they can persist. Making an astrochemical model which contains enough chemistry to form all of the molecules in Table 1 is at the bleeding edge of research in this discipline, but the underlying principles which make it possible to have a rich chemistry in the gas-phase have been understood for about 30 years and are fairly simple. Nearly all of the gas in a dark cloud is in the form of molecular hydrogen and because of this, we often measure the abundance of molecules *relative* to H_2. The next most abundant molecule, CO, is present at a level of about 0.0001 relative to H_2. It has a *fractional abundance* of 0.0001. Helium atoms are present in interstellar gas too, typically at a fractional abundance of 0.1. Cosmic rays are high energy protons which pervade space - they can penetrate the interior of dark clouds where other radiation cannot go, and ionise H_2, H and He. Due to the low densities and temperatures in the interstellar medium, the only efficient chemical reactions between molecules are two-body processes which have small or no activation barriers. These are mainly *ion-neutral* reactions, which proceed at rates of around 10^{-9} cm^3 s^{-1}. This rate is independent of temperature, so that room-temperature experimental data can be directly applied to low-temperature astrochemistry. Reactions between neutrals can also occur, but *neutral-neutral* reactions are about a hundred times slower than

Table 1. Detected interstellar, circumstellar and cometary molecules, by number of constituent atoms and ordered by mass. *Source:* `astrochemistry.net`.

2	3	4	5	6	7	8	≥ 9
H_2	H_3^+	CH_3	CH_4	C_2H_4	CH_3NH_2	CH_3CH_3	CH_3CH_2OH
CH	CH_2	NH_3	NH_4^+	CH_3CN	CH_3CCH	CH_2OHCHO	CH_3OCH_3
CH^+	NH_2	H_3O^+	CH_2NH	CH_3NC	$c\text{-}C_2H_4O$	CH_3COOH	CH_3CH_2CN
NH	H_2O	C_2H_2	H_3CO^+	NH_2CHO	CH_3CHO	$HCOOCH_3$	CH_3COCH_3
OH	H_2O^+	H_2CN	CH_3OH	CH_3SH	CH_2CHCN	CH_3C_3N	CH_3C_4H
OH^+	C_2H	$HCNH^+$	SiH_4	H_2CCCC	C_6H	C_6H_2	NH_2CH_2COOH
HF	HCN	H_2CO	$c\text{-}C_3H_2$	$HCCCCH$	HC_5N	HC_6H	C_6H_6
C_2	HNC	$c\text{-}C_3H$	H_2CCC	$H_2C_3N^+$		C_7H	C_8H
CN	HCO	$l\text{-}C_3H$	CH_2CN	HC_2CHO			HC_7N
CN^+	HCO^+	$HCCN$	H_2CCO	C_5H			HC_9N
CO	N_2H^+	$HNCO$	NH_2CN	C_5N			$HC_{11}N$
CO^+	HOC^+	$HNCO^-$	$HCOOH$				
N_2^+	HNO	$HOCO^+$	C_4H				
SiH	H_2S	H_2CS^+	HC_3N				
NO	H_2S^+	C_3N	$HCCNC$				
HS	C_3	C_3O	$HNCCC$				
HS^+	C_2O	$HNCS$	C_5				
HCl	CO_2	SiC_3	SiC_4				
SiC	CO_2^+	C_3S					
SiN	N_2O						
CP	HCS^+						
CS	$NaCN$						
SiO	$MgCN$						
PN	$MgNC$						
AlF	$c\text{-}SiC_2$						
NS	$AlNC$						
SO	C_2S						
SO^+	OCS						
$NaCl$	SO_2						
SiS							
$AlCl$							
S_2							
FeO							
KCl							

ion-neutral reactions, and they often possess activation barriers. In general it is not easy to determine whether a neutral-neutral reaction will proceed under interstellar conditions, so experimental or theoretical data is required in this case. One rapid reaction type is *dissociative recombination*, where an ion (e.g. HCO^+) is broken apart by an electron. These reactions typically occur at a rate of 10^{-6} cm^3 s^{-1}, but they are temperature dependent and so in general experimental data is required. Experiments are particularly important for this class of reactions because they can determine the *branching ratios* of the reaction. As an example, the dissociative recombination of N_2H^+ has three possible exit channels: N + N + H, N_2 + H and N + NH. Before the reaction was measured in the laboratory, astrochemists assumed that the reaction always produced N_2 + H. Recent measurements have found that this only happens 35% of the time, whereas 65% of the time the exit channel is N + NH, so our models have to be adjusted accordingly, and the consequences examined. There is a close interaction between observation, theory and laboratory experiment in astrochemistry, and the cycle between them is one of the field's most attractive features.

For many molecules, their formation in the gas-phase is via the dissociative recombination of a precursor ion. That is, if you want to form a particular molecule by gas-phase chemistry, you first form its *protonated* counterpart, and recombine it. Molecular hydrogen is so abundant that anything that can react with it does so very efficiently, so for simple examples involving species like water, ammonia or methane, the problem reduces to initiating the chemistry, or, creating the initial ion. As an example, consider the gas-phase formation of water. The last step is the recombination of H_3O^+ (protonated water) with an electron, producing H_2O and H. We can form H_3O^+ by successive reaction of O^+ with H_2: $O^+ \longrightarrow OH^+ \longrightarrow OH_2^+ \longrightarrow OH_3^+$, where each reaction also produces a hydrogen atom. The problem therefore reduces to forming either O^+ or OH^+. Since H_2 is the dominant form of hydrogen in a dark cloud, the product of its cosmic ray ionisation, H_2^+, will react straight away with H_2 again to form H_3^+, the simplest molecular ion. A given ion will transfer a proton to anything which has a *proton affinity* greater than it. For H_3^+, that is most things, the notable exceptions being atomic nitrogen and molecular oxygen. For the formation of water, this means that OH^+ can be produced by reaction of H_3^+ with atomic oxygen. The fact that H_3^+ is so abundant explains why molecular ions like HCO^+ and N_2H^+ are readily observed - they are the products of proton transfer between H_3^+ and CO and N_2. In essence it is the low proton affinity of H_2 which drives interstellar chemistry. The interstellar detection of H_3^+ was finally made in 1996 with the United Kingdom Infrared Telescope (UKIRT) on Mauna Kea, Hawai'i. This was an important event because it confirmed that our understanding of the chemistry in dark clouds is largely correct, as the molecule was observed at an abundance predicted by models. Subsequently, however, it was detected in diffuse clouds with an unexpectedly high abundance, indicating that our understanding of the chemistry in those objects is incomplete.

Of course, molecules are not only formed, they are also destroyed. Reaction with H_3^+ partly returns a species to its protonated form from where it can recombine again, but the reaction can also dissociate the molecule. The reaction of a species with He^+ is even more destructive, because it isn't possible for a species to react with He^+ to produce the protonated form and therefore the reaction is always dissociative (incidentally, He^+ does not react with H_2). This means that the cosmic ray ionisation of helium (which produces He^+) is a critical reaction for interstellar chemistry. If the rate of this reaction is changed even by a factor of 2 - a small error for astronomers - the calculated abundances of nearly all species are significantly affected. In fact studies have shown that this is the reaction to which the overall abundances of species are most sensitive. As well as destruction by chemical reactions, molecules can also be destroyed by radiation, and they can even be removed completely from the gas-phase by sticking to a grain surface, a process known as adsorption or more informally, *freeze-out*.

To build a model of gas-phase chemistry in a dark cloud, one obviously needs a source of reactions and their *rate coefficients*. The main source used by the community is the publicly available UMIST Database for Astrochemistry, which resides online at `udfa.net`. This database contains reaction data for around 4200 chemical reactions involving around 400 atomic and molecular species, and includes all the processes discussed above and many more besides. Once the reaction network is specified for the given problem, software exists which can translate the network into a system of coupled ordinary differential equations describing the rate of change of the abundance of each species. Since the rate coefficients depend on temperature, density, the amount of UV radiation (for photoprocesses) and on the cosmic ray ionisation rate, these parameters must be given. Once the initial abundances of species are determined, the system of ODEs can be integrated in time using library routines, and the chemical evolution of the object be found. It is a remarkable feature of such a system that after around a million years (for a dark cloud model at least) the chemistry approaches a *steady state*, where the abundances of all the species in the model remain constant. If one is only interested in these abundances, the integration can be done away with and a non-linear system of algebraic equations solved instead. This is usually done using a generalisation of Newton's root finding algorithm.

2.3. *Gas-grain and solid state astrochemistry*

Some of the molecules in Table 1 are difficult to form via gas-phase reactions. H_2, in fact, is one such molecule, but there are other more complex ones too. These include formaldehyde (methanal; H_2CO) and methanol (CH_3OH). The observed abundances of these molecules in certain regions of the interstellar medium are greater than those that can be produced by gas-phase chemistry alone, and so reactions on grain surfaces are needed to form these species. We know for certain that these species are present in ice mantles, because they have been observed directly. It has been demonstrated experimentally that they can form by successive hydrogenation of CO molecules, so that $CO \longrightarrow HCO \longrightarrow H_2CO \longrightarrow CH_3O \longrightarrow CH_3OH$. This process is most efficient at a temperature of 15 K, while at 20 K, the process is inefficient. Hence, in regions where larger abundances of methanol and formaldehyde are observed, we assume that they were made on grains and then some process returned the ice mantle species to the gas phase. This process can be thermal desorption, or a non-thermal process like cosmic-ray heating or grain mantle explosions. The simultaneous calculation of realistic solid-state and gas-grain chemistry is currently slightly out of reach - we expect such models to be forthcoming within a few years. The reason for this is that in order to model this chemistry properly, the stochastic nature of gas-grain interaction has to be considered. In pure gas-phase models, the chemical abundances are found in a deterministic way using differential equations, as described above. In grain surface chemistry though, the limiting factor is the accretion of gas-phase species onto grains, which is inherently stochastic. Initial attempts to include this chemistry in gas-phase models forced the gas-grain interaction into the framework of deterministic kinetics, which is unrealistic. Of course, it may turn out that this is a reasonable approximation to the full calculation in some cases.

2.4. *Interstellar organic chemistry*

Many complex organic molecules are observed in regions known as *hot molecular cores*. These objects are associated with high-mass star formation, and are hot enough (around 200 K) that the species in ice mantles on dust grains have been returned to the gas by sublimation. In regions where ice mantles have been removed from their parent dust grains, we have conditions where we can expect both gas-phase and solid-state chemistry to contribute to the organic complexity in the gas. Instead of relying on solid-state astrochemistry to produce *every* species we can't explain by pure gas-phase chemistry, we can rely on the former to produce a few simpler *precursor* species which may then react in the warm gas to produce the more complex molecules. These simpler species are precisely those which are known to exist and be efficiently formed on grain surfaces, including methanol and formaldehyde. Models which evaporate appropriate quantities of these species from dust grains produce much higher abundances of more complex organics (e.g. ethanol) than pure gas-phase models, bringing the models into closer agreement with observations. Methanol, when protonated, can transfer an alkyl cation to other simple evaporated neutral species to form species like methyl formate, acetic acid and dimethyl ether, all of which are observed in hot cores. The proposed formation mechanism for interstellar glycine is that the precursor molecule NH_2OH (hydroxylamine) is desorbed from grain surfaces and, when protonated, reacts with acetic acid to form protonated glycine, which then recombines. Of course, to confirm the theory, the precursor molecule should be observed as well, and these observations are currently underway.

2.5. *Isotopic fractionation chemistry*

We also see many isotopically-substituted species in interstellar gas. For example, hydrogen cyanide (HCN) has been detected in its ^2H (deuterium), ^{13}C and ^{15}N forms as well as the main isotopomer $^1H^{12}C^{14}N$. Such observations can provide further information about the physical conditions in the interstellar medium, but are also useful for identifying chemical processes via isotopic labelling. Deuterium fractionation is especially useful in this regard due to its high sensitivity to temperature. The cosmic abundance of deuterium is about 0.00002 relative to hydrogen, but some molecules in interstellar clouds are observed to have D/H ratios orders of magnitude bigger than that. For example, the DCN/HCN ratio in a typical dark cloud may be around 0.01. This enhancement is a chemical effect which relies on low temperature to work. The deuterium is brought out of its reservoir in HD (the deuterated form of molecular hydrogen) by the reaction of HD with H_3^+. The products of that reaction are H_2D^+ and H_2. From there, the deuterium is propagated around the chemical network via ion-neutral reactions. The crucial fact is that the back-reaction between H_2D^+ and H_2, which returns deuterium to HD, has an activation barrier. This means that at low temperatures ($T < 30$ K), the deuterium is permanently out of its reservoir and so the chemistry is *fractionated*.

Recently, multiply-deuterated species have also been detected. Amazingly, the fractionation ratio of fully-deuterated ammonia (ND_3/NH_3) was measured to be 0.0009 - almost 12 orders of magnitude higher than the expected value based on the cosmic D/H ratio! This, together with observations of D_2CO and various isotopomers of methanol and ammonia, has put real pressure on astrochemical models.

3. Prestellar and Protostellar Cores

Prestellar cores are regions of interstellar clouds which are in the first stage of low-mass star formation. They are collapsing, and so the density of the gas increases towards the centre, but there is as yet no star, so they are cold, still at the temperature of the cloud, around 10 K. The rate of 'freeze-out' of molecular species from gas to dust grain surface depends on the number of collisions of gas with dust, which depends on the velocity of the gas (and therefore the gas temperature) and the density. So as the density of the gas increases, the freeze-out rate increases too, and molecular material is removed from the gas-phase, in a specific order which depends on the binding energy of each species (see also the chapter by Malcolm Gray in this volume). In prestellar cores therefore, the opportunity exists to directly observe different stages of the core collapse by using observations of different species, and thereby work out how the collapse proceeds. Regions in the centres of such cores have been found where the CO abundance indicates that it has frozen out, and likewise CS, but there is still N_2 (inferred from N_2H^+) and ammonia.

One interesting aspect of observations of prestellar cores is that the level of deuterium fractionation in D_2CO, for example, is correlated with the amount of CO which is frozen out of the gas-phase. The D_2CO/H_2CO ratios observed in these sources range from 0.01 to 0.1 (remember, the cosmic D/H ratio is about 0.00001), with the higher ratios observed towards sources with greater CO depletion. It has been known theoretically since 1989, when the first deuterium fractionation models were constructed, that the fractionation increases with increasing gas-phase CO depletion, i.e. with increasing density. (In the following discussion, it may help to refer to Figure 2). This is because H_3^+ and H_2D^+ are mainly destroyed by neutral molecules like CO, but in cold, dense regions, this destruction diminishes

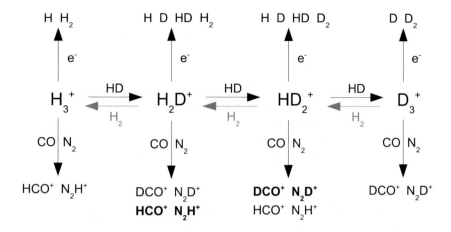

Fig. 2.   Reactions involving H_3^+ and its deuterated isotopomers. They are formed by successive reaction with HD. At temperatures less than 30 K, the back-reactions with H_2 (grey arrows) do not happen, so the gas becomes fractionated..The main destructions of the ions are through dissociative recombination with electrons and ion-neutral reactions with abundant species.

due to the freeze-out of these species. The conversion from H_3^+ to H_2D^+ does not change however, and so there is relatively more deuterium in molecules. The problem is that the enhancements in multiply-deuterated species observed in prestellar and protostellar sources are too large for conventional models to reproduce even in light of this consideration. This is exacerbated in prestellar cores because they are not hot enough for ice mantles to have been evaporated, so grain surface chemistry cannot be relied upon to 'save us' by producing the high ratios in the solid-state. This prompted some researchers to reconsider the theory of gas-phase deuterium fractionation by adding the multiply-deuterated forms of H_3^+ (HD_2^+ and D_3^+) to their models. The results were impressive: the D_2CO ratios now agree much better with observations, while some ratios in other molecules now exceed those observed! Although it is no doubt still not the end of the story, this change represents a leap in our understanding of deuterium chemistry in cold gas. Given the importance of H_3^+ to interstellar chemistry in general, it may seem obvious in hindsight that the inclusion of these ions makes such a difference. Why then were they not included in models before? Normally, that is under normal interstellar cloud conditions, the dominant destruction of all the forms of H_3^+ is by recombination with electrons. We have to add a deuterium atom to an ion by reacting HD with it, and normally we can expect these reactions to be uncompetitive compared to the dissociative recombination. In the case of dense gas however, the electron fraction is lower, and eventually the gas will reach a density where the recombination reaction is not dominant. In that special case, a cascade occurs from H_3^+ to D_3^+ (see Figure 2). Model calculations show that in fact D_3^+ will be the dominant (most abundant) ion in these dense regions. In the same way that H_3^+ readily donates a proton to most other species, D_3^+ donates a deuterium atom and this is how the fractionation in molecules gets so large. The fact that the deuterated forms of H_3^+ are abundant offers observational astrophysicists a unique probe of an otherwise hidden part of star formation - the very centre of the prestellar core. The density is so high that all molecules are completely frozen-out, but H_2D^+ and HD_2^+ will still be abundant in the gas (D_3^+ is harder to detect), and can be used to trace the physical conditions.

Another consequence of including these ions in models is that they increase the gas-phase atomic D/H ratio. This is important because of the deuterated forms of methanol and formaldehyde observed in protostellar sources. These are objects which are in the next stage of low-mass star formation after the prestellar phase. There is now a young star at the centre of the object, and the heat from it removes the ice mantles from dust grains. Since the deuterated forms of methanol observed are highly fractionated, and these species are not formed efficiently in the gas-phase (see above), we need a way of fractionating them on the grain surface. The same is true of hydrogen sulphide (D_2S) which has also been observed with a high ratio. It turns out that the critical number for achieving high fractionation on grain surfaces is the accreting (freezing) gas-phase atomic D/H ratio. The accretion occurs during the prestellar phase, and those models with multiply-deuterated H_3^+ produce the ratio which models of the solid-state fractionation require to make the ratios in methanol agree with those observed in protostellar sources.

4. Protoplanetary Discs

As the molecular cloud core undergoes gravitational collapse, it is rotating. The enhanced centripetal effects flatten the protostellar object into a disc. The observational evidence is that most pre-main sequence objects have discs, and it is now widely accepted that planetary systems form from these discs by the agglomeration of particles. A simple study of our Solar System lends credence to the argument that planetary systems form out of discs: first, all the planets' orbits are in roughly the same plane, and second, all the orbits are in the same sense. It is difficult to imagine how these criteria could be satisfied if not because the planets formed out of a rotating disc.

The life of a protoplanetary accretion disc can be divided into three phases. In the *formation phase*, the disc is built up by the infall of matter from the cloud core on a timescale of around 100 000 years. The mass of the disc at this stage is comparable with the mass of the resulting stellar system, because not much accretion onto the central object has taken place. In this phase all the infalling material encounters the accretion shock as it falls onto the disc. Next comes the *viscous phase*, where the supply of external matter is greatly reduced and mass accretion proper, that is, accretion onto a central object, starts. Angular momentum is expelled and the disc spreads. Now the mass of the disc is less than the mass of the central object, and a small amount of mass carries a large proportion of the angular momentum. This phase lasts for a few hundred thousand years, and it is in this phase that planet formation starts by agglomeration of microscopic interstellar particles into macroscopic objects. What remains in the disc is nearly unprocessed material from the molecular cloud (except any processing by the accretion shock). In the final *clearing phase*, the gaseous component of the nebula is dispersed, perhaps by strong pre-main sequence stellar winds.

Detailed high-resolution observations of discs around stars are at least a few years away yet, and the community awaits results from the SMA and ALMA interferometers to provide more data. Several molecules have been detected in discs by current telescopes though - these include CO, CN, CS, H_2CO, HCN and HCO^+. The only deuterated molecules detected so far are DCO^+ and HDO. These observations are sensitive to the outer regions of the discs, at radii of a few hundred A.U. from the centre (an Astronomical Unit is the mean distance between the Earth and the Sun, and is equivalent to about 149598 million metres).

Originally, the chemistry in protoplanetary discs was investigated using equilibrium calculations, which just consider thermodynamics and do not explicitly include chemical reactions. Equilibrium is a valid approximation for the region extremely close to the central object where the temperature and density are high, but it is unrealistic in the cooler, less dense outer disc. The chemistry occurring there has to be calculated kinetically. Figure 3 shows schematically some of the processes affecting chemistry which should be included in a model. These include sources of ionising radiation as well as mass transport processes. Most researchers working in this field adopt an approach where a complex dynamical code is obtained and values of density, temperature and so on are used as inputs to a stand-alone chemical model. The disadvantage with this approach is that the dynamics and chemistry are treated as if they are independent, which is certainly not the case, but the (significant) advantage is that both sides of the model (dynamics, chemistry) can be as complex as possible. Recently, some progress has been made in the direction of coupling chemistry and dynamics in disc models - in a model where diffusive transport is included, significant effects on the distribution of molecules are found.

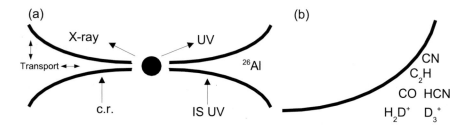

Fig. 3.   (a) Schematic showing the important physical processes affecting chemistry in a protoplanetary disc. These include ionisation by UV, X-rays and cosmic rays, ionisation due to the decay of extinct radionuclides and mass transport processes. (b) Blow up of one quadrant of the disc, showing the positions of molecules: C_2H, CN and dissociation products are most abundant near the surface, while molecular ions like H_2D^+ dominate the gas in the cold midplane where other species are frozen out. The warm intermediate layer contains most of the observed abundant molecules.

One-dimensional chemical kinetic models have been calculated to study the chemistry in the midplane of the disc. In this region, irrespective of the physical model chosen, the temperature is low (10-15 K) in the outer part of the disc ($r > 100$ A.U.). There is also not much ionisation due to the amount of material above the midplane which blocks radiation. Molecular species are therefore totally frozen onto grain surfaces. This is a problem, because the observations clearly show a rich gas-phase chemistry in the outer regions of the discs, but in midplane models there isn't any at these radii. This situation changed when models were calculated in 2D using physical models which included the UV irradiation of the disc by the central star. This radiation creates a warm layer of gas above the midplane, and it is therefore from that layer that the molecular line emission comes. Due to the high level of ionisation at the surface, this region is abundant in molecules which are photodissociation products. These include CN and C_2H, which are the products of HCN and C_2H_2 photodissociation respectively.

The fact that the molecular emission comes from the warmer upper layers of the disc means that we have to look for a different tracer of the conditions in the midplane. If we could measure the dominant ion there, we could use that information to deduce the electron fraction. Furthermore, if the ion contained deuterium, we could also use it to probe the temperature profile. Since H_3^+ doesn't freeze out, we could use H_2D^+ or HD_2^+ for this purpose. The same arguments presented in Section 3 apply though, and D_3^+ is likely to be the dominant midplane ion at some radii. Our best chance to probe the midplane may therefore be to observe H_2D^+ and/or HD_2^+ and interpret the results with a chemical model.

5. The Solar System

Between the evolution of the protoplanetary disc and the formation of the Solar System, the key question for astrochemistry is: how much processing did pristine interstellar material undergo in the disc before it became incorporated into Solar System bodies like comets and meteorites? Since Hale-Bopp, our knowledge of molecular abundances and particularly deuterium fractionation ratios in comets (specifically cometary *comae*) has increased significantly. Comets are made of dust and ice, but as a comet nears the Sun, the ice starts to sublime which produces the coma of gas which we see easily with the naked eye in some cases. As the desorbing *parent* molecules are photodissociated by radiation from the Sun, chemistry occurs which produces many *daughter* species. Using models of the chemistry in the coma we can attempt to work backwards from the observed molecules and deduce the abundances of solid-state molecules in the comet nucleus. To date, there are around 50 molecules detected in cometary comae, and they are almost all seen in interstellar clouds as well, which although not proof that the nucleus is composed of pristine interstellar material, is certainly suggestive. The types of molecules observed, coexisting reduced and oxidised species, and the D/H ratios observed are similar to those seen in interstellar hot cores - gas which also originated from the sublimation of ice mantles. By comparing cometary gas with interstellar gas we can at least conclude that the molecules formed under interstellar *conditions*. Approximately the same story is true for meteoritic material in the Solar System as well. Many organic molecules have been found in meteorites, including but not limited to amines and amides, alcohols, aldehydes, ketones, aliphatic and aromatic hydrocarbons, amino acids, carboxylic acids, purines and pyrimidines. Enrichments in deuterium and ^{15}N fractionation are also observed, again signifying at least similar formation conditions for the molecules, if not that they are in fact unprocessed interstellar material. Comets are split into two families based on their period. The short period, Kuiper Belt comets are thought to have formed at roughly the same radius from the protosun as where they are today ($\sim$ 40 A.U.), while the long period, Oort Cloud comets formed closer to the protosun ($\sim$ 10 A.U.) and were ejected into much larger orbits ($\sim$ 50 000 A.U.) possibly by gravitational interaction with the giant planets. There are some chemical differences observed between the two families, and one can suppose that this is either because they were formed differently or that they have had different processing histories. In fact, using a midplane protoplanetary disc model, it was found that over the 5-40 A.U. comet-forming region the solid-state chemical composition changes very little, suggesting the latter explanation.

Astrobiology is a relatively new field of research which is basically concerned with the origins of life on Earth and in the Universe. Observations of biologically significant molecules in the interstellar medium attract much attention largely because of the implied connection between interstellar and Solar System material. The detection of glycine (the simplest amino acid) in the ISM for example, strengthens the thesis that amino acids were already present in the early Solar System and that perhaps the first *prebiotic* material was delivered to Earth by comets. The detection of glycine in the coma of a comet would be even more significant.

Bibliography and Further Reading

Ehrenfreund P., Charnley S.B., *Organic molecules in the interstellar medium, comets, and meteorites*, Annual Reviews of Astronomy & Astrophysics **38**, 427 (2000).
Markwick A.J., Charnley S.B., *Chemistry in protoplanetary disks*, in Astrobiology: Future Perspectives, Kluwer (2004).
Kuan Y.-J. *et al.*, *Interstellar glycine*, The Astrophysical Journal **593**, 848 (2003).
Roberts H., Herbst E., Millar T.J., *Enhanced deuterium fractionation in dense interstellar cores resulting from multiply-deuterated H_3^+*, The Astrophysical Journal (Letters) **591**, 41 (2003).
For copies of any of the above email `ajm@ajmarkwick.com`.

Andrew Markwick

Fig. 4. Andrew Markwick (centre), with Helen Roberts (left) and Tom Millar (right) above Zermatt, Switzerland, September 2003.

Andrew Markwick was born in Middlesbrough in north-east England. He was educated at Stokesley School and Sixth Form College, following which he got a first in Mathematics with Astrophysics at UMIST in Manchester. After that he stayed at UMIST to do a PhD in Astrophysics supervised by Prof. Tom Millar, which was finally entitled *"Chemistry in dynamically evolving astrophysical regions"*. He received his doctorate in October 2000, by which time he had already taken up a PPARC funded post-doc (PDRA) position. In January 2003 he decided it was time to leave UMIST and so he took a 2 year NRC (National Research Council) Research Associateship to NASA Ames Research Center in California. His research interests are varied - he has published papers describing aspects of the chemistry in protoplanetary discs, AGB circumstellar envelopes and dark interstellar clouds, and a few observational papers using infrared and radio data. His many other interests include walking (especially in the North York Moors), playing music (guitar, keyboards, drums), football (he supports Middlesbrough FC) and all aspects of technology. You can contact him by email (ajm@ajmarkwick.com) or find out more by visiting www.ajmarkwick.com.

CHAPTER 15

EXTRASOLAR PLANETS

Hugh R. A. Jones and James S. Jenkins

Astrophysics Research Institute, Liverpool John Moores University,
Twelve Quays House, Egerton Wharf, Birkenhead CH41 1LD, UK
E-mail: hraj@astro.livjm.ac.uk

More than one hundred planets are known outside our solar system. These extrasolar planets have been found in orbit around stars ranging from spectral type F to M with semi-major axes ranging from 0.02 to 6 au (1 au = Earth-Sun distance). Their masses range from around 0.1 to 10 Jupiter masses, though mass values are usually somewhat ambiguous because of the unknown inclination of their orbits. Migration theory suggests that planets of Jupiter-like mass are to be found at Jupiter-like distances, though so far we have primarily detected an important minority at smaller radii. The extrasolar planets discovered thus far are found around primary stars with a particularly high metal content, that may increase towards shorter periods and thus correlate with greater migration. One of the most enigmatic results is the high eccentricities of extrasolar planets in comparison with the Solar System. As time goes by our searches become ever more sensitive to a wider range of primaries with lower mass and longer period extrasolar planets. We are now at the beginning of a new adventure: the search for planetary systems that are more like our own.

1. Introduction

It was the renaissance philosopher Giordano Bruno who first suggested there might be other worlds orbiting the stars of the night sky. Bruno's heretical philosophising came to a fiery end, when in 1600 he was burned at the stake. However, his musings set the stage for one of astronomy's 'Holy Grails' — the search for planets around other stars. Bruno's sad death at the hands of the Holy Inquistion set the stage for fruitless searches over the following 395 years. In 1991 the first definitive extrasolar planet (exoplanet) was discovered around a pulsar using timing measurements

(Wolzcan & Frail 1992). A further three planets have now been discovered around PSR 1257+12 as well as a planet around PSR B1620-26 (Backer, Foster & Sallmen 1993; Sigurdsson *et al.* 2003). While these are landmark discoveries, the planets' location, next to a stellar remnant, helps little in understanding our own Solar System. However it does lead us to believe that planets are common throughout the Galaxy, and possibly the Universe.

In 1995 the long search ended when Mayor & Queloz (1995) announced the detection of the first exoplanet around a Sun-like star. The radial velocity of the G2V star 51 Pegasi was used to infer the presence of a Jupiter mass planet in a 4.2 day orbit. The discovery was quickly confirmed independently (Marcy & Butler 1996) and also corroborated by Doppler evidence for Jupiter mass planets in orbit around a number of other nearby stars. Although very exciting, the indirect nature of the detection meant that the planet around 51 Pegasi was controversial for some time. Evidence was put forward for asymmetric variations in line profiles indicating that the radial velocity variations arose within the photosphere of the star. Whilst such measurements proved incorrect, today the proof of each individual exoplanet still requires substantial scrutiny to ensure that radial velocity variations are not caused by variations in the photosphere or chromosphere of the target star. There is a rather long list of objects for which planetary status has been retracted or is at least uncertain (Schneider 2004). Nonetheless the number of bona fide objects increases rapidly and today we have an inventory of more than 100 exoplanets. To those outside the field this may sound like a large number, however this is not a sufficient number to understand even the main features of exoplanets. In fact we have only just begun to probe the parameter space. The most pertinent measure of the infancy of this field is that we do not yet have the sensitivity to detect the presence of our own Solar System if it were present around any other nearby star. The detection of Solar Systems like our own seems the vital scientific stepping-stone to the search for other life forms in the Universe.

2. Finding Exoplanets

The word planet means 'wanderer' in Greek, thus planets move on the sky relative to the fixed background stars. This definition needs updating since, with modern equipment, even quite distant galaxies can be seen to move. Discussions about the status of Pluto and 'free-floating planets/brown dwarfs' are described in McCaughrean *et al.* (2001). These discussions led the International Astronomical Union (IAU), in February 2003,

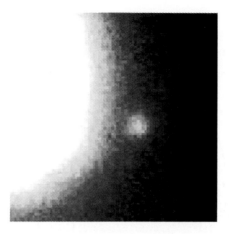

Fig. 1. The difficulty of direct exoplanet imaging can be seen in this image (Oppenheimer 1999) of the 15-70 M_{JUP} (Saumon *et al.* 2000) brown dwarf Gl229B (centre) 7 arcsec from its M2V companion Gl229A (left). By contrast Jupiter is approximately ten times closer in and a million times fainter.

to create a working definition of what constitutes a planet and what does not. The three main criteria they settled on were as follows:

• Objects with true masses below the limiting mass for thermonuclear fusion of deuterium (currently calculated to be 13 Jupiter masses for objects of solar metallicity) that orbit stars or stellar remnants are 'planets' (no matter how they formed). The minimum mass/size required for an extrasolar object to be considered a planet should be the same as that used in our Solar System.

• Substellar objects with true masses above the limiting mass for thermonuclear fusion of deuterium are 'brown dwarfs', no matter how they formed nor where they are located.

• Free-floating objects in young star clusters with masses below the limiting mass for thermonuclear fusion of deuterium are not 'planets', but are 'sub-brown dwarfs' (or whatever name is most appropriate).

Planetary mass objects may have already have been imaged in young star forming regions (e.g. Tamura *et al.* 1998). Apart from the masses of these objects being very dependent on poorly constrained theoretical models, the 'free-floating' nature of these objects means they fall outside the currently accepted notion of exoplanet. An important strand in most definitions is the concept that to be a 'planet' an object must be in orbit around a 'star'. This proximity to a much brighter object as well as their relative

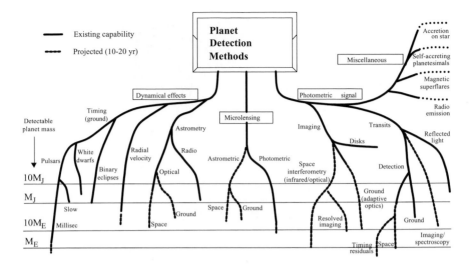

Fig. 2.   There are many plausible ways to detect exoplanets. This schematic has become known as the 'Perryman-tree' and been adapted from Perryman (2000).

faintness makes planets difficult to find, e.g. Fig. 1. Discovery would be simplified were it possible to directly image exoplanets. The best opportunity so far is perhaps the controversial planet around Epsilon Eridani. At only 3.2 pc, it should soon become feasible to image this object, although with a separation of 1 arcsecond and a magnitude difference of 15 (a factor of 1,000,000 in brightness) this observation is at the limit of current technology. This difficulty in direct imaging means that a wide range of innovative planet searching ideas have been devised, see Fig. 2.

2.1. *Radial velocity – Doppler wobble*

So far radial velocity or Doppler wobble has been the most important technique for finding exoplanets, e.g., Fig. 3. While the velocities of stars have been measured for many years, the breakthrough in making exoplanet detections by this method has been to observe a stable reference source at the same time as the target star (Butler *et al.* 1996; Baranne *et al.* 1996). The current best long-term precision is around 2 m s^{-1}. For comparison Jupiter causes the Sun to wobble with a velocity of 12.5 m s^{-1} over a 12 year period. Planets one tenth the mass of Jupiter, albeit with much shorter periods, are now detected by the radial velocity technique. The next generation of radial velocity searches can be expected to improve efficiency and

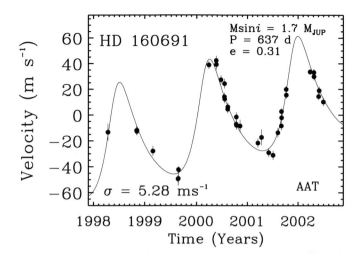

Fig. 3. The figure shows the detection of a radial velocity exoplanet around the star HD 160691 by the Anglo-Australian planet search (Jones *et al.* 2002). Assuming 1.08±0.05 $M_\odot$ for the primary, the minimum (M sin i) mass of the companion is 1.7±0.2 M_{JUP} and the semi-major axis is 1.5±0.1 au. In addition the figure also shows a trend indicating a second companion.

bring long-term precisions down to below 1 m s^{-1} by using a new generation of specifically designed spectrometers along with robotic telescopes and automated scheduling.

2.2. *Astrometric*

Since a star and a planet orbit around their common centre-of-mass, a large planet will cause movement in the position of a star. Jupiter induces such a motion on the Sun. At a distance of 10 parsecs, Jupiter could be seen to give an angular motion of 0.5 milliarcseconds to the Sun. This motion is tantalisingly close to limits of current technology. The last major astrometry mission HIPPARCOS managed a precision of 2 milliarcseconds. Its successors GAIA and SIM will measure angular motion of microarcseconds and will thus be sensitive to Jupiter mass objects in orbit around all local stars. Fig. 4 shows astrometric confirmation of the exoplanet around Gl876.

2.3. *Transits*

Around 1 in 3000 stars (e.g. Horne 2003) are expected to have a planet in orbit around them which moves into the line of sight between the star and

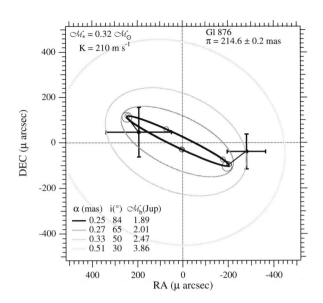

Fig. 4.   The figure shows the four orbits permitted for the astrometric perturbation of Gl876A due to Gl876b (Benedict *et al.* 2002). The innermost orbit is determined from the simultaneous astrometry-RV solution. Plotted on this orbit are the phases of the astrometric observations, with the two primary phases indicated by large circles. Also plotted (+) are the astrometry-only residual normal points at $\phi = 0.26$ (periastron, lower right) and $\phi = 0.72$. These normal points are connected to the derived orbit by residual vectors.

the Earth. The drop in brightness of a star can be detected as a cool planet transits across it and blocks some of the light. The detection of planetary transits requires the highest quality photometry as well as frequent sampling to avoid many of the numerous false positives, e.g., grazing/partial eclipses from two stars, reddened A-type stars eclipsed by M dwarfs, giants eclipsed by dwarfs, brown dwarfs and triple systems. One planet has been seen in transit as predicted from its Doppler wobble measurements, see Fig. 5. Konacki *et al.* (2003) has made the first direct transit detection (confirmed by Doppler follow-up Torres *et al.* 2003) and there are many new candidates (Udalski *et al.* 2002). Two of these new candidates have also been confirmed by Doppler follow-up (Bouchy *et al.* 2004). Thus we have 3 planets discovered independently by the transit method. However these planets are located very far away around stars in the Galactic bulge and are thus no good for follow-up analysis. This does not make them any less important, on the contrary, planets in the Galactic bulge tell us a

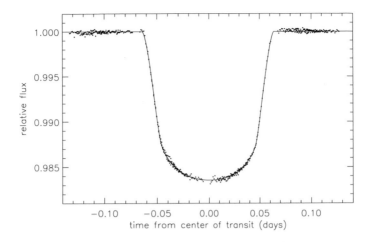

Fig. 5. Light curve from Brown *et al.* (2001) obtained by observing four transits of the planet of HD209458 using the STIS spectrograph on the Hubble Space Telescope. The folded light curve can be fitted within observational errors using a model consisting of an opaque circular planet transiting a limb-darkened stellar disk. In this way the planetary radius is estimated as 1.347 ± 0.060 R$_{\rm JUP}$, the orbital inclination 86.6 ± 0.14, the stellar radius 1.146 ± 0.050 R$_\odot$. Satellites or rings orbiting the planet would, if large enough, be apparent from distortions of the light curve or from irregularities in the transit timings. No evidence is found for either satellites or rings, with upper limits on satellite radius and mass of 1.2 R$_{\rm JUP}$ and 3 M$_{\rm JUP}$, respectively. Opaque rings, if present, must be smaller than 1.8 planetary radii in radial extent. The high level of photometric precision attained in this experiment confirms the feasibility of photometric detection of Earth-sized planets circling Sun-like stars.

lot about how robust planetary systems actually are. In addition to more than 20 ground-based projects a number of space missions are also planned (Horne *et al.* 2003). It is nonetheless feasible to search for planetary transits without access to Space Agency budgets (e.g., transitsearch.org).

2.4. *Gravitational microlensing*

The presence of planets can also be inferred by their focusing and hence amplification of light rays from a distant source by an intervening object. Microlensing (Paczynski 1986) describes the gravitational lensing that can be detected by measuring the intensity variation of a macro-image made of any number of micro-images that are generally unresolved to the observer. Microlensing is capable of probing the entire mass range of planets from Super-Jupiter mass to Earth mass. Fig. 6 shows the first detection of a planetary microlensing event (Bond *et al.* 2004). The universality of

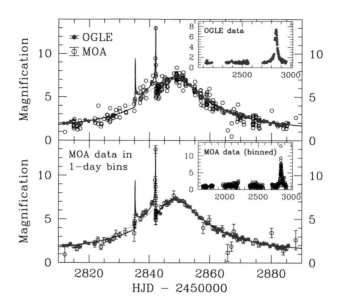

Fig. 6.   The first observed planetary microlensing light curves with best fitting and
single lens models are shown (Bond *et al.* 2004) . The OGLE and MOA measurements are
shown as filled circles and open circles, respectively. The top panel presents the complete
dataset during 2003 (main panel) and 2001–2003 OGLE data (inset). For clarity, the error
bars were not plotted, but the median errors in the OGLE and MOA points legend are
indicated. The lower panel is as the top panel but with the MOA data grouped in 1 day
bins except for the caustic crossing nights, and with the inset showing MOA photometry
during 2000–2003. The binary and single lens fits are plotted in flux units normalized to
the unlensed source star brightness of the best planetary fit, and they are indicated by
the solid black and grey dashed curves, respectively.

gravity guarantees that gravitational microlensing is also sensitive to unbound planetary mass objects as well as brown dwarfs. One conventional
thought is that planets are formed through accretion and brown dwarfs are
formed through fragmentation during star formation. Abundant statistics
of microlensing planets can provide an important clue to test this scenario.
Microlensing planets are expected to be distant and so follow-up will be
limited.

3. Properties of Exoplanets

We will consider the properties of exoplanets and their host stars announced
up until the end of March 2004. Our information comes primarily from
Marcy *et al.* (2004) and Schneider (2004). We use metallicity values [Fe/H]

from Santos *et al.* (2004). It is important to note that these catalogues consist primarily of exoplanet *candidates*. They should be considered as *candidates* because apart from HD 209458b (Fig. 5) and Gl876b (Fig. 4), their orbital inclinations are not well constrained.

It is expected that as the frequency and quality of radial velocity data continues to improve the number of multiples will rapidly increase. While it is premature to assess the fraction of multiple systems it is important to note that the long running Lick sample of 51 stars has so far found two triples and one double in addition to a further five so-far single planets (Fischer *et al.* 2003). This means that the multiple fraction is then 3/51 and the current fraction bearing any exoplanet is then 8/51 or 15%. It is vital to note that this is a *lower* limit to the fraction of exoplanets around *suitable* solar-type stars. The original Lick sample numbered 100, though was pared down to 51 by the removal of spectroscopic binaries, chromospherically active, rapidly rotating and evolved stars.

3.1. *51 Pegasi's exoplanet is not typical*

While the discovery of 51 Pegasi was a landmark, as with the earlier discovery of planets around pulsars, it was met with scepticism in part because of the difficulties of the measurement but as much because 51 Pegasi b seemed to have nothing in common with our Jupiter (apart from mass). Its orbit seemed to require radical new ideas about the formation of planets; though in fact all that was required was the rediscovery of the robust theoretical concept of inward planetary migration driven by tidal interactions with the protoplanetary disk (Goldreich & Tremaine 1980; Lin & Papaloizou 1986; Ward & Hourigan 1989). Thus, while such a large mass planet could not form in the glare of radiation from its Sun, it was entirely plausible that it had migrated into position through the disk of material around 51 Pegasi (Lin, Bodenheimer & Richardson 1996). Although 51 Pegasi-like objects dominated the early discoveries, other types of planets are much more common. Of the 100 or so exoplanets that have been discovered the 51 Pegasi-like planets (3-5 day orbital periods) comprise 12 and represent a class of planets circling about 0.5% of stars. The 51 Pegasi class were found first because they are easiest to find. Relatively heavy and close to their stars they exert the largest force on their stars and are thus by far the easiest to detect by the radial velocity method. As more planets are discovered other types of biases in our understanding of exoplanets resulting from our experimental sensitivity are starting to reveal themselves.

Although the exoplanets discussed in this review were all found using the radial velocity technique they are not discovered from a single well documented and quantified methodology. The compilation relies on a number of different ongoing surveys operating with different samples, sensitivities, instruments, scheduling, strategies and referencing techniques. Cumming *et al.* (1999, 2003) has thoroughly investigated the observational biases inherent in the Lick and Keck surveys but has yet to report findings for the bulk of detected exoplanets. So far none of the surveys have the 3 m s^{-1} precision over 15 years necessary to detect Jupiter and thus do not yet constrain the frequency of Solar System analogues. Nonetheless a very wide range of properties have already been found.

3.2. *The planetary mass function rises towards lower masses*

Fig. 7 shows that the number of exoplanets rises exponentially towards lower masses. Also over-plotted on this graph for comparison is a normalised curve for the mass-function index with an exponential power of -1.7. As the sensitivity of the radial velocity technique increases, the number of low mass planets will continue to increase very steeply, meaning the actual mass function will be increasingly masked. However, the Baysian approach by Cumming *et al.* (2003) promises to incorporate a knowledge of detection sensitivities for a single sample. With the assumption that the mass function

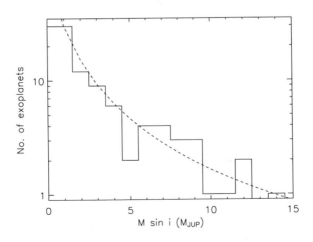

Fig. 7.   The number of exoplanets per unit mass are shown as a solid line. The dotted line depicts the number of exoplanets proportional to M_{JUP} sin i$^{-1.7}$.

is constant with period it appears that a steep mass function is favoured. The simulations of Tabachnik & Tremaine (2002) and Zucker & Mazeh (2002) using around 60 planets favour a much flatter mass distribution. So far, the relatively small numbers of objects as well as the selection biases preclude much confidence in a particular value for the mass function. Zucker & Mazeh (2002) note the probable lack of high mass exoplanets at short periods. Based on Fig. 7 it seems that the expected lower average mass of close-in planets persists to periods beyond the 51 Pegasi-type objects.

3.3. *The parent stars of exoplanets are very metal-rich*

An important characteristic of a star is the fraction of 'metals' (elements heavier than Helium) it contains - the metallicity. Gonzalez (1997) found exoplanet host stars to be metal-rich. This conclusion has been confirmed by many authors with different samples, methodologies and spectral synthesis codes (e.g., Reid 2002; Santos *et al.* 2003). Fig. 8 shows just how metal-rich the exoplanet primaries are. Only a single bin is anywhere close to solar metallicity. All other bins are at least 0.1 dex above the solar; whereas the Sun and other solar type dwarf stars in the solar neighbourhood have an average metallicity of 0 or even slightly less (Reid 2002). Above solar metallicities, where most of the known exoplanets are hosted, the probability of

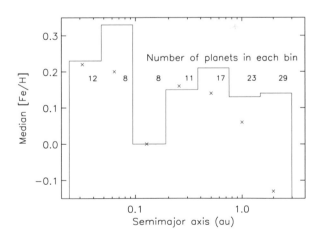

Fig. 8. Median spectroscopic metallicities of the primaries of exoplanets plotted as a function of semimajor axis. The metal-rich nature of exoplanets is evident from this plot, with only one bin approaching solar metallicity. The crosses represent the low-mass third of exoplanets (0.1-1.1 M_{JUP}sini). The final bin has been omitted because there are only two planets in this bin, neither of which are in the low-mass third of exoplanets.

detecting an exoplanet is proportional to its metallicity. By a metallicity of
+0.3 dex the frequency of stars with exoplanets is greater than 20% (Fischer *et al.* 2004; Santos *et al.* 2004). This result, representing the only link
between the presence of planets and a stellar photospheric feature, has been
given two main explanations. The first, is based on the classical view that
giant planets are formed by runaway accretion of gas on to a 'planetesimal'
having up to 10 Earth masses. In such a case, we can expect that the higher
the proportion of dust to gas in the primordial cloud (i.e. the proportion
of metals), and consequently in the resulting protoplanetary disk, the more
rapidly and easily may planetesimals, and subsequently the observed giant
planets be built.

Opposing this view, it had been proposed that the observed metallicity
'excess' may be related to 'pollution' of the convective envelope of the star
by the infall of planets and/or planetesimals. So far evidence for this has
only been found in one of the many systems investigated. While such pollution may play a small role, giant stars with planets all have high metallicities
and show no evidence for pollution (Santos *et al.* 2003). These are crucial
systems because they are well mixed and not subject to the mixing and
surface uncertainties that limit our confidence in measuring 'pollution' in
Sun-like stars. Besides the [Fe/H] differences, there is currently some debate
about possible anomalies concerning other elements. But the relatively low
number of exoplanets known, and possible systematics with respect to the
samples do not permit firm conclusions to be reached (Santos *et al.* 2003).
So far, it seems that the higher metallicity of most planet-harbouring stars
arises because high metallicity environments have a higher probability of
planet formation (Livio & Pringle 2003). The relatively low metal content
of the Solar System may be consistent with the relative lack of migration (Lineweaver 2001). Thus it appears that like the early detections of
51 Pegasi-type exoplanets we are finding a surfeit of exoplanets around
metal-rich stars because they are easier to detect.

3.4. *Why don't exoplanets have circular orbits like our Solar System?*

Apart from the short-period exoplanets whose orbits are circularised by the
tidal pull from their parent star, Fig. 9 shows the eccentricity of exoplanet
orbits is much higher than in the Solar System. Fischer *et al.* (2003) find
that it is rather close to that observed for stellar binaries. According to our
paradigm of planetary formation a planet (formed in a disk) should keep a

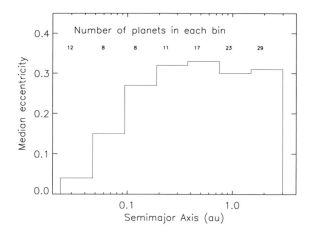

Fig. 9.   Eccentricity versus semimajor axis for exoplanets. In the Solar System Pluto is the only planet with an eccentricity of greater than 0.1.

relatively circular (low eccentricity) orbit. In order to boost exoplanet eccentricities it is necessary to imagine interactions between multiple planets in a disk and between a planet and a disk of planetesimals and perhaps the influence of a distant stellar companion. In fact dynamical interactions between planets seem inevitable since even with the fairly poor sampling of known exoplanets, 10 multiple systems have already been found. Of these a number are in resonant orbits. Thus 'dynamical fullness' is probably important and suggests that interactions play a vital role in determining the properties of many exoplanets. These orbital complexities mean that, to understand exoplanets more generally, it will be necessary to find all the main components in planetary systems. This will require using results from different techniques, particularly radial velocity and astrometric, to disentangle the various planetary components in orbit around nearby stars.

Various mechanisms have been proposed to explain the orbital eccentricities, namely planet-planet interactions, planet-disk interactions, planet migration leading to resonance capture and phenomenological diffusion processes (e.g. Tremaine *et al.* 2003). A comprehensive model for orbital eccentricities may include more than one process. Two or more planets will migrate in a viscous disk at different rates, allowing them to capture each other in mean-motion resonance. Such resonances can then pump the eccentricities of both planets up to values as high as 0.7 (Chiang 2003) as observed. The growth in eccentricities, especially as the disk dissipates, can render the two orbits unstable as close passages occur. Ford *et al.* (2003)

show that such close passages often lead to ejection of the less massive planet, leaving one behind. This sequential set of processes provides a natural explanation for the resonances and eccentricities observed among single and multiple planets. Models combining dynamical scattering and tidal interactions can now naturally account for the observed range of periods and eccentricities (Adams & Laughlin 2003).

3.5. *Is the semimajor axis distribution consistent with migration theory?*

The distribution of exoplanets with semimajor axes in Figs. 8 and 10 suggest that exoplanets show key differences with period. Fig. 10 (solid line) indicates two separate features in the exoplanet semimajor axis distribution. A peak of close-by exoplanets is seen in the 51 Pegasi-type objects, then a dearth, followed by a smooth rise in the number of exoplanets toward longer semimajor axes. The close-by peak has been an important motivation in the development of migration theories for exoplanets (Lin *et al.* 1996; Murray *et al.* 1998; Trilling *et al.* 1998; Ward 1997; Armitage *et al.* 2002; Trilling, Lunine & Benz 2003). This peak can be explained by a stopping mechanism such as Lindblad resonances between the planet and disk (Kuchner & Lecar 2002) and photoevaporation (Matsuyama, Johnstone & Murray 2003) limiting exoplanet migration to periods of less than 3 days.

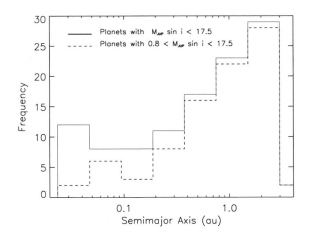

Fig. 10.   The semimajor axis distribution for exoplanets. The solid line represents the entire exoplanet sample and may suggest a number of interesting features. The dashed line represents all planets above 0.8 M$_{\rm JUP}$sini, this was chosen because it is the mass of HD 114729b the lowest mass long-period planet yet found.

Once there exists a sample of short-period exoplanets derived from a wider range of primary masses it should be possible to distinguish between different stopping mechanisms by comparing the orbital characteristics of the exoplanet with those of the primary. For example, Kuchner & Lecar predict that planets around early A-type main sequence stars will collect at a radius much further from the star (0.3 au) than the radius where solar-type stars collect. The rise in the number of exoplanets towards longer periods is becoming more apparent as more planets are discovered and is very well reproduced by exoplanet migration scenarios which envisage planets migrating inwards (e.g., Armitage *et al.* 2002; Trilling *et al.* 2003) as well as outwards (Masset & Papaloizou 2003). Indeed when a completeness correction is made to Fig. 10 (dashed line) the close-by peak does not seem as pronounced as first thought, and suggests that the peak may well be a selection effect.

3.6. *Exoplanet atmospheres*

Future space missions such as DARWIN and TPF should enable direct exoplanet spectra to be acquired rather than making inferences about exoplanets from studying their host stars. Synthetic spectra for exoplanet atmospheres have made rapid enhancements through modelling of brown dwarfs and the incorporation of stellar irradiation leading to detailed predictions for observable exoplanet polarisations. Their first direct test has come from HD209458. In addition to the exquisite light curve (Fig. 5), Charbonneau *et al.* (2002) have made a detection of the atmosphere of HD209458b. They detected additional dimming of sodium absorption during transit due to absorption from sodium in the planetary atmosphere. The observed dimming is reasonably well modelled by planetary atmosphere models that incorporate irradiation and allow for sodium to be out of thermal equilibrium (Barman *et al.* 2002). Vidal-Madjar *et al.* (2003, 2004) have also detected atomic hydrogen, carbon and oxygen in the extended upper atmosphere of HD209458b. Combining this with models for HD209458b, which put the upper atmosphere at a temperature of 10,000 K, this gives strong evidence for atmospheric evaporation. This evaporation confirms the conclusions by Hebrard *et al.* (2003) and may lead to new types of planets being discovered with hydrogen poor atmospheres or even with no atmospheres at all (Trilling *et al.* 1998). However Winn *et al.* (2004) found no evidence for an excess of Balmer Hα absorption during a transit of HD209458. This puts constraints on the excitation temperature, $T_{ex} < 8000$ K, whereas Lecavelier & Etangs *et al.* (2004) find an excitation temperature > 8000 K (in agreement with Vidal-Madjar *et al.* 2003). Therefore even though models for the atmosphere of HD209458b explain some of the transit properties further work is required.

4. The Hunt for Terrestrial Planets

Currently the radial velocity technique is only sensitive to gas giant planets around solar-type stars in the solar neighbourhood. Earth mass planets do exist, yet so far they have only been found around pulsars and are of little consequence when dealing with terrestrial planets. The radial-velocity detection method is extremely mass dependent, its signal is proportional to the ratio of the planet and stellar mass. In order to detect terrestrial planets around stars via the Doppler technique, we either increase the sensitivity of the detection method or we decrease the mass of the target stars we are observing. With an amplitude stability of 1 ms^{-1} over long timescales it should be possible to probe well below the mass of the Earth around low-mass stars such as M-dwarfs.

One of the primary goals of these searches is to find an inhabited planet outside our own solar system. Therefore a region known as the Habitable Zone (HZ) is a useful concept; planets in this region can sustain liquid water at their surface. This means that the planets need a surface temperature approximately 270–370 K in order for water not to freeze or boil away. The more luminous the star, the further out the HZ is located. While the radial-velocity technique is far from sensitive to the HZs around solar-type stars, M dwarfs on the other hand have very low luminosities and therefore their HZs are located at much smaller radii. Searches for terrestrial planets around M dwarfs have already begun in earnest. Endl *et al.* (2003) have ruled out the presence of any planets with minimum masses between 4-6 Earth masses in the HZ around our closest neighbour Proxima Centauri.

5. The Future of Planet Searches

The impetus of discovery and characterisation means that exoplanet discovery should continue to increase as objects are found from a wide range of techniques. The power of characterisation using several techniques has already been proven for Gl876 and HD209458. A much deeper understanding of exoplanets should become apparent once such data exists for a large sample. The 'dynamical fullness' seen for our own Solar System and the multiple exoplanets, suggests that the characterisation of exoplanets will increasingly be made using a variety of techniques. In the short-term the combination of radial velocity, astrometry, transit and interferometric imaging measurements seem most promising. Notwithstanding an increasing rate of discovery and characterisation of an even broader spectrum of exoplanets, the next few years should continue to bring dramatic improvements in the realism of exoplanet formation and evolution simulations.

The revolution in exoplanet science is just beginning. At the moment we are sensitive to a small fraction of the planet parameter space and in particular are not yet sensitive to our own Solar System orbiting a nearby star. Despite having plenty of examples of exoplanets and of higher mass brown dwarfs, we are still challenged by the definition of a planet. We need a better idea of planets as a function of host star properties, planet mass, composition (gas, ice, rock, metal) and orbit parameters and the intercorrelation of these parameters. It will be fascinating to see the importance of environment on exoplanets, not only on the major planets that we detect over the next few years but also the minor constituents such as comets and asteroids. Substantial research into chemical differentiation will be necessary; do other terrestrial planets have non-equilibrium atmospheres?

The current 'Holy Grail' of exoplanet discovery, is to obtain a direct image of a terrestial exoplanet. A number of instruments are in the pipeline at the moment to make this concept a realisation, such as the DARWIN and the OWL (OverWhelmingly Large telescope). Many of the precursor technologies are now being developed and realised by the current generation of interferometers, e.g. the VLTI. The range of ground and space based endeavours underway now (e.g. Schneider 2004) suggests that by the time the current generation of astrophysically inclined school leavers have finished their PhDs, the research edge may have moved to the characterisation of more Earth-like planets. By then the technical and scientific resources should exist to start to constrain a range of scenarios for exolife.

References

1. Adams F.C., Laughlin G., 2003, Migration and Dynamical Relaxation in Crowded Systems of Giant Planets, Icarus, 163, 290
2. Armitage P.J., Bonnell I. A., 2002, The Brown Dwarf Desert as a Consequence of Orbital Migration, MNRAS, 330, L11
3. Armitage P.J., Livio M., Lubow S.H., Pringle J.E., 2002, Predictions for the Frequency and Orbital Radii of Massive Extrasolar Planets, MNRAS, 334, 248
4. Backer D. C., Foster R. S. Sallmen S., 1993, A Second Companion of the Millisecond Pulsar 1620-26, Nature, 365, 817

5. Baranne A. *et al.*, 1996, ELODIE: A Spectrograph for Accurate Radial Velocity Measurements, A&AS, 119, 373

6. Barman T. *et al.*, 2002, Non-LTE Effects of Na I in the Atmosphere of HD 209458b, ApJ, 569, 51

7. Benedict *et al.*, 2002, A Mass for the Extrasolar Planet Gl 876b Determined from Hubble Space Telescope Fine Guidance Sensor 3 Astrometry and High-Precision Radial Velocities, ApJ, 581, L115

8. Bouchy F. *et al.*, 2004, Two New 'Very Hot Jupiters' Among the OGLE Transiting Candidates, A&A, in press

9. Bond I.A. *et al.*, 2004, OGLE 2003-BLG-235/MOA 2003-BLG-53: A Planetary Microlensing Event, ApJ, 606, 155L

10. Brown T., Charbonneau D., Gilliland R., Noyes R., Burrows A., 2001, HST Time-Series Photometry of the Transiting Planet of HD 209458, ApJ, 552, 699

11. Butler R.P., Marcy G.W., Williams E., McCarthy C., Dosanjh P., Vogt S.S., 1996, Attaining Doppler Precision of 3 m s^{-1}, PASP, 108, 500

12. Charbonneau D., Brown T.M., Noyes R.W., Gilliland R.L., 2002, Detection of an Extrasolar Planet Atmosphere, ApJ, 568, 377

13. Chiang E.I., 2003, Excitation of Orbital Eccentricities by Repeated Resonance Crossings: Requirements, ApJ 584, 465

14. Cumming A., Marcy G.W., Butler R.P., 1999, The Lick Planet Search: Detectability and Mass Thresholds, ApJ, 526, 896

15. Cumming A., Marcy G.W., Butler R.P., Vogt S.S., 2003, The Statistics of Extrasolar Planets: Results from the Keck Survey, ASP Conference Series 294: Scientific Frontiers in Research on Extrasolar Planets, eds. D. Deming and S. Seager, 27

16. Endl M., Kurster M., Rouesnel F., Els S., Hatzes A. P., Cochran W. D., 2003, Extrasolar Terrestrial Planets: Can We Detect Them Already?, ASP Conference Series 294: Scientific Frontiers in Research on Extrasolar Planets, eds. D. Deming and S. Seager, 75

17. Fischer D.A. *et al.*, 2003, A Planetary Companion to HD 40979 and Additional Planets Orbiting HD 12661 and HD 38529, ApJ, 586, 1394

18. Fischer D., Valenti J. A., Marcy G., 2004, Spectral Analysis of Stars on Planet Search Surveys, ASP Conference Series, in press

19. Ford E.B., Rasio F.A., Yu K, 2003, Dynamical Instabilities in Extrasolar Planetary Systems, ASP Conference Series 294: Scientific Frontiers in Research on Extrasolar Planets, eds. D. Deming and S. Seager, 181

20. Goldreich P., Tremaine S., 1980, Disk-Satellite Interactions, ApJ, 241 , 425

21. Gonzalez G., 1997, The Stellar Metallicity-Giant Planet Connection, MNRAS, 285, 403

22. Hebrard G., Etangs A. Lecavelier des., Vidal-Madjar A., Desert J. -M., Ferlet R., 2003, Evaporation Rate of Hot Jupiters and Formation of Chthonian Planets, astroph, 12384

23. Horne K., 2003, Status and Prospects of Planetary Transit Searches: Hot Jupiters Galore, ASP Conference Series 294: Scientific Frontiers in Research on Extrasolar Planets, eds. D. Deming and S. Seager, 361

24. Jones H.R.A., Butler R.P., Tinney C.G., Marcy G.W., Penny A.J., McCarthy C., Carter B.D., 2002, Extrasolar planets around HD 196050, HD 216437 and HD 160691, MNRAS, 337, 1170

25. Konacki M., Torres G., Jha S., Sasselov D., 2003, A New Transiting Extrasolar Planet, Nature, 421, 507

26. Kuchner N., Lecar M., 2002, Halting Planet Migration in the Evacuated Centres of Protoplanetary Disks, ApJL, 574, 87

27. Lecavelier des Etangs A. *et al.*, 2004, Atmospheric Escape from Hot Jupiters, A&A, 418, L1

28. Livio M., Pringle J., 2003, Metallicity, Planetary Formation and Migration, MNRAS, 346, 42

29. Lineweaver C.H., 2001, An Estimate of the Age Distribution of Terrestrial Planets in the Universe: Quantifying Metallicity as a Selection Effect, Icarus, 151, 307

30. Lin D.N.C., Bodenheimer P., Richardson D.C., 1996, Orbital Migration of the Planetary Companion of 51 Pegasi to its Present Location, Nature, 380, 606

31. Lin D.N.C., Papaloizou J.C.B., 1986, On the Tidal Interaction Between Protoplanets and the Protoplanetary Disk. III Orbital Migration of Protoplanets, ApJ, 309, 846

32. Marcy G.W., Butler R.P., Vogt S.S., Fischer D.A., McCarthy C. *et al.*, 2004, The California & Carnegie Planet Search, http://exoplanets.org

33. Marcy G.W., Butler R.P., 1996, First Three Planets, SPIE, 2704, 46M

34. Masset F.S., Papaloidzou J., 2004, Runaway Migration and the Formation of Hot Jupiters, ApJ, in press

35. Matsuyama I., Johnstone D., Murray N., 2003, Halting Planet Migration by Photoevaporation from the Central Source, ApJL, 585, 143

36. Mayor M., Queloz D., 1995, A Jupiter-mass Companion to a Solar-type Star Nature, 378, 355

37. McCaughrean M.J., Reid I.N., Tinney C.G., Kirkpatrick J.D., Hillenbrand L.A., Burgasser A., Gizis J., Hawley S.L., 2001, What is a planet?, Science 291, 1487

38. Murray N., Hansen B., Holman M., Tremaine S., 1998, Migrating Planets, Science, 279, 69

39. Oppenheimer B., 1999, Brown Dwarf Companions of Nearby Stars, PhD, California Institute of Technology

40. Paczynski B., 1986, Gravitational Microlensing at Large Optical Depths, ApJ, 301, 503

41. Perryman M., 2000, Extra-Solar Planets, Rep. Prog. Phys., 63, 1209

42. Reid I.N., 2002, On the Nature of Stars with Planets, PASP, 114, 306

43. Santos N.C., Israelian G., Mayor M., Rebolo R., Udry S., 2003, Metallicity, Orbital Parameters, and Space Velocities, A&A, 398, 363

44. Santos N.C., Israelian G., Mayor M., 2004, Spectroscopic [Fe/H] for 98 Extrasolar Planet Host Stars. Exploring the Probability of Planet Formation, A&A, 415, 1153

45. Saumon D. *et al.*, 2000, Molecular Abundances in the Atmosphere of the T Dwarf Gl229B, ApJ, 541, 374

46. Sigurdsson S., Richer H.B., Hansen B.M., Stairs I.H., Thorsett S.E., 2003, A Young White Dwarf Companion to Pulsar 1620-26, Science, 301, 193

47. Schneider J., 2004, The Extrasolar Planets Encyclopaedia, http://www.obspm.fr/encycl/encycl.html

48. Tabachnik S., Tremaine S., 2002, Maximum-likelihood method for estimating the mass and period distributions of extrasolar planets, MNRAS, 335, 151

49. Tamura *et al.*, 1998, Isolated and Companion Young Brown Dwarfs in the Taurus and Chamaeleon Molecular Clouds, Science, 282, 1095

50. Torres G., Konacki M., Sasselov D., Jha S., 2003, The transiting planet OGLE-TR-56b, astroph, 0310114

51. Tremaine S., Zakamska N.L., 2003, Extrasolar Planet Orbits and Eccentricities, astroph, 12045

52. Trilling D.E., Benz W., Gulliot T., Lunine J.I., Hubbard W.B., Burrows A., Orbital Evolution and Migration of Giant Planets: Modelling Extrasolar Planets, ApJ, 500, 428

53. Trilling D., Lunine J.I., Benz W., 2003, Orbital migration and the frequency of giant planet formation, A&A, 394, 241

54. Udalski A. *et al.*, 2002, The Optical Gravitational Lensing Experiment. Planetary and Low-Luminosity Object Transits in the Carina Fields of the Galactic Disk, Acta Astronomica, 52, 317

55. Vidal-Madjar A. *et al.*, 2003, An extended upper atmosphere around the extrasolar planet HD209458b, Nature, 422, 143V

56. Vidal-Madjar A. *et al.*, 2004, Detection of Oxygen and Carbon in the Hydrodynamically Escaping Atmosphere of the Extrasolar Planet HD209458b, ApJ, 604L, 69

57. Ward W.R., Hourigan K., 1989, Orbital Migration of Protoplanets,: The Inertial Limit, ApJ, 347, 490

58. Ward W.R., 1997, Protoplanet Migration by Nebula Tides, ICAR, 126, 261

59. Winn J.N. *et al.*, 2004, A Search for Hα Absorption in the Exosphere of the Transiting Planet HD 209458b, PASJ, in press

60. Wolszczan A., Frail D., 1992, A Planetary System around the Millisecond Pulsar PSR1257+12, Nature, 255, 145

61. Zucker S., Mazeh T., 2002, On the Mass-Period Correlation of the Extrasolar Planets, ApJ, 568, 113L

Hugh R. A. Jones

Dr Hugh Jones is a Reader / Academic Manager at Liverpool John Moores University. Hugh lecturers at Liverpool John Moores and Liverpool Universities as well as developing a suite of distance learning courses (astronomy.ac.uk), serving on the advisory board for learning and teaching subject network in the Physical Sciences and developing electronic gadgets (madlab.org). He is the UK principal investigator for the Anglo-Australian planet search which has found many new extra-solar planets including the first such planet to be found in an Earth-like orbit, the 100th extra-solar planet and the system most like our own. Some of the editing for this review was done on a train journey to visit his new nephew Rory (the image shows them together).

James S. Jenkins

James Jenkins was born on the 6th of September 1980 in Falkirk Royal Infirmary and was brought-up in a town called Bo'ness. During this time he attended Bo'ness Primary School and Bo'ness Academy, eventually completing an HND in Measurement & Control Engineering at Falkirk College of Further & Higher Education.

After his HND was completed James turned his attention to astronomy and graduated from Liverpool University with a 1st Class honours degree (2003) in Astrophysics. He is currently a Ph.D student at Liverpool John Moores University working with Hugh Jones on the search for planets outside our Solar System. He was a keen amateur footballer, until a knee injury put paid to that, and is an avid supporter of Falkirk Football Club.

Solar System and Climate Change

CHAPTER 16

OUR SOLAR SYSTEM

Andrew Coates

Mullard Space Science Laboratory,
University College London

Up until the dark ages, humankind knew of six planets including our own. The invention of the telescope, and the beginnings of scientific thought on orbits and planetary motion, were in the seventeenth century. The next three centuries added Uranus, Neptune and Pluto to the known list as well as the many moons, asteroids and comets that we know today. It is only in the latter part of the 20th century that we have been privileged to carry out in-situ exploration of the planets, comets and the solar wind's realm and to begin to understand the special conditions on Earth which meant that life started here. This is leading to a detailed view of the processes which have shaped our solar system.

Here, we briefly review our current knowledge of the solar system we inhabit. We discuss the current picture of how the solar system began. Important processes at work, such as collisions and volcanism, and atmospheric evolution, are discussed. The planets, comets and asteroids are all discussed in general terms, together with the important discoveries from space missions which have led to our current views. For each of the bodies we present the current understanding of the physical properties and interrelationships and present questions for further study. The significance of recent results, such as proof that there were one standing bodies of water on Mars, and the discovery of what appears to be an Oort cloud comet, are put into context.

What is in store for planetary exploration and discoveries in the future? Already a sequence of Mars exploration missions, a landing on a comet, further exploration of Saturn and the Jovian system and the first flyby of Pluto are planned. We examine the major scientific questions to be answered. We also discuss the prospects for finding other Earth-like planets elsewhere, and for finding extraterrestrial life both within and beyond our own solar system.

1. Introduction – The Solar System in the Last Four Millennia

Astronomy is one of the oldest observational sciences, existing for about four millennia. This time is estimated on the basis that names were given to those Northern constellations of stars which were visible to early civilisations. Two millennia ago, a difference between the more mobile planets and comets on the one hand and the fixed stars on the other was realised by Ptolemy but not understood. At the dawn of the present millennium six planets (Mercury, Venus, Earth, Mars, Jupiter and Saturn) were known. Aurora were seen on Earth and recorded but understanding still eluded us. In the first decades of the last millennium comets were seen as portents of disaster. Little further progress was made during the Dark Ages.

Science began to take giant leaps forward in the sixteenth century. Copernicus realised that Earth was not at the centre of the solar system. Tycho Brahe's planetary observations enabled Kepler to formulate laws of planetary motion in the seventeenth century. In the same century, Galileo invented the telescope and Newton developed his theory of gravitation. Newton also invented the reflecting telescope. The eighteenth and nineteenth centuries saw increasing use of this new technology, which deepened and increased understanding of the objects in the sky. Amongst many other discoveries were the periodicity of Halley's comet and the existence of the planets Uranus, Neptune and Pluto – one planet per century for the 18th, 19th and 20th. The observations were all made in visible light to which our eyes are sensitive and which is transmitted through our atmosphere. The nineteenth century also gave us the basics of electromagnetism.

In the twentieth century we observed many important scientific and technological advances. In terms of astronomy we have gained the second major tool for the exploration of the universe in addition to the telescope – namely the spacecraft. Space probes have not only opened up the narrow Earth-bound electromagnetic window which only allows us to detect visible light and some radio waves from the ground, but it has also allowed in-situ exploration and sampling of our neighbouring bodies in the solar system. Using the techniques of remote sensing to look back at the Earth has added a new perspective. For the first time we can now begin to understand our place in the universe and the detailed processes of the formation of the universe, our solar system and ourselves. We are truly privileged to be able to use these techniques to further this scientific understanding.

Our sense of wonder in looking at the night sky has not changed over the millennia. As we use better ground and space based telescopic techniques and more detailed in-situ exploration one can only feel a sense of excitement at the discoveries that the beginning millennium may bring. In this paper we will review our solar system in general terms. We consider scientific questions for the future and speculate a little on how they will be answered.

2. Origin of the Solar System

The objects in the solar system now reflect the history of its formation over 4.5 billion years ago. Because the planets are confined to a plane, the ecliptic, it is thought that the Sun and planets condensed from a spinning primordial nebula. The heavier elements in the nebula are thought to be present due to earlier nearby supernova explosions. As the nebula collapsed and heated, the abundant hydrogen fuel ignited and fusion reactions started in the early Sun. Gas further from the centre of the nebula became progressively cooler and condensation occurred onto dust grains. This caused differences in composition due to the progressively cooler temperatures away from the Sun. Gravitational instabilities then caused the formation of small, solid planetesimals, the planetary building blocks. Accretion of these bodies due to collisions then formed the objects familiar to us today.

This model predicts different compositions at different distances from the Sun, and this is seen in the different classes of solar system objects today. In the inner solar system, for example, the temperatures were of order 1000K, near to the condensation temperature of silicates, while in the outer solar system temperatures were of order 100K, where water and methane ice condense. The forming planetesimals therefore have different compositions in the inner and outer solar system. The cores of the outer planets are associated with their much colder planetesimals, some of which remain as comets. The inner, rocky planets are associated with their own planetesimals, of which the asteroids are the partially processed survivors in the present solar system.

The formation process can be seen going on elsewhere, for example in the Orion nebula. Protoplanetary disks have been imaged there using the Hubble space telescope, and structures which look like forming solar systems have been seen edge-on. Such patterns were also seen in the mid-1980s near the star Beta Pictoris, in that case the dust cloud was inferred to be due to pieces from forming or collided comets. Several other examples have been seen using infrared telescopes in space, including the recent Spitzer telescope detection of a new planet in Taurus just 1 million years old, and the presence of ice and simple organics in forming planetary systems. It is thought that the basic formation of solar systems is widespread throughout the Universe.

3. Processes: Collisions, Accretion and Volcanism

Clearly, collisions played an important role in the early solar system in the formation of planetesimals and larger bodies, and in collisions between these bodies. Accretion of material onto the forming bodies also played a key role. In the case of the outer planets, a rocky/icy core formed and if the core was big enough, accretion of gas and dust from the solar nebula occurred. Since temperatures were cold in the outer solar system it was possible for the large cores to retain solar nebula material. This is why the current composition of the atmospheres of Jupiter, Saturn, Uranus and Neptune are close to solar proportions. For the inner planets, temperatures were higher and it was possible for light atoms and molecules such as hydrogen and helium to escape.

The Apollo missions to the Moon provided information on the cratering rate with time since the solar system's formation. Cratering density was combined with dating of the returned samples from known positions on the surface, revealing the density of craters with time since 4.6 billion years ago. The cratering rate was high from 4.6 to 3.9 billion years ago, the 'late heavy bombardment' period, following which it slowed down between about 3.9 and 3.3 billion years ago and it has been at an asymptotic 'steady' level since about 3 billion years ago. From this we can infer that there are two populations of Earth-crossing bodies. The first reduced with time as collisions happened or the bodies were ejected from the solar system. The second group is the present population of Earth-crossing bodies, and the population must be being replenished as well as lost.

The cratering process clearly depends on the size of the impacting object and its impact speed. Clearly the important parameter is the kinetic energy of the object. In general impactors are travelling at tens of km per second, making impact kinetic energies large. There are two physical shapes of crater seen – simple craters (for small impactors) and complex craters (for large impact energy). For larger energies, the complex craters may include wall collapse and a central peak. Both crater types produce ejecta, the distance these move from the crater depends on the gravitational force on the target body. The Moon contains a well-preserved cratering record as there is no erosion by atmosphere or oceans; both crater types are seen there.

In contrast, the Earth's cratering record is affected by erosion by the atmosphere, by oceans and by volcanic and tectonic processes. Over 150 impact craters have been discovered on Earth. Of these about 20% are covered by sedimentation. Previously undiscovered craters are occasionally found, such as a new structure under the North Sea discovered by oil exploration geophysicists in 2002. However, in general only the remnants of relatively recent craters remain, though one of the large structures (Vredefort, South Africa) is 2.02 billion years old.

As well as the Moon and Earth, cratering is widespread in the solar system and is a feature of all solid bodies – planets, moons, asteroids and even comets – visited by spacecraft. Analysis of the cratering record on each body can be used in studies of a body's age and geological history.

Some extreme examples of cratering are worth mentioning. First, on Saturn's moon Mimas is the large (130km) crater Herschel. The impact which produced this crater may have almost torn Mimas apart, as there are fissures on the other side of Mimas which may be related. Mimas is primarily icy according to albedo measurements, so this and the many other craters seen on Mimas shows that craters are well-preserved even on icy bodies.

A second extreme example is the Hellas impact basin in the southern hemisphere of Mars. This massive crater is some 2100km across and 9km deep. A ring of material ejected from the crater rises about 2km above the surroundings and stretches to some 4000km from the basin centre, according to recent data from the Mars Orbiter Laser Altimeter on Mars Global Surveyor. As a rough rule of thumb, the diameter of any resulting crater is approximately 20 times the diameter of the impactor, so the original impactor in this case must have been $\sim$ 100km. The actual relationship between impactor size and crater size involves the kinetic energy of impact, the gravity of the impactor, the strength of the impacting material (relevant for small impacts), the angle of incidence and the composition of the target.

As a third example we can consider the formation of Earth's Moon. Based on Apollo sample composition, and solar system dynamics calculations, it is now thought that the most likely explanation for the formation of the Moon was a large (Mars-sized) object hitting the Earth and the coalescence of material from that impact. Clearly collisions have a very important role in the solar system's history.

Collisions clearly still occur too: some 500T of material from meteor trails and other cosmic junk hit the top of Earth's atmosphere every year. In 1994, we witnessed the collision of comet Shoemaker-Levy 9 into Jupiter

– the first time such an event could be predicted and observed. Currently there are great efforts on Earth to locate and track the orbits of near-Earth objects, as sooner or later a large impact will happen. The comfort we can take is that such collisions are very rare.

Another major, fundamental process in the evolution of solar system objects is volcanism. While only two bodies in the solar system have currently known active volcanism (Earth and Io) there is evidence on many bodies for past volcanism (including on Mercury, Venus and Mars), and for cryovolcanism on Europa and Triton, and there are many other locations where evidence of resurfacing is seen.

In general, an energy source is needed to drive volcanism. This may be from the heat of formation, from radioactive decay of the constituents, from tidal forces, or from angular momentum. How much heat is left from formation depends on the size of the object. In particular, the area to volume ratio is larger for small bodies, so it is expected that these lose heat more quickly. Therefore, for example, Mars had active volcanism early in its history but this has now died out, while Earth still has active volcanism. On Io, the power for the volcanic activity comes from tidal forces. Venus on the other hand had rampant volcanism some 800 million years ago, when the planet was effectively resurfaced, but current activity has not yet been directly detected. This is surprising as the size is similar to Earth, but perhaps the slow current spin rate plays a role. However, earlier missions detected indirect evidence for current volcanism, and if present it would help to explain the high (90 bar) current atmospheric pressure despite billennia of atmospheric erosion by the solar wind. Venus Express will seek evidence of current activity.

4. Evolution of Atmospheres

The main sources of atmospheric material are:

- Solar nebula
- Solar wind
- Small body impact
- Outgassing of accreted material

The clues for the relative strengths of these sources come from the relative abundances of the inert gases, of primordial (^{20}Ne, ^{36}Ar, ^{38}Ar, ^{84}Kr, ^{132}Xe) or radiogenic (^{40}Ar [from ^{40}K], ^{4}He [from ^{238}U, ^{235}U, ^{232}Th]) origin. The former must have been there since the beginning of the solar system as they are not products nor do they decay, and the latter build up with time. The inert gases take no part in chemical processes, and most are too heavy to escape.

Minerals in the forming grains and planetesimals in the solar nebula contained trapped gases – volatile components. These included iron nickel alloys (which contained nitrogen), serpentine (which contained water, ammonia and methane) and several others. On accretion into a larger body, these cores would have attracted a 'primary atmosphere' of gases from the surrounding solar nebula (principally hydrogen, and otherwise reflecting the solar abundance).

As we saw earlier, in the first 0.5 by the impact rate by planetesimals was very high, bringing more material to the planetary surfaces (a 'veneer' of volatiles), the surfaces were heated by impacts, and the atmosphere was heated. There was also higher heating at that time, from gravitational accretion, from radioactive decay of ^{40}K, ^{238}U, ^{235}U and ^{232}Th. Also, heating would have occurred by 'differentiation' as heavier substances dropped towards the centre of the recently formed bodies.

All this early heat would have increased tectonic activity at the surfaces, at least of the solid bodies. Larger planets have hotter cores, so gas should be released faster. Earth's gas would be expelled faster than Venus (0.8 M_{earth}) and much faster than Mars (0.1 M_{earth}). Isotope ratios at present show that both Earth and Venus are effectively fully 'processed' (although there is still an active cycle involving volcanism and solution in the oceans), while Mars is much less so and it is estimated that only about 20% of its volatiles have been released.

Gases trapped in the interior can be released by several processes. These are internal heating, volcanic activity, and impacts of large meteors – asteroids or comets.

Once released, gases would remain gravitationally bound assuming that (1) the body is big enough and (2) the gas is cool enough to remain gravitationally bound.

Other chemical reactions would start and the atmosphere would evolve over billions of years. Many changes have occurred since the formation of the solar system 4.6 billion years ago.

For the terrestrial planets, the isotope abundance ratios for the primordial isotopes are quite different to solar abundances, and more similar to those found in meteorites. This rules out 1 and 2 above as potential mechanisms. For Venus, Earth and Mars, the abundance ratios for the three planets is about the same, but the absolute abundances decrease with distance from the Sun. This apparently rules out asteroid impacts as a dominant mechanism as the impact rates should be similar for these three planets.

However, this picture is modified due to the mixing of the isotopes ^{36}Ar/^{132}Xe and ^{84}Kr/^{132}Xe. For Venus the values indicate a quite different origin, much more solar, in that atmosphere. For Earth and Mars however the proportions show that the two planets are either end of a mixing line: beyond the Earth end of that line conditions are like those in comets, while beyond the Mars end they are more like the primitive chondrite meteorites. This seems to indicate that a proportion of Earth's current atmosphere is cometary in origin whereas in Mars case the source was asteroids.

These are now the expected proportions at the end of the bombardment about 4 billion years ago. Much of the water in Earth's atmosphere and oceans may have come from comets.

The surface temperature of the planets at this stage plays a role – controlled by the distance to the Sun. The temperature of a body (ignoring internal sources and atmospheric effects) can be estimated from the balance between solar input and radiation (assumed black body). Under that assumption, the temperature T can be written $T = 279(1\text{-}A)^{0.25}R^{-0.5}$ K, where A is the albedo and R in AU. If the kinetic energy of gas molecules is high enough it can escape the gravitational pull of the planet, if not it will be trapped to form an atmosphere. Putting in figures it can be shown that Jupiter, Saturn, Neptune, Uranus, Venus, Earth, Mars should all easily have atmospheres, with Io, Titan and Triton are on the borderline. Clearly the real picture is much more complex in terms of the energy balance, presence of an atmosphere, greenhouse effect, internal heating, etc. However these simple considerations do remarkably well. The atmospheric evolution is then different for terrestrial planets and for the outer planets; this forms part of the next sections.

5. Terrestrial Planets

Due to the temperature variation in the collapsing primordial nebula, the inner regions of the solar system contain rocky planets (Mercury, Venus, Earth and Mars). The condensation temperatures of the minerals forming these planets were higher than the icy material in the outer solar system. While we can treat the inner planets as a group, the diversity of the planets and of their atmospheres is remarkable. The processes of impacts, volcanism and tectonics are vital in the evolution of planets. Our understanding also depends on the sources and dissipation of heat.

The origin of the atmospheres of the inner planets is an important topic in itself. It is now thought that the origin is due to outgassing of primitive material from which the planets were made. As was pointed out before, however, the cometary volatile content of our own atmosphere seems to be higher than that for Mars at least. The subsequent evolution of the atmospheres depends on two major factors: the distance from the Sun (which controls radiation input) and the mass of the body (controlling firstly heat loss rate from the initial increase due to accretion, secondly heating rate due to radioactivity and thirdly atmospheric escape speed). The presence of life on Earth has also played an important role in determining atmospheric composition here.

Mercury is a hot, heavily cratered planet which is difficult to observe because of its proximity to the Sun. Only one spacecraft, Mariner 10, has performed three fast flybys of the planet. Mercury is remarkable because of its high density, second only to the Earth. The other terrestrial planetary bodies, the Earth, Venus, Mars and the Moon, all fit on a straight line which relates density and radius, but Mercury is way off that line. It is unclear why this is – but may be associated with the origin of the planet or the stripping of its outer layers by an early impact or the intense T-Tauri phase solar wind.

Another remarkable and unexpected discovery from Mariner-10 was the presence of a strong magnetic field and a magnetosphere. It is likely that the planet has a larger iron-rich core in relation to its radius than the others. It is strange that there is so much iron, at this distance in the forming solar nebula FeS (one of the major expected condensing iron-rich minerals) should not have even condensed. Mercury has an 'exosphere' rather than an atmosphere, since the pressure is so low that escape is as likely as a gas collision. An atmosphere was not retained because of the low mass (size) and high temperature (proximity to the Sun). The planet has an eccentric, inclined orbit. The rotation period is in a 2:3 ratio with its rotation. This indicates that the slightly non-spherical shape of the planet was important during its formation.

The surface of Mercury contains a well-preserved, little disturbed cratering history. During the early bombardment in the first 0.8 billion years since formation, towards the end of which a large impact produced the 1300km Caloris Basin feature, there was important tectonic activity. There is also evidence of significant shrinkage of the planet due to cooling, causing 'lobate scarp' structures; this shrinkage was also early in the planet's history.

It appears likely from radar measurements that some of Mercury's polar craters, which are never heated by the Sun, may contain condensed ice from cometary impacts, similar to the deposits inferred from Lunar Prospector measurements of hydrogen at the Moon. It would be interesting to estimate the cometary impact rate here in comparison with that at lunar orbit, to understand more about the collision history. But one of the fundamental unknowns about Mercury is what lies on its unseen side, not imaged by Mariner-10. Radar measurements indicate some structure, which could be a large volcano, but the Messenger and BepiColombo missions should explore this in detail.

The Venus-Earth-Mars trio is particularly important for us to understand because we know that life evolved at least on the Earth. In the case of Venus the planet, although similar in size to the Earth, was closer to the Sun; water evaporated early because of the higher temperature and caused a greenhouse effect, causing further evaporation of water and eventually a runaway greenhouse effect. Lack of surface water meant that fixing of evolved carbon dioxide in the rocks via as at Earth was impossible. Carbon dioxide evolved into the atmosphere over the 4 billion years since formation continued the greenhouse effect after the hydrogen (from water dissociation) has escaped to space and the oxygen had oxidised rocks. The surface temperature is now 750 K, the atmosphere is thick and supports sulphuric acid clouds, and our sister planet is completely inhospitable to life – though there have been recent suggestions that the Venus atmosphere might be a possible site. There is a total of only 10 cm equivalent depth of water in the Venus atmosphere, compared to 3km in the atmosphere and oceans of Earth. The clouds are observed to rotate much faster (~ 4 days) in the equatorial regions than the planetary rotation rate (longer than a Venus year, and retrograde); this 'super-rotation' is one of the aspects of the Venusian atmosphere that is not yet well understood.

In terms of the space programme there have been missions to study the atmosphere (Mariner, Pioneer, Venera, Vega), to map below the clouds using radar (Magellan) and landers (Venera) have transmitted pictures briefly from the inhospitable surface. For the future, ESA's Venus Express will study the atmosphere, surface and solar wind interaction. It will use new windows in the infrared, discovered since the Pioneer Venus and Magellan missions. It will study (1) the global structure and dynamics of the atmosphere below the clouds for the first time, and the cause of super-rotation (2) aspects of surface science including the search for current volcanism and tectonics, and (3) the escape of the Venus atmosphere to space via solar wind scavenging.

Mars on the other hand is smaller than the Earth and further from the Sun. Isotopic ratios between radiogenic ^{40}Ar and primordial ^{36}Ar and indicate that only about 20% of the gas in the rocks has been evolved into the atmosphere because of reduced tectonic activity (small size). Also due to the size much of the early atmosphere was lost to space since the loss of a magnetic field 3.8 by ago, and the present atmospheric pressure is less than 1% that of Earth's. Substantial carbon dioxide ice deposits are present at the poles but there is not enough in the atmosphere to cause a greenhouse effect.

Martian oxygen isotopic ratios show that there must be a source of oxygen, perhaps frozen sub-surface water, which increases the proportion of the lighter isotope which would otherwise preferentially be lost to space, to the observed level. This has been partially confirmed recently by Mars Odyssey, which sees hydrogen, probable evidence of water, near the poles. However the neutron analysis technique, used also for the Moon by Lunar Prospector, only works for the top 1m of the Martian soil due to absorption, and it is hoped that the Mars Express ground-penetrating radar will survey down to 5km. On the other hand, images from Viking and Mars Global Surveyor contain evidence, in the form of outflow channels, that liquid water flowed on the Martian surface about 3.8 billion years ago. More recently, NASA's Opportunity spacecraft has shown the first solid evidence for rock formation in water which was flowing at some point in the past. There is also spectacular geological evidence that Opportunity landed near an ancient shoreline where a fair sized body of water once stood. This appears to prove conclusively that Mars was once indeed warmer and wetter.

The present Mars is much colder, dryer and much less hospitable to life than it once was. The equivalent depth of water in the atmosphere is only 15 microns. There is no possibility of finding liquid water on the surface now due to the low pressure and average temperature. Any life on Mars was probably primitive and lived over 3.8 billion years ago. At that time, in addition to being warmer and wetter, Mars had a magnetic field, as shown by evidence of remanent magnetism in ancient rocks, by Mars Global Surveyor. This would have played a role in protecting any life at that time from galactic and solar cosmic radiation. In addition it would have protected the Mars atmosphere against scavenging by the solar wind. That protection is no longer there, and for the last 3.8 billion years the Martian atmosphere has been gradually stripped away. The possibility for current life on Mars could remain as sub-surface organisms near any water deposits, possibly similar to primitive organisms found in the Antarctic ice.

There is some recent evidence from Mars Global Surveyor, of water seepage from underground deposits to the surface, where it would immediately sublime. This discovery is significant – from the age of the surface this may have happened within the last million or so years.

At the time of writing, new results are pouring in from NASA's Spirit and Opportunity (see above), as well as from ESA's Mars Express. Highlights from Mars Express include the direct detection of infrared spectral features of water ice covering the (mainly CO_2) Southern polar icecap (from the OMEGA and PFS experiments). The confirmation of small amounts of methane in the Martian atmosphere is also significant as it must be the product of either current volcanism or of present primitive, presumably sub-surface life. Resolving this dichotomy will be one of the key areas of research in the next few years, with potentially astounding consequences.

Unfortunately the UK-led Beagle 2 spacecraft was not heard from since its December 2003 release from Mars Express. This mission was significant in that it was looking for direct evidence of past or present life on the planet. It is hoped that future missions will carry on this important quest: at present, the first to specifically address this is ESA's Exomars rover, proposed for launch in 2011 as part of the Aurora programme. Other future Mars missions are NASA's Mars Climate Orbiter (looking for signs of water) and Phoenix (polar lander to study icecap region) missions.

In the longer term, some serious scientists propose that greenhouse gases could be introduced into the Martian atmosphere to warm the planet and release some of the trapped water and carbon dioxide, ultimately giving a hospitable environment for humans. This is an interesting idea in theory and a good target for computer simulations. In the opinion of this author terraforming would be the ultimate in cosmic vandalism if implemented.

The Earth was at the right place in the solar system and was the right size for life to evolve. The presence of liquid water on the surface meant that carbon dioxide could be dissolved in rocks as carbonates, some of this is recycled due to volcanism. As life developed, starting at the end of the bombardment phase, photosynthesis became important, leading to the production of oxygen and the fixing of some of the carbon in the biomass. Enough oxygen in the atmosphere rapidly led to production of stratospheric ozone, which allowed the protection of land-based life forms from harmful EUV radiation.

The Earth is also the planet we know the most about. Looking at the Earth from space gave us a new perspective: an enhanced feeling that the Earth is special and indeed fragile. The average temperature of the Earth's surface is close to the triple point of water where solid, liquid and vapour may all exist. That is part of how we came to be here.

Our Moon is the first planetary satellite in terms of proximity to the Sun. Its density is much lower than the Earth's and there is effectively no atmosphere. The cratering record is therefore well preserved but the maria show that volcanic activity was important after the early bombardment and up to about 3.2 billion years ago. Despite intensive study by spacecraft (Luna, Ranger, Surveyor, Apollo, Clementine, Lunar Prospector) the origin of the moon has not yet been determined completely from the competing theories (simultaneous formation, catastrophic impact on early Earth, capture). However the current favourite which fits most of the evidence is that a Mars-sized body hit and coalesced with the Earth, and the moon formed from the impact ejecta.

As was already mentioned, the recent Clementine and especially Lunar Prospector missions returned evidence for hydrogen, perhaps in the form of water mixed with regolith, in craters towards the pole of the Moon. If this is water it would be from cometary impacts, and one of the exciting aspects of this is that there could be cometary material available for scientific study at a relatively close distance. Though studying this would be a large technological challenge the scientific benefits of examining this pristine material from the early solar system are significant.

The other missions going to the Moon soon, including SMART-1 and Selene, will have the possibility of producing better mineral maps and composition information over the whole lunar surface, adding significantly to the information gained from the Apollo and Russian lunar samples.

The satellites of Mars, namely Phobos and Deimos, may be captured asteroids based on their physical characteristics. However, the understanding of the dynamics of their capture is by no means solved.

6. Our Planetary Neighbours

Despite the proximity of our planetary neighbours and the many space missions which have explored them, many important questions remain. Why is Mercury's core so large? Might a collision early in its life explain this, and its orbital eccentricity and inclination? How is seismic activity affected? Is there ice at Mercury's poles? Why is there super-rotation in the Venus atmosphere? What is the surface composition of Venus and Mercury? What is the geological history? How oxidised is the Venusian surface and what is the history of water in the Venus atmosphere? Is there current volcanism on Venus? Why is Venus' axial rotation retrograde? What changes is humankind making to the Earth's climate and do these need to be ameliorated? To what extent was Earth's magnetic field a 'cradle for life'? Might Mars have included such a cradle up to 4 billion years ago when the Martian field was lost? What is the origin of the Moon? Do the Martian atmospheric loss rates to space support the models? Where is the water on Mars now? What is the history of other volatiles? Was there, and is there, life on Mars? Could and should we terraform Mars?

7. Outer Planets

The outer planets group contains the gas giants (Jupiter, Saturn, Uranus and Neptune) and the icy object Pluto. The gas giants are heavy enough and were cold enough when they were formed to retain the light gases hydrogen and helium from the solar nebula, and these constituents form most of the mass of the planets reflecting the early composition. The visible disk for telescopes and space probes is ammonia- and water-based clouds in the atmosphere. At Jupiter, the largest planet and closest gas giant to the Sun, the cloud structure shows a banded and colourful structure caused by atmospheric circulation. The detailed cloud colours are not fully understood. There is no solid surface as such, but models of the internal structure of the gas giants show increasing pressure below the cloud tops, ultimately the pressure becomes so high that a metallic hydrogen layer forms at about 80% and 50% of the radius respectively. Dynamo motions in this layer, assuming it must be liquid, power the powerful planetary magnetic fields. A rocky/icy core is thought to be present at about 25% of the planetary radius.

Jupiter rotates rapidly, providing some energy via the Coriolis force for atmospheric circulation. However, both Jupiter and Saturn have internal heat sources which mean that they emit 67% and 78% more energy than they receive from the Sun respectively. This gives most of the energy for the atmosphere, but the origin of the internal heat source is not fully understood; nor is it clear why Saturn's internal heat flux is higher than Jupiter's. Models suggest that helium precipitation within the metallic hydrogen core, in which helium is insoluble, may be responsible. There are also strong zonal (east-west) winds near the equator on Jupiter and Saturn, stronger on Saturn where they reach two thirds of the speed of sound. Their origin is not fully understood. The planets also have important long-and short-lived atmospheric features, of which the most prominent is the great red spot on Jupiter. This long-lived feature, seen for at least 300 years, is surprisingly stable and so far there is no adequate model to describe it. Seasonal spot features appear in Saturn's atmosphere. An apparently long-lived spot feature appears on Neptune.

In-situ results at Jupiter have recently been enhanced by data from the Galileo orbiter and probe. While the orbiter discovered unexpected dipole magnetic fields on Ganymede, and weaker fields at Io and Europa, the probe sampled the atmospheric composition, winds, lightning and cloud structures at a point which turned out to be non-typical in the Jovian atmosphere. One of the discoveries was a lower than solar helium abundance which provides support for the idea of helium precipitation in the metallic hydrogen layer; a similar conclusion was arrived at based on Voyager data at Saturn. Also there is less water than expected.

The gas giants each have important and fascinating moons. At Jupiter, Io has the only known active volcanoes other than Earth, providing sulphur-based gases for the Jovian magnetosphere; Europa may have a liquid water ocean under its icy crust; Ganymede has its own 'magnetosphere within a magnetosphere'; and Callisto has a very old, cratered surface. Our knowledge of these has been revolutionised by in situ observation, before this only the albedos and orbital periods were known.

At Saturn, Titan is a tantalising planet-like object, and the only moon with a significant atmosphere – 1.5 times the Earth's pressure at the surface. However, its face was shrouded from Voyager's view by organic haze in its thick nitrogen-methane atmosphere. The atmosphere may hold clues about Earth's early atmosphere; there may be methane or ethane based precipitation systems; and the ionosphere forms a significant source for Saturn's magnetosphere. Cassini-Huygens will study Titan and some of Saturn's 30 known icy satellites in detail starting in July 2004.

At Uranus, the moon Miranda graphically indicates the accretion theories as it appears to be made up of several different types of structure seen elsewhere. Also the moon system is out of the ecliptic because the spin axis of Uranus at 98° inclination is almost in the ecliptic itself. At Neptune, Triton is an icy satellite with a very thin atmosphere but it is in a retrograde orbit and is spiralling closer to Neptune; in tens of millions of years it may break up to produce spectacular rings. It may be similar in characteristics to Pluto and Charon. Uranus and Neptune have another curious property – their magnetic field axes are significantly different from their rotation axes. This has recently been shown to be consistent with motions in a shell of conducting core material further towards the surface than at other planets.

Ring systems are present at all the gas giants but spectacularly so at Saturn. Saturn's rings were discovered by Galileo, found to be separate from the planet by Huygens, found to have gaps by Cassini who also suggested that they were composed of separate particles, this idea was mathematically proved two centuries later by James Clark Maxwell. Detailed exploration was begun by the flyby missions Pioneer and Voyager which found remarkable structures including warps, grooves, braids, clumps, spokes, kinks, splits, resonances, and gaps. Whole new areas of solar system dynamics were opened up, including the study of electromagnetic forces which may be important in spoke formation. The rings are less than a kilometre thick, as low as tens of metres in places and composed of billions of chunks of ice and dust ranging from microns to tens of metres in size. But the main question has not yet been satisfactorily answered: how did the rings form? Was it break-up of a smaller satellite or cometary capture?

And then there is Pluto, with its moon Charon. Following an elliptical, inclined orbit and currently the furthest planet, Pluto is an icy body rather than a gas giant. It may be closely related to but larger than the icy Kuiper belt objects, the outer solar system planetesimals, and it may also be related to Triton. Much will be learned by the first spacecraft reconnaisance of the Pluto-Charon system. But as Pluto goes towards aphelion its tenuous and interesting methane-based atmosphere will condense and become much less dense. In 2010-2020 it is expected that a rapid atmospheric collapse occurs. There is a good case to get to Pluto as soon as possible and another excellent case for a visit near the next perihelion in 2237.

8. Questions for the New Millennium on Outer Planets and Their Satellites

What causes the cloud colours in the gas giants? Are the internal structure models correct? What causes the internal heat source? Why are the zonal winds so high on Jupiter and, particularly, Saturn? Why are atmospheric features, such as the Jovian great red spot, so stable? Does Europa have water oceans and is life a possibility there? Will the Titan atmosphere evolve further when the Sun becomes a red giant? Why is the Uranian spin axis so tilted and what are the effects? Are there strange, shell-like dynamos in action at Uranus and Neptune? Can the study of Saturn's rings give us more information about the radiation-plasma-dust mixture in the early solar system? What are the basic characteristics of Pluto-Charon?

9. Comets

Comets are the building blocks of the outer solar system. Their formation was at the low temperatures prevalent at these distances in the primordial nebula and they retain volatile material from the early solar system. They are relatively pristine bodies making their study important. Some of them collided and coalesced to form the cores of the gas giants. From the orbits of comets we find that many ($\sim 10^{12}$) were formed in the Uranus-Neptune region but were expelled to form the spherical (and theoretical) Oort cloud with radius approximately 50,000 AU. The orbits of these distant members of the solar system may be disturbed by passing stars. They may then plunge into the inner solar system where their orbits have random inclinations. Others ($\sim 10^8$) form the Kuiper belt just beyond Pluto's orbit; their inclinations are close to the ecliptic plane. Whatever their origin, comet nuclei are dirty snowballs which, when they near the sun, emit gas and dust which form the plasma and dust tails (seen in all comets, e.g. Hale-Bopp in 1997). Halley's comet, for example, loses about a metre of material per orbit and has orbited the Sun about 1000 times, but activity varies significantly from comet to comet.

The recent discovery of Sedna, with an elliptical, highly inclined orbit outside the Kuiper Belt and yet well inside the postulated Oort Cloud, is something of a challenge to existing theories. Its orbital properties suggest that it may be more closely related to the Oort cloud. Its size, at 0.75 the diameter of Pluto, has rekindled the debate about whether Pluto should be classed as a planet. Its discovery follows that of over 700 Kuiper belt objects in the last few years. These outer solar system objects have features which are more reminiscent of comets, rather than being either rocky or gas giant planets.

The space missions in the mid-1980s confirmed that comets reaching the inner solar system have a nucleus. They measured gas, plasma and dust and led to an understanding of tail formation. The main surprises were the darkness of the nucleus (Albedo $\sim 4\%$) and jet activity rather than uniform gas and dust emission. This is consistent with the idea of a cometary crust developing after multiple passes through the inner solar system, and fissures form in this as a result of phase changes below, due to the temperature variations over the comet's orbit. This idea also plays a role in the sudden flaring of comets, as seen at Halley when beyond the orbit of Neptune in 1990, and may be important in the breakup of cometary nuclei as seen in Schwaschmann-Wachmann 3 in 1998.

The composition measurements showed that the abundance relative to silicon of volatiles like C, N and O are more similar to solar abundances, so relatively pristine, compared to more processed bodies like meteorites and (even more so) the Earth.

The Deep Space 1 encounter with comet Borrelly in 2001 showed a slightly darker surface (albedo < 3%) even than comet Halley. This may indicate even less volatiles in the crust as the comet, a Jupiter class object which has visited the inner solar system many times, is thought to be 'older' than Halley.

Cometary missions which will provide unique data in the future include the ambitious ESA Rosetta Orbiter and Lander, NASA's Stardust currently bringing back dust samples to Earth, and Deep Impact which released pristine material from the heart of a comet for study.

10. Cometary Missions

Given their importance in the early solar system, what is the detailed composition of several comets? Can we bring an icy sample back to Earth for analysis? Is there an Oort cloud? What is the relation to planets? Might comets have brought volatiles to the inner planets?

11. Asteroids

In some sense the asteroids belong with the inner planets. Many asteroids occur in the main belt between Mars and Jupiter. Some are in other orbits, including orbits which cross the Earth's. A wide variation of eccentricities and inclinations of the orbits are also present. Spectral studies allow the classification of asteroids into types: C-type, dark, rich in silicates and carbon, mainly outer main belt; S-type, rocky bodies, mainly inner main belt and Earth crossing; M-type, iron and nickel. A few other asteroids do not fit this scheme. It seems likely that asteroids are the remains of inner solar system planetesimals rather than due to the destruction of a larger body. However, there have been collisions between bodies since the early bombardment leading to fragmentation. There is evidence for craters from Gaspra, Mathilde and Ida (and Ida's moon Dactyl, may itself have been due to a collision), and from Eros. Dust and boulders on Eros, seen during the NEAR-Shoemaker landing sequence, were a surprise but are perhaps debris from collisions. Collisions with the Earth and other planets may have been important, and some asteroids are the nearest remaining objects to the inner solar system planetesimals. In future, some workers think that commercial mining of asteroids for minerals may become economically feasible.

Table 1. Missions (Stage: 1 = flyby, 2 = orbiter, 3 = lander, 4 = sample return)

Object	Past missions	Stage	Current and future missions
Mercury	Mariner 10	1	Bepi-Colombo ESA-Japan, Messenger NASA
Venus	Mariner, Pioneer Venus, Venera, Vega, Magellan	3	ESA Venus Express, Japan Planet-C
Earth	Many	n/a	Many
Moon	Luna, Ranger, Surveyor, Zond, Apollo, Clementine, Lunar Prospector	4	Japan Lunar-A, Selene
Mars	Mars, Mariner, Viking, Phobos, Pathfinder, MGS, Odyssey	3	NASA Mars Climate Orbiter 05, Phoenix 07, Mars Science Lab 09, ESA ExoMars 11, Aurora? Mars Express, Spirit, Opportunity
Jupiter	Pioneer, Voyager, Galileo	3	NASA Juno Polar Orbiter, Ulysses
Saturn	Pioneer, Voyager	1	NASA-ESA Cassini-Huygens
Uranus	Voyager	1	–
Neptune	Voyager	1	–
Pluto	–	0	New Horizons (Pluto)
Asteroids	Galileo, NEAR, DS1	1, 2, (3)	Muses-C, Dawn
Comets	ICE, Sakigake, Suisei, VEGA, Giotto, DS1	1	Stardust, Rosetta, Deep Impact
Sun + i/p medium	Yohkoh, Cluster, IMAGE,...	n/a	Solar-B, STEREO, MMS, Solar Orbiter, SDO, LWS missions, WIND, ACE, SOHO, TRACE, Cluster, Image, NASA IBEX

12. Asteroid Missions

Why are asteroid types diverse? What is the composition? Which, if any, are planetesimals? How pristine? Do they contain interstellar grains from before the solar system? Is there any water? What is their origin? Which asteroids do meteorites come from? Might they be a future source of raw materials?

Table 1 (above) shows a list of the past and future missions approved over the next decade or so. The natural sequence of solar system missions involves four stages: (1) initial reconnaissance by flyby, (2) detailed study by orbiter, (3) direct measurement of atmosphere or surface via entry probe and (4) sample return. The stage we have reached for each body is also shown in Table 1.

We are at a particularly exciting time for solar system exploration. Recently, the first proof that there was once at least one standing body of liquid water on Mars has been found be NASA's Opportunity probe. Europa, Jupiter's Moon, has a liquid ocean under an icy crust. We are about to explore Titan, which has an atmosphere similar to that of the early Earth, in detail. And new bodies are being found in the Kuiper belt and beyond. At the same time, extremophile life has been found under Earth's surface. The environments for life will be one of the key areas for future research and space missions within the next several decades. Currently, Earth is the only body on which life is known. I suspect that evidence for life elsewhere will be found within the next decade.

13. Planetary Mission Example – Cassini-Huygens

As an example of a planetary mission we consider the Cassini-Huygens mission to Saturn. The Cassini orbiter, with a mass of 5.5 tonnes at launch and as large as a bus, represents the last of NASA's big planetary missions at the present time, while Huygens fitted nicely as a medium-sized ESA mission.

The scientific objectives once Cassini-Huygens arrives at Saturn after its long journey, performing gravity assists at Venus (twice), Earth and Jupiter can be split into five categories:

   (i) Saturn – atmosphere
  (ii) Saturn – rings
 (iii) Titan
 (iv) Icy satellites
  (v) Magnetosphere

Clearly Huygens will contribute substantially to (iii), but the Cassini orbiter during its four year tour will contribute to all five. There are many exciting questions in each area such as:

- What is the structure of the planet? Why does Saturn emit 80% more heat than incident on it from the Sun? Why do winds blow at 2/3 the speed of sound at the equator? What colours the clouds and why is this less marked than Jupiter?
- What is the composition of the ring particles? How do the dynamics work, what is the role of gravity, shepherding, electrostatics, collisions, etc? How did the rings form?
- Do the cold temperature and the atmospheric constituents mean that there is methane-ethane precipitation on Titan? What is the state of the surface? What atmospheric layers does this moon have and why is the surface pressure so high at 1.5 bar? How does the photochemistry work in the hydrocarbon haze? What is the role of magnetospheric electrons in the temperature balance in the ionosphere and atmosphere? Does Titan have a magnetic field?
- Why are Saturn's satellites so diverse? Why are some of the icy surfaces craters, others reformed? Why are some surfaces, such as half of Iapetus, dark? Do any have a magnetic field?
- Is Saturn's magnetosphere intermediate between Jupiter (rotation-driven) and Earth (solar wind-driven)? What difference does size make in the convection system? What is the interaction with the rings, may it explain the ring spokes? What is the interaction with Titan, in the magnetosphere and when in the solar wind?

To answer these questions and others, the orbiter carries 12 instruments: 6 remote sensing and 6 in-situ. The Huygens probe also carries 6 instruments. There is UK participation in 6 of the orbiter instruments and 2 on the probe.

Since Cassini-Huygens' arrival at Saturn in July 2004, a wealth of data is being provided to the world-wide planetary community. The results are revolutionising our knowledge of Saturn and its system, and will tell us more about the origin of the solar system.

14. Other Solar Systems

One of the most fundamental questions facing mankind as the new millennium begins is – Has there been or could there be life elsewhere? The question has been haunting us for at least two millennia, but the answer is now closer given recent developments in technology which are overcoming the vast difficulties for our observations. The answer will have profound and exciting implications for scientific, cultural, philosophical and religious thinking. Most scientists in the field now believe there was insufficient evidence for NASA's 1996 announcement of early Life on Mars. Nevertheless, the announcement of the result was introduced by the President of the United States, generated huge interest and drew comment from leaders in many fields in addition to science. This illustrates the importance of the answer to humankind. Without life elsewhere, we feel alone and isolated in the Universe. Whether life has or has not evolved elsewhere yet, we would like to know why.

As we look into the night sky, at the 3,000 or so visible stars, it is natural for us to speculate on life elsewhere. The knowledge that our galaxy alone is some 100,000 light years across, contains some 200 billion stars, and that the Universe contains some ten times as many galaxies as the number of stars in our Milky Way, makes the presence of life elsewhere seem possible and indeed likely. Attempts have been made to quantify this, and the Drake Equation is still the best way of writing down the number of contemporaneous civilizations in our galaxy as the product of seven terms: the formation rate of stars, the fraction of those with planets, the number per solar system that are habitable, the fraction where life has started, the fraction with intelligence, the fraction with the technology and will to communicate, and the lifetime of that civilisation. Reasonable estimates for all of these parameters from astronomy, biology and sociology lead to values between zero and millions of civilizations within our own galaxy. Although we only know the first term with any accuracy as yet, my guess would be towards the higher end. But without firm evidence it is impossible that we will ever be able to determine most of these terms accurately enough and the final answer will remain indeterminate by this approach. Unfortunately it seems that we are in the realm of speculation rather than science, creating a ripe area for science fiction.

Attempts to detect life elsewhere have included passive searches for organized signals and active approaches including the transmission of powerful microwave bursts and attaching plaques to the Pioneer and Voyager spacecraft leaving the solar system. Despite recently enlisting the assistance of much computing power around the world, the passive search approach feels forlorn, but is still worth a try. As for our own electromagnetic transmissions they have reached a very small part of our galaxy in the 100 years since radio was invented and in the 26 years since the Arecibo telescope was used to shout to the cosmos 'we are here'. But we cannot get around the inverse square law for intensity of electromagnetic waves, making detection of our signals highly challenging – and in any case it would take a significant time to get an answer. Discounting an enormous stroke of luck, a more scientific and systematic approach is likely to be via the detection and analysis of extrasolar planets by remote sensing. The technology for this is just becoming available.

In the last few years we have seen the first tantalising evidence for other solar systems than our own, in different stages of formation. In 1984, the star Beta Pictoris was seen to be surrounded by a cloud of gas and dust reminiscent of early solar system models. Other examples of dust-gas disks around stars have been detected recently using the Hubble Space Telescope. Have planetesimals, or comets, formed near these stars? Are observed structures in the cloud evidence for forming planets? We should begin to be able to answer these questions soon, and we should find many other examples.

So far, planet-hunters have found firm evidence for over 160 planets around Sun-like stars, and several additional candidates. There are even a handful of multiple planet systems. The main technique used for this is to detect the anomalous motion, or 'wobbling', of the companion star, using large ground-based telescopes, although in one reported case to date an occultation, or reduction of the companion star's light due to the object crossing its disk (a transit, as those of Venus and Mercury which are visible from Earth), was detected. The planets discovered so far are all inferred to be massive gas giants, most are several times heavier than Jupiter and the smallest so far is the size of Saturn. All orbit closer to the star than the giant planets in our own solar system, the main hypothesis for this being that they form further out and lose orbital energy via friction with the remaining gas and dust. Multiple large gas planets around another star have also been inferred from observations. One problem in the identification of any extrasolar planets is to distinguish them from 'brown dwarf' companions, failed stars which were too light for fusion to start. Telescope technology is advancing, and dedicated observational space missions using the transit method (COROT, Kepler, Eddington), using the radial velocity or wobble method (Eddington, Gaia) later, using interferometry (Darwin, TPF) have been proposed and will be flown within the next decade. We must choose the most promising stars to observe, starting with stars like our own Sun. We can expect that the present catalogue will increase soon and significantly, the observable planetary size will decrease becoming closer to the Earth's size, accurate distinction techniques between planets and brown dwarves will be developed and statistics will be built up on size distributions and orbits. We will ultimately know whether our own solar system is unique.

Given the, in my opinion, likely existence of small rocky planets elsewhere, and the fact that we are most unlikely to have the technology to make *in situ* measurements, we will need to establish methods for detecting the presence of life on these objects using remote sensing. Using data from the Galileo spacecraft during its Earth swingbys, Carl Sagan and colleagues detected the presence of life on Earth using atmospheric analysis and radio signal detection. The challenge is enormously greater for planets of remote stars. However, we may speculate that spectroscopic observations of ozone, oxygen and methane may hold the key, at least to finding clues for life as we know it. But it is unlikely we will be able to prove conclusively that life exists there. There is an optimistic view, shared by many scientists including myself. Surely amongst the many planets elsewhere there must be at least some which have similar conditions to the Earth? The accepted critical planetary properties for life to emerge are size, composition, stellar heat input and age. To this list we should add the presence of a magnetic field to protect against stellar winds and radiation.

Within our own solar system, where *in situ* exploration is possible, we must search for evidence of life on Mars during its early, warmer, wetter history. Following the loss of Beagle 2, several new missions are planned with this in mind, and ESA's Inspirations initiative and Aurora programme, and NASA's Exploration Initiative, provide the context. In addition, two other possible sites for life within our own solar system sites should be explored. First, Europa may be a possible present site for life, in its liquid water ocean underneath an icy crust, deep enough to be shielded from Jupiter's powerful radiation belts. Second, Titan is a potential future site for life when the Sun exhausts its hydrogen fuel, becomes a red giant and warms the outer solar system. But as well as looking for tangible clues within our own solar system it is clear that we must compare with other solar systems and broaden the search.

15. Conclusions

In summary there are many exciting and challenging ways we can explore solar systems in the new millennium: our own with *in situ* studies and beyond with remote sensing. A compelling, cross-cutting theme is whether there may be life beyond the Earth. The answers to be gained from all these explorations are fundamental to a better understanding of our place in the universe.

16. Latest News

Cassini-Huygens has been orbiting Saturn since 1 July 2004. Discoveries have been made in several areas including:

Saturn: rotation period changed, and equatorial winds weaker than Voyager saw

Icy satellites: the retrograde moon Phoebe is a captured Kuiper-belt comet; a bulge at Iapetus' equator; an atmosphere at Enceladus

Magnetosphere: aurorae depend more on solar wind pressure than magnetic field orientation; dynamic injection events; inner magnetosphere is water dominated

Rings: dynamics and composition, and their interaction with moons and plasma; ring atmosphere and ionosphere dominated by molecular oxygen

Titan: hydrocarbons, neutral up to benzene and charged up to C8 series in the upper atmosphere; active cloud systems; cryovolcanism; no nitrogen torus

Huygens landing revealed dendritic river-like channels, dried up lake beds, ethane clouds, sub-surface methane, ice pebbles, atmospheric turbulence during descent

The successful Deep Impact mission removed pristine material from below the surface of comet Tempel-1 on 4 July 2005, watched by space- and ground-based observatories. Results will determine the strength of the cometary surface and interior, and the composition of the pristine material.

ESA's SMART-1 arrived at the Moon and started mapping the surface composition; NASA's Genesis parachute failed to open but some science with the collected solar wind samples will be possible; ESA's Rosetta is on the way to its target comet Churyumov-Gerasimenko in 2014; NASA's Messenger was launched towards Mercury, arrival 2009.

Further Reading

Lewis, J.S. 1997, *Physics and chemistry of the solar system*, London: Academic press.
Beatty, J.K., Petersen, C.C. & Chaikin, A. 1999, *The New Solar System*, Cambridge: Cambridge University Press.
Stern, S.A. 1992, The Pluto-Charon system, *Ann. Rev. Astron. Astrophys.*, **30**, 185–233.

Andrew Coates

Andrew Coates, born at Heswall, Cheshire, gained a BSc (first) in Physics from UMIST in 1978, and MSc (1979) and D.Phil. (1982) in plasma physics from Oxford University. He has been at UCL's Mullard Space Science Laboratory since. He was Royal Society University Research Fellow and is now Reader in Physics and Head of Planetary Science. He was guest scientist at MPAe, Germany and University of Delaware, and was a media fellow at BBC World Service. Space mission involvements include AMPTE, Giotto, Meteosat, Polar, Cluster, Double Star, Mars Express, Beagle 2, Venus Express, Rosetta and Cassini-Huygens where he leads the team which provided the electron spectrometer for the orbiter (shown in the photograph). Aged 47, he was recently on several PPARC committees and on the ESA Solar System Working Group. Scientific interests include the solar wind interaction with planets and comets and space instrumentation; he has over 100 publications. Interests include popularising science and helping bring up twin daughters.

CHAPTER 17

PLANETARY UPPER ATMOSPHERES

Ingo Müller-Wodarg

Space and Atmospheric Physics Group,
Imperial College London,
Prince Consort Road,
London SW7 2BW, U.K.
E-mail: i.mueller-wodarg@imperial.ac.uk

Earth and most planets in our solar system are surrounded by permanent atmospheres. Their outermost layers, the thermospheres, ionospheres and exospheres, are regions which couple the atmospheres to space, the Sun and solar wind. Furthermore, most planets possess a magnetosphere, which extends into space considerably further than the atmosphere, but through magnetosphere-ionosphere coupling processes closely interacts with it. Auroral emissions, found on Earth and other planets, are manifestations of this coupling and a mapping of distant regions in the magnetosphere into the upper atmosphere along magnetic field lines. This article compares planetary upper atmospheres in our solar system and attempts to explain their differences via fundamental properties such as atmospheric gas composition, magnetosphere structure and distance from Sun. Understanding the space environment of Earth and its coupling to the Sun, and attempting to predict its behaviour ("Space Weather") plays an important practical role in protecting satellites, upon which many aspects of todays civilisation rely. By comparing our own space environment to that of other planets we gain a deeper understanding of its physical processes and uniqueness. Increasingly, we apply our knowledge also to atmospheres of extrasolar system planets, which will help assessing the possibility of life elsewhere in the Universe.

1. Introduction

Earth, like most planets in our solar system, is surrounded by a permanent layer of gas, an atmosphere. When following the changes of temperature with altitude, as shown in Figure 1, we may identify regions of negative

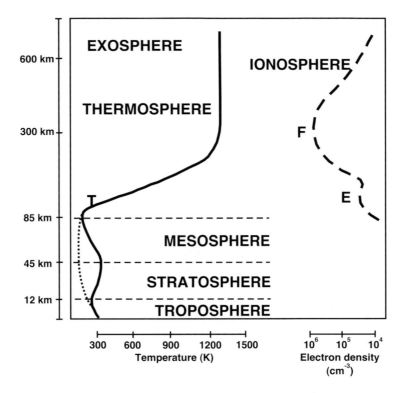

Fig. 1. The vertical thermal structure of the Earth's atmosphere and density structure of its ionosphere. Note that electron densities increase from right to left.

and positive temperature gradient which define layers in the atmosphere. From the surface to top these are the *troposphere* (0 - ~12 km), *stratosphere* (~12 - 45 km), *mesosphere* (~45 - 85 km) and *thermosphere* (above ~85 km). Part of the thermosphere is ionized, forming an *ionosphere*. The upper boundary of any "*~sphere*" is denoted by the similar word ending in "*~pause*", so the top of the mesosphere and bottom of the thermosphere is the *mesopause*. The upper thermosphere/ionosphere regime is the *exosphere*, a region (above ~700 km on Earth) where collisions between gas particles become rare and upward moving atoms can escape. The lower boundary of the exosphere is the *exobase*. While the bottom of the thermosphere is characterized by a sharp rise of temperature with height it becomes isothermal above a certain altitude. This height-independent temperature value is often referred to as the *exospheric temperature*.

What generates this distinct layer structure in our atmosphere? The main reason lies in the fact that our atmosphere consists of a variety of gases which not only each absorb different parts of the solar spectrum, but are also distributed non-uniformly with altitude, thus giving a height structure to the solar absorption and thereby heating rates. This along with the ability of some gases (eg., CO_2) to radiate in the infrared, cooling their surroundings, gives rise to the distinct atmospheric layer structure. As an example, the occurrence of a stratosphere on Earth is due to the presence of ozone which absorbs solar radiation between around 200 and 300 nm wavelength, heating that region. So, without ozone the vertical temperature profile would in that altitude regime instead adopt the dotted line (see Figure 1) and there would be no stratosphere. While we may identify the atmospheric layers of Figure 1 in many atmospheres of our solar system (Earth, Gas Giants, Titan), others (Venus, Mars, Triton, Pluto) are believed not to have a stratosphere, due to the lack of an "ozone-equivalent" constituent heating that region and causing a temperature profile inversion. Dominant physical processes differ between layers and one can at times study one region in isolation, but it is important to remember that all atmospheric layers are coupled through vertical transport of gases, energy or momentum. This adds considerable complexity to the study of atmospheres.

On the basis of dominant physical processes scientists generally distinguish between two broad domains, the lower and upper atmospheres, and the studies of these regions also carry separate names. For the case of Earth, the lower atmosphere consists of the troposphere and stratosphere, and the discipline exploring these is *meteorology*, while the upper atmosphere consists of the mesosphere, thermosphere, ionosphere and exosphere, which are studied in the discipline called *aeronomy*. The name "aeronomy", created by Sidney Chapman, was officially introduced in 1954, a few years before the launch of the first artificial satellite. The purpose of this interdisciplinary field is to study any atmospheric region (earth, planet, satellite, comet) where ionization and photodissociation processes play a role. Aeronomy examines processes that couple the atmospheres of planets and moons to the solar wind and ultimately to the Sun itself. This implies that any concept, method or technique developed for the terrestrial atmosphere can be adapted to other bodies of the solar system. Sometimes in the literature the stratosphere is seen as part of the upper atmosphere and thus included in aeronomy studies, but we will not consider stratospheric processes here. This article will highlight some of our current understanding of aeronomy, with an emphasis on comparison between different solar system bodies (*"comparative aeronomy"*). By seeing Earth or any other planet or moon in context with the rest of the solar system, we understand better their differences and may identify any features unique to a particular atmosphere. Testing our theories of atmospheric behaviour under the different environments of other planets not only enables us to learn about those worlds, but also urges us to revise some of our understanding of our own planet.

Recently, Mendillo *et al.*[1] identified the following as some of the key questions posed in solar system aeronomy: (1) What are the constituents of each atmosphere? (2) How do they absorb solar radiation? (3) What thermal structures result from heating versus cooling processes? (4) What types of ionospheres are formed? (5) What are the roles of atmospheric dynamics? (6) Does a planetary magnetic field shield the atmosphere and ionosphere from solar wind impact? (7) How do trapped energetic particles and electrodynamics affect the atmospheric system? Some of these questions will be reviewed in this article.

2. Methods of Exploring Planetary Atmospheres

The techniques for acquiring information about our own as well as other bodies' atmospheres have reached a high level of ingenuity and accuracy and merit a discussion far beyond the scope of this brief review. Readers are therefore referred to more comprehensive descriptions in the literature, discussing both historical and current techniques.[2-7] On a broad scale, we may distinguish between *remote sensing* and *in situ* observations, where an instrument either measures from a distance or within the environment itself. While in-situ measurements often give us more detailed information about specific processes, they are confined to the location of the instrument (path of the spacecraft carrying it), whereas remote sensing techniques give information on a broader spatial and temporal scale, and may often be easier and cheaper to accomplish. Most remote sensing measurements are made by detecting and spectrally analyzing radiation, whereas in-situ measurements can sample atmospheric particles directly in addition to detecting radiation. Instruments can measure in an *active* or *passive* way, in which they either detect emissions from the region of interest or send out a signal (eg., radar) and measure its return. By spectrally analyzing the radiation we learn about the atmospheric gas composition, its dynamics and the planet's magnetic environment.

Once observations have been made, sophisticated *data analysis* needs to be carried out since in most cases the parameters of interest (eg., temperature or composition of atmospheric gases) are not detected directly, but derived from instrument signals by making assumptions about the atmosphere or environment. Often, data analysis is a complex procedure which may take years and cause much controversy since different analyses make different assumptions and may obtain diverging results. One recent example for the range of possible results is the analysis of Voyager observations of Saturn's exospheric temperature; while the analysis of a stellar occultation experiment yielded a value of 800 K,[8] a solar occultation experiment on board the same spacecraft gave 400 K.[9]

Space probes which made important contributions to aeronomy in the solar system include Mariner (Venus, Mars), Venera (Venus), Pioneer Venus, Pioneer (Jupiter, Saturn), Voyager (Jupiter, Saturn, Uranus, Neptune), Galileo (Jupiter, Venus) and the on-going or future missions Ulysses (Sun, Jupiter), Mars Global Surveyor, Cassini (Saturn, Jupiter, Venus), Mars Express and Venus Express. Thanks to these spacecraft and a vast range of ground based observations exploring our own and other aeronomic systems over the past 3 decades, today we have an unprecedented understanding of atmospheres in our solar system. Observations are accompanied by detailed theoretical modelling work which takes advantage of the increasing computing capabilities to simulate the physics of planetary atmospheres. Such numerical models play a crucial role in helping us understand the observations and predict other, yet unobserved, features.

3. Atmospheres in the Solar System

In order for a planet or moon to possess an atmosphere, its solid body or dense inner core need to bound gravitationally the gas particles, preventing the bulk of them from escaping into space. By contrast, bodies such as comets possess what is called a coma, where gas particles escape into space continuously but are replenished, thus giving these bodies a transient atmosphere which will disappear as soon as the gas source vanishes. Other examples of bodies with transient atmospheres are our Moon, the planet Mercury and Jupiter's moons Io, Ganymede, Europa and Callisto. The exact composition of an atmosphere depends on the available gases at the time when the solar system was formed as well as the evolution under the influence of geology (via outgassing), impacts of other bodies, solar radiation and the solar wind. On Earth, the presence of life has also considerably influenced the atmospheric composition. Broadly, we may on the basis of their composition divide the non-transient atmospheres in our solar system into three categories, the Nitrogen atmospheres (Earth, Titan, Triton, Pluto), Carbon dioxide atmospheres (Venus, Mars) and Hydrogen atmospheres (Jupiter, Saturn, Uranus, Neptune). Mercury is the only planet not to have a permanent atmosphere, while Saturn's Titan and Neptune's Triton are the only moons with permanent atmospheres. Titan's atmosphere is particularly thick, with surface pressure values comparable to those on Earth. Pluto, the outermost planet, is thought to have an atmosphere similar to that of Triton, which partly condenses out when Pluto on its elliptical orbit is sufficiently far from the Sun and temperatures are low enough.

3.1. *Thermospheres*

As illustrated in Figure 1, a thermosphere is one of the outermost layers of an atmosphere, characterized by a steep vertical temperature gradient in its bottom region and height-independent temperatures further up. These large temperatures are a result of the highly energetic (Ultraviolet and X-ray) parts of the solar spectrum being absorbed by gases in a low density environment. Another feature of the thermospheric environment is the importance of vertical molecular conduction, as first pointed out for Earth in 1949 by Spitzer.[10] Energy is transferred by collisions between molecules, a process which is particularly important in regions of large temperature gradients, as that in the lower thermosphere. Energy deposited in the thermosphere by solar absorption or magnetospheric sources (see Sections 4, 5) is conducted down into the mesosphere. From there it is largely lost back into space by radiative cooling through vibrational transitions in polyatomic molecules, giving the mesopause its cold temperatures.[11]

Infrared cooling is also found to be important within the thermospheres of some planets, in particular Venus, Titan and Jupiter. On Venus, the high abundances of CO_2 causes strong cooling at 15 μm[12] and explains the cold exospheric dayside temperatures of below 300 K.[13] While Venus is the planet with an atmosphere located closest to the Sun, the energy is thus effectively radiated back into space, keeping exospheric temperatures relatively low. Infrared cooling on Titan is caused by HCN, a byproduct of ionospheric chemistry which has strong rotational bands and despite its low fractional abundance of less than ~0.1 % causes significant thermospheric cooling.[14] Molecular conduction on Titan only becomes important in the upper regions of the thermosphere.[14,15] In Jupiter's thermosphere an effective radiative coolant is H_3^+, the main ionospheric constituent. Since its

Table 1. Key properties of upper atmospheres on planets and moons in our solar system.

Body	Heliocentric distance [AU]	Siderial rotation period	Main neutral gases	Exospheric Temperature [K]	Main ions	Peak ion density [cm^{-3}]	Magnetic dipole moment (Earth=1)/tilt	Solar EUV, particle/Joule heating [10^9 W]
Venus	0.72	-243 days	CO_2, CO, N_2, O	100-300	O_2^+, CO_2^+, O^+	10^5	$<10^{-4}$/?	300, 0
Earth	0.98-1.02	23^h22^{min}	N_2, O_2, O	800-1400	O^+, O_2^+, NO^+	10^6	$1/10.8^o$	500, 80
Mars	1.4-1.7	24^h15^{min}	CO_2, CO, N_2, O	200-350	O_2^+, CO_2^+, O^+	10^5	$<10^{-4}$/?	25, 0
Jupiter	5.0-5.5	9^h55^{min}	H, H_2, He*	940	H^+, H_3^+	10^5	$20,000/9.6^o$	800, 10^5
Saturn	9.0-10.1	10^h39^{min}	H, H_2, He	420	H^+, H_3^+	10^4	$600/<1^o$	200, 200
Titan	9.0-10.1	15.95 days	N_2, CH_4	180	$C_xH_y^+$	10^3	0?	3, < 0.2
Uranus	18.3-20.1	17^h15^{min}	H, H_2, He	800	H^+, H_3^+	10^4	$50/58.6^o$	8, 100
Neptune	29.8-30.3	16^h7^{min}	H, H_2, He	600	H^+, H_3^+	10^3	$25/47^o$	3, 1
Triton	29.8-30.3	-5.87 days	N_2, CH_4	102	N^+?	10^4	?	0.05, 0.1
Pluto	29.7-49.3	-6.38 days	N_2, CH_4	100?	$HCNH^+$?	10^3?	?	0.05, ?

Values either unknown or estimated are marked with a "?". Negative rotation periods denote retrograde rotation. The table is adopted from Strobel.[31]

discovery on Jupiter in 1989,[16] H_3^+ has been imaged on several occasions with ground based telescopes, such as NASA's Infrared Telescope Facility on Mauna Kea, Hawaii,[17] and emissions were found to be concentrated in the auroral regions. H_3^+ emissions have also recently been discovered near the poles of Saturn,[18] but are less pronounced than on Jupiter. The principal gases in the Earth's thermosphere are O, O_2 and N_2, and the small amounts of CO_2 and NO there are insufficient on a global scale to cause significant infrared cooling under normal conditions, but locally and during geomagnetic storm conditions on Earth they have to be taken into account.

Table 1 summarizes key properties of upper atmospheres in the solar system, such as exospheric temperatures, major gas composition and heating rates. Apart from the variations in composition between the planets, note the exospheric temperatures and how their values change with distance from the Sun. The average exospheric temperature values, as well as their diurnal and solar cycle ranges, are also plotted in Figure 2 versus location in the solar system. While one might initially expect planets located closer to the Sun to be hotter, the trends in Figure 2 and Table 1 show that this is not generally the case. One important current topic of research in aeronomy is to understand this unexpected trend, and this will be discussed further in Section 5. The earlier discussion of infrared cooling in Venus' thermosphere already gave a reason for the low exospheric temperatures there.

Solar heating of the dayside thermosphere sets up day-night pressure gradients, which drive horizontal winds and thereby a system of global circulation. The nature of thermospheric winds depends amongst other things on the magnitude of day-night pressure gradients as well as the planet's rotation rate. On fast rotating planets (such as Gas Giants, see Table 1), there is too little time to build up substantial day-night temperature gradients, and the direction of thermospheric winds is furthermore influenced by the well known *Coriolis forces*. Thermospheric winds can furthermore be affected by the motion of ions, which respond to the presence of electrical fields (see Section 3.2), adding considerable complexity to thermospheric circulation. Thermospheric winds, in turn, transport gases, affecting the distribution of individual constituents. Thermospheric dynamics are studied in detail using *General Circulation Models*, and we now have such models for most planets in the solar system.

As noted previously, any atmospheric region should be seen not as an isolated regime, but coupled to its surroundings. What makes the thermosphere particularly complex is the variety of its coupling to surrounding areas. Forming one of the the outermost regions of an atmosphere, it "sits"

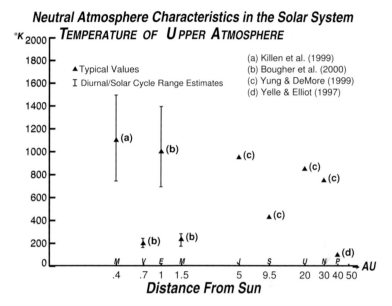

Fig. 2. Exospheric temperatures in the solar system versus distance from the Sun (from Mendillo *et al.*[20]).

on the entire atmosphere below and is subject to phenomena such as upward propagating waves. These waves have a profound influence on the Earth's lower thermosphere, accelerating winds there and thus also affecting the structures of temperatures and densities. Recently, observations have shown the possibility of phenomena such as global warming, which appear primarily in the lower atmosphere, to be detectable in the Earth's thermosphere and ionosphere as a result of vertical chemical and dynamical coupling.[19] The thermospheric environment is also coupled to interplanetary space through the magnetosphere which can generate substantial winds and heating (see discussions in 4 and 5). The coupling of thermospheres to regions as vastly apart as the lower atmosphere and solar wind or Sun itself via simultaneously occurring processes like chemistry, neutral gas and plasma dynamics as well as electromagnetism make them some of the most complex and fascinating environments to study.

3.2. *Ionospheres*

Energetic photons entering the upper atmosphere of Earth or other planets not only heat neutral gases (see Section 3.1), but are capable of ionizing them as well, generating a layer of plasma embedded in the thermosphere, called the ionosphere. The same process which causes the neutral atmosphere's vertical thermal structure is often also responsible for the creation of ionospheres with several separate layers of enhanced plasma density. Depending on the distribution of neutral gases with height, different parts of the solar spectrum ionize at varying altitudes, creating separate layers of ionization. The dashed curve in Figure 1 shows the structure of the Earth's ionosphere in terms of electron densities. Two distinct layers can be identified, namely the "*E-layer*" and "*F-layer*" which peak near 90 and 300 km altitude, respectively. Historically, these names derive from the fact that the first ionospheric layer to be discovered was named the "E-layer", where "E" stands for "electric", and other layers discovered subsequently below and above the E-layer were given names of letters before or after the letter "E" in the alphabet. The basic difference between these regions, apart from their electron densities and altitudes of occurrence, is that E-layer plasma is largely in photochemical equilibrium, whereas the F-layer is controlled both by photochemistry and plasma transport and thus not in photochemical equilibrium.[20] Dominant ions in the terrestrial E-layer are molecular ions such as O_2^+ and NO^+, whereas the F-layer near its main peak consists mainly of the atomic ion O^+. Molecular ions in the E-layer are in an environment of larger neutral particle densities than F-layer plasma, so they rapidly recombine after sunset and the E-layer largely disappears during the night. In the F-layer recombination of ions is far more inefficient and additional plasma is supplied from other regions by transport, so the F-layer survives through the night. Peak ion densities in the F-layer are usually less than 0.1 % of the ambient neutral gas densities, but we will show later (Section 4) that the ionosphere often plays a crucial role in thermospheric properties and provides the coupling between the thermosphere and magnetosphere. Note that not all ionospheres in the solar system contain all the layers identified on Earth.

Figure 3 along with Table 1 summarize some key ionosphere properties in the solar system, and we may identify two important features. Firstly, the Earth's is according to our current knowledge possibly the only ionosphere in which the main peak density (F-layer) ion is atomic (O^+), whereas on Venus and Mars it is O_2^+, on the Gas Giants H_3^+ and on Titan a hydrocarbon molecule ($C_xH_y^+$). The dominant ion on Triton may also be atomic (N^+ or C^+), but there is as yet too little observational evidence to confirm this unambiguously. Secondly, peak ion densities tend to decrease with distance from the Sun, except for the value on Earth, which is a factor of 5 larger than that of any other ionosphere in the solar system. It seems, therefore, that the Earth's ionosphere stands out amongst the planets. This difference appears unrelated to properties of its magnetic field, as listed in Table 1.

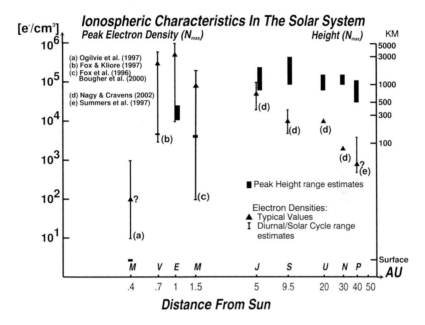

Fig. 3.   Ionospheric characteristics in the solar system versus distance from the Sun. On the Gas Giants (Jupiter, Saturn, Uranus, Neptune), "surface" is defined as the 1 bar pressure level in the atmosphere. (from Mendillo *et al.*[20]).

Recent observations have shown day-to-day variations in the Earth's E-layer to correlate well with those of the main ionospheric peak on Mars, and both these to reflect changes in the solar EUV flux during the same period.[21] The significance of this observation is that it demonstrated similarities between the Martian ionosphere and terrestrial E-layer, indicating that Mars has an "E-layer type" ionosphere which is controlled by photochemical equilibrium, as expected from an ionosphere dominated by molecular ions. The importance of molecular ions (Table 1) indicates that most other ionospheres in the solar system, like that of Mars, are likely to be of "E-layer type" and in photochemical equilibrium, with Earth being a special case where the E-layer is in photochemical equilibrium, but the main peak, the F-layer (being atomic), is not. While this statement is a reasonable description of the global behaviour of ionospheres in the solar system, it does not capture more detailed properties. Locally, winds or other magnetospheric coupling processes may disrupt photochemical equilibrium, and on most planets the ionosphere above the peak (the *topside ionosphere*) is not in photochemical equilibrium.

4. The Space Environment of Magnetized Bodies

4.1. *What is a magnetosphere?*

The thermospheres and ionospheres, as described in Sections 3.1 and 3.2, are coupled to space, the Sun and solar wind by vast regions beyond the atmospheres which surround magnetized bodies, called *magnetospheres*. Caused by the presence of magnetic fields, they form cavities which protect the planet or moon from the direct influence of the solar wind. In fact, life on Earth may not have evolved without the presence of our protecting magnetosphere. Figure 4 illustrates schematically the structure of the Earth's space environment. The internal magnetic dipole field without the presence of the solar wind would form field lines similar to those of a bar magnet. However, the solar wind, a stream of high energetic particles ejected from the Sun, affects the global shape of magnetic field lines surrounding Earth, compressing them on the sunward side (left half of Figure 4) and streching them out on the anti-sunward side into a *tail* (right half of Figure 4). The boundary layer where solar wind pressure and magnetic field pressure balance is the *magnetopause* (dashed line in Figure 4). The environment inside the magnetopause is the magnetosphere, so the magnetic field shields Earth from the bulk of the solar wind particles. While Figure 4 highlights some features specific to Earth, the schematic equally applies in principle to any magnetized body in the solar system. The eighth column in Table 1 compares magnetic dipole moments ("strengths") and magnetic field tilts between the various bodies in the solar system. A tilt of 0° would imply the "bar magnet" to be aligned with the rotational axis of the planet. We see from the values in Table 1 that the bodies in the solar system with significant magnetic fields, and thus magnetospheres, are Earth, Jupiter, Saturn, Uranus and Neptune. Titan and Triton are special cases - they do not possess intrinsic magnetic fields, but their orbits lie within the magnetospheres of Saturn and Neptune, respectively, so they have *induced* magnetospheres.

4.2. *Aurora*

One particular manifestation of the Earth's coupling to the solar wind via the magnetosphere are the spectacular Northern/Southern Lights or *aurora*

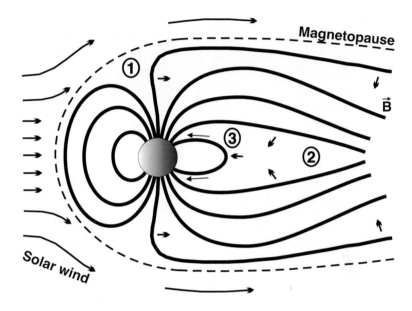

Fig. 4.   Schematic of the Earth's magnetosphere and its interaction with the solar wind.

borealis/australis which, seen at night from the ground, are vast curtains of light changing rapidly their shape and intensity (Figure 5). These auroral emissions appear to the naked eye in green and red colours and seem so violent and mysterious that they have captured human curiosity and fear throughout history, being given an almost mythological status and interpretation. Overall, aurora appear in the form of ring shaped bright regions roughly centred around the north and south magnetic poles of a planet, the *auroral ovals*. A scientific exploration into the origins and nature of the Earth's aurora began in the 18th and 19th centuries,[22,23] and more powerful observational techniques led to the discovery of auroral emissions also on other planets, such as Jupiter[24] (Figure 6) and Saturn.[25] Other forms of atmospheric emissions that appear less spatially defined and less bright than the polar, highly variable aurora are summarized as *airglow*. Based on differences in the creation mechanisms one may define an aurora as being *any optical manifestation of the interaction of extra-atmospheric energetic electrons, ions and neutrals with an atmosphere.*[26] What distinguishes airglow from aurora is the primary process causing them: airglow is generated by solar photons, whereas aurora is generated by energetic particles from outside the atmosphere (which excludes photoelectrons, a by-product of

Fig. 5.   Auroral emissions, as seen from the ground (Photo courtesy of Jouni Jussila).

some chemical processes inside the atmosphere). Auroral emissions can be observed at visible, infrared, UV and, for Jupiter and Earth, at X-ray wavelengths.

4.3. *Magnetosphere-Ionosphere coupling*

The processes which lead to auroral emissions on Earth and other magnetized bodies form part of a field referred to as *Magnetosphere-Ionosphere coupling*. What distinguishes planets are the atmospheric composition and the processes accelerating the energetic particles entering an atmosphere. So, the study of auroral emissions gives us important information about a planet's atmosphere and magnetosphere, which will be reviewed below. This comparison will show that aurora on Earth originates from coupling to the solar wind, while on Jupiter they originate primarily from the fast rotation of the planet and the effect this has on the magnetosphere. Not enough is known yet about the origins of Saturn's aurora, but the planet may be an "in-between case" between Earth and Jupiter.[27] Realizing differences such as these is an important step towards a more comprehensive understanding of the planets in our solar system.

4.3.1. *Earth*

On Earth, changes in auroral brightness and shape are related to disturbances in the geomagnetic field which are induced by its interaction with the solar wind, triggered for example by solar flares or coronal mass ejections. At times, these disturbances are particularly violent and lead to *geomagnetic storms*, which manifest themselves as periods (from below an hour to a day) of particularly bright auroral displays and an expansion of the auroral oval, making it visible at times from mid latitudes. The nature of solar wind-magnetosphere interaction on Earth during such events is illustrated in Figure 4. The solar wind "drags" magnetic field lines across the poles (marked as "1" in Figure 4) towards the tail region, where these contract into the tail ("2") and recombine into loops (closed field lines), which convect inward towards the nightside of the planet, accelerating plasma along the field lines ("3") mapping into the high latitude regions of the atmosphere and causing auroral emissions. Enhanced aurora on Earth is thereby a direct result of Sun-Earth coupling via the solar wind.

Understanding this response of the Earth's upper atmosphere to solar disturbances is an active field of research called *Space Weather* and has important practical applications. The strong currents, electric fields and fluxes of incident energetic particle in the upper atmosphere during geomagnetic storms are capable of severely disrupting satellites and even power grids on the ground, so it has become desireable to be able to predict these events in order to switch off satellites or otherwise protect sensitive electrical equipment.

4.3.2. Jupiter

Jupiter's magnetosphere is the largest structure in our solar system after the Sun's own atmosphere, extending out by up to 100 Jupiter radii (R_J) into space. If it were visible with the naked eye in its entirety from Earth, it would have a radius in the sky larger than the Moon, despite its much larger distance from Earth. Earth based observations of Jupiter's aurora as well as in-situ measurements of field strength by the Pioneer, Voyager, Ulysses, Galileo and, most recently, Cassini spacecraft have given many clues about its structure and processes, suggesting Jupiter's magnetosphere to be different in many ways to that of the Earth. While plasma in the Earth's magnetosphere originates mainly from the atmosphere and partly from the solar wind, that in Jupiter's magnetosphere originates primarily from the Galilean moons, in particular the volcanic moon Io which injects > 1 tonne of material (mainly SO_2) per second into space. This is then ionized by solar EUV radiation or energetic particle collisions. The moon Europa, through bombardment of its surface ice by energetic magnetospheric particles, ejects around 50 kg of atomic oxygen per second, while only the light ions (H^+, He^{++}) originate from Jupiter's upper atmosphere. The dominant mechanism accelerating plasma in Jupiter's magnetosphere is not the solar wind (like for Earth), but a break-down of the magnetosphere's co-rotation and the disturbance that this breakdown generates.[28] Jupiter rotates once every 10 hours and up to a distance of around 20 Jupiter radii the plasma co-rotates at the same angular velocity. Beyond that distance, however, co-rotation cannot be maintained and the outer regions of Jupiter's magnetosphere "lag behind" the inner ones. The co-rotation break-down leads to acceleration of plasma which along magnetic field lines travels towards the planet and enters the atmosphere, generating auroral emissions. The exact morphology of these complex interactions is currently an active subject of study, with numerous theoretical models being developed to understand the latest observations.[28,29] Figure 6 shows an artist's impression of Jupiter's magnetosphere and illustrates the plasma flow along magnetic field lines between the Galilean satellites, in particular Io, and Jupiter's upper amosphere, where it creates "foot prints" of the moons in the auroral emissions.

Both on Earth and Jupiter, as well as other magnetized bodies, it is important to realize that the horizontal structures observed in auroral emissions in the atmosphere along magnetic field lines map to different regions in the magnetosphere. As can be seen from the schematic in Figure 4, field lines crossing the equatorial plane closer to the planet map into more

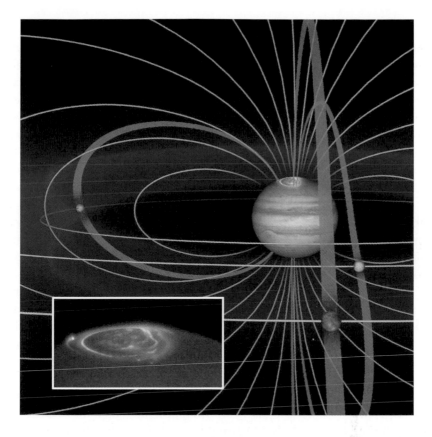

Fig. 6. Jupiter's space environment, with moons Io, Europa and Ganymede (insert: Jupiter's aurora, as seen in the UV). (Illustration courtesy of John Clarke.)

equatorial regions, whereas those extending out to larger distances map into higher latitudes. The auroral emissions are thus a direct projection of magnetospheric processes into the atmosphere. One striking example can be seen in the auroral maps of Jupiter, such as those derived from Hubble Space Telescope (HST) observations in the UV (insert in Figure 6): we can clearly identify bright patches in the aurora which are magnetic footprints of the Galilean satellites Io (above the left hand limb), Europa and Ganymede (bright dots near the central meridian).[30] These are striking examples of how seemingly uncoupled regions, such as moons and the planet's atmosphere, are indeed closely linked through the presence of the seemingly invisible magnetosphere.

5. Energy Crisis in the Solar System

Another important on-going field of research in solar system aeronomy is to understand the exospheric temperatures on planets. As discussed earlier (see Section 3.1) and shown in Figure 2 and Table 1, exospheric temperatures do not generally decrease with distance from the Sun. Venus is colder than expected from its location close to the Sun, whereas Gas Giants are far hotter than we would expect them to be. It is evident that the intensity of solar radiation alone is insufficient to understand this. In Section 3.1, CO_2 cooling was found to explain Venus' cold temperatures, but what processes explain the hot thermospheres of Gas Giants, in particular Jupiter? This is an outstanding problem in our understanding of the outer planets. The two currently most favoured processes which could additionally heat the upper atmospheres of Gas Giants are (1) deposition of energy from the solar wind and magnetosphere into the upper atmosphere, (2) heating due to upward propagating waves.[33]

Magnetosphere-Ionosphere coupling on Earth (Section 4.3.1) not only creates the spectacular auroral emissions, but also deposits significant amounts of energy into the upper atmosphere through the precipitating particles themselves and, secondly, a process called *Joule heating*. Essentially, Joule heating is equivalent to the creation of thermal energy by a current flowing through a resistor. The currents on Earth are driven by electric fields generated by the interaction between the solar wind and magnetosphere (see Section 4.3.1) and mapped into the auroral latitudes. Joule heating is particularly effective during geomagnetic storms, at times doubling local temperatures. Furthermore, the collisions of fast ions with slower ambient neutrals act to accelerate the neutral gas, creating strong thermospheric winds. Could similar processes play a role on other planets?

The right column in Table 1 compares heating rates in the upper atmospheres of Earth and other planets due to solar EUV absorption, particle precipitation and Joule heating. The given values are global averages and the balances of these heating processes may vary locally, in particular in the regions of precipitation and Joule heating. The absence of induced electrical fields, and thereby Joule heating on unmagnetized planets (Venus, Mars) makes solar heating the dominant energy source. Titan and Triton with their induced magnetospheres are special cases and magnetospheric energy sources are thought to play a role primarily on Triton. The moon is exposed to Neptune's magnetosphere which, due to the large tilt of 47° between magnetic and rotational axes, is particularly dynamic.[34] The interaction between Titan's atmosphere and Saturn's magnetosphere will be examined in detail with the forthcoming Cassini observations between 2004 and 2008. The planets in Table 1 where particle precipitation and Joule heating clearly play an important role from a global perspective are Jupiter and Uranus, followed by Saturn, Neptune and Earth. On Jupiter, magnetospheric energy sources are thought to exceed solar EUV heating rates by a factor of $\sim$100, on Saturn the two rates are comparable in magnitude and on Earth, being closer to the Sun, solar EUV heating is on average $\sim$6 times stronger. The outer planets where magnetospheric heating processes are most important are also those with the largest exospheric temperatures, so this trend clearly suggest one important "missing link" in our understanding of the temperature trends of Figure 2 throughout our solar system to be these magnetospheric energy sources. Many detailed questions remain unresolved, in particular how this magnetospheric energy, deposited at high magnetic latitudes, can be globally distributed on planets like Jupiter and Saturn, whose fast rotation hinders such transport by winds.

Heating from below by upward propagating waves would solve the problem of energy transport into equatorial regions, and much work has gone into investigating the possibility of waves as an energy source for the upper atmosphere. While for planets Earth, Venus and Mars the importance of waves as a source of momentum in their upper atmospheres has been demonstrated, wave dissipation on these planets releases only insignificant amounts of thermal energy. For Jupiter, some studies have suggested wave heating to play a key role, while others have contradicted this finding[33]. Further observations and more accurate calculations are necessary to solve this issue.

6. Looking Beyond: Extrasolar Planets

One fundamental question that has long intrigued humans on Earth is whether other forms of life are present in our solar system and beyond. Within our solar system, the search for life has concentrated on Mars and will advance to Europa an possibly Titan. The first discoveries in 1992 of a planet orbiting a pulsar, and in 1995 of a planet orbiting a "normal" star (51 Pegasi) have given the question of life outside our solar system a new dimension. If other planets exist in other solar systems, what is the likelihood of finding life on these? In this context much effort has gone into determining whether some of the now more than 100 observed extrasolar planets possess atmospheres and, if so, then how these compare to the atmospheres in our solar system, and whether they could support forms of life. In 2000, a pioneering observation discovered for the first time spectro-scopically lines of Na and H during the transit of planet HD209458b in front of its star, and thereby detected for the first time directly the atmosphere of an extrasolar planet.[35] Most of the extra-solar planets which we can currently detect, including HD209458b, are Jupiter-like, but orbiting close to the star, with semimajor axes below 0.1 A.U. These properties not only pose new challenges to our theories of solar system formation, but also to our understanding of their atmospheres. Will these atmospheres be similar to those of Gas Giants in our solar system? What exospheric temperatures will they have, and what controls their energetics? What chemical and dynamical processes will occur, and what is the nature of their coupling to internal or induced magnetospheres? While these questions cannot yet be answered through observations, first calculations are now being carried out to understand the aeronomy of such planets.[36] While still struggling to understand the aeronomy of planets in our own solar system, we may already apply our current knowledge and experience to extrasolar planets, predicting some key properties and working towards more accurate observations to help us understand not only Earth in context with other planets of our solar system, but our solar system in context with others. Aeronomy thus promises to capture our curiousity, inspire our imagination and demand our efforts for many decades to come.

Acknowledgments

This work was supported by the British Royal Society as part of my University Research Fellowship. My special gratitude goes to Marina Galand for her detailed and valuable comments on the draft. I would also like to thank Michael Mendillo for his great advice on the manuscript and for providing me with two of the figures, and finally I also thank John Clarke and Jouni Jussila for providing me with further figures.

References

1. M. Mendillo, A. Nagy and J. H. Waite, Introduction, in Atmospheres in the Solar System: Comparative Aeronomy, *Geophys. Monogr. Ser.* **130** edited by M. Mendillo, A. Nagy, and J. H. Waite, AGU, Washington, D. C. (2002).
2. H. Rishbeth and O. K. Garriott, Introduction to Ionospheric Physics, *Academic Press*, New York (1969).
3. M. H. Rees, Physics and Chemistry of the Upper Atmosphere, *Cambridge University Press*, New York (1989).

4. T. Encrenaz, J.-P. Bibring, M. Blanc, M.-A. Barucci, F. Roques and Ph. Zarka, The Solar System, 3rd edition, *Springer*, Berlin (2004).

5. M. Mendillo, F. Roesler, C. Garnder and M. Sulzer, The application of Terrestrial Aeronomy Groundbased Instruments to Planetary Studies, in Atmospheres in the Solar System: Comparative Aeronomy, *Geophys. Monogr. Ser.* **130** edited by M. Mendillo, A. Nagy, and J. H. Waite, AGU, Washington, D. C. (2002).

6. J. T. Clarke and L. Paxton, Ultraviolet Remote Sensing Techniques for Planetary Aeronomy, in Atmospheres in the Solar System: Comparative Aeronomy, *Geophys. Monogr. Ser.* **130** edited by M. Mendillo, A. Nagy, and J. H. Waite, AGU, Washington, D. C. (2002).

7. D. T. Young, Mass Spectrometry for Planetary Science, in Atmospheres in the Solar System: Comparative Aeronomy, *Geophys. Monogr. Ser.* **130** edited by M. Mendillo, A. Nagy, and J. H. Waite, AGU, Washington, D. C. (2002).

8. M. C. Festou and S. K. Atreya, Voyager ultraviolet stellar occultation measurements of the composition and thermal profiles of the Saturnian upper atmosphere, *Geophys. Res. Lett.* **9**, 1147 (1982).

9. G. R. Smith, D. E. Shemansky, J. B. Holberg, A. L. Broadfoot, B. R. Sandel, and J. C. McConnell, Saturns upper atmosphere from the Voyager 2 EUV solar and stellar occultations, *J. Geophys. Res.* **88**, 8667 (1983).

10. L. Spitzer, The Atmospheres of the Earth and Planets, G. P. Kuiper, ed., *Univ. Chicago Press* 211 (1949).

11. J. W. Chamberlain, Upper atmospheres of the planets, *Astrophys. J.* **136**, 582 (1962).

12. S. W. Bougher, D. M. Hunten and R. Roble, CO_2 cooling in terrestrial planet thermospheres, *J. Geophys. Res.* **99**, 14609 (1994).

13. J. L. Fox and S. W. Bougher, Structure, luminosity and dynamics of the Venus thermosphere, *Space Sci. Rev.* **55**, 357 (1991).

14. R. V. Yelle, Non-LTE models of Titan's upper atmosphere, *Astrophys. J.* **383**, 380 (1991).

15. I. C. F. Müller-Wodarg, R. V. Yelle, M. Mendillo, L. A. Young and A. D. Aylward, The thermosphere of Titan simulated by a global three-dimensional time-dependent model, *J. Geophys. Res.* **A9**, 20833 (2000).

16. P. Drossart (and 11 others), Detection of H_3^+ on Jupiter, *Nature* **340**, 539 (1989).

17. J. E. P. Connerny and T. Satoh, The H_3^+ ion: a remote diagnostic of the jovian magnetosphere *Phil. Trans. R. Soc. Lond.*, **A358**, 2471 (2000).

18. T. Stallard, S. Miller, G. E. Ballester, D. Rego, R. Joseph and L. Trafton, The H_3^+ latitudinal profile of Saturn, *Astrophys. J. Lett.* **521**, L149 (1999).

19. M. J. Jarvis, B. Jenkins and G. A. Rodgers, Southern hemisphere observations of a long-term decrease in F region altitude and thermospheric wind providing possible evidence for global thermospheric cooling, *J. Geophys. Res.* **103**, 20774 (1998).

20. A. Nagy and T. Cravens, Solar system ionospheres, in Atmospheres in the Solar System: Comparative Aeronomy, *Geophys. Monogr. Ser.* **130** edited by M. Mendillo, A. Nagy, and J. H. Waite, AGU, Washington, D. C. (2002).

21. M. Mendillo, S. Smith, J. Wroten, H. Rishbeth and D. Hinson, Simultaneous ionospheric variability on Earth and Mars, *J. Geophys. Res.* **108** 1432, doi:10.1029/2003JA009961 (2003).

22. J. J. D. de Mairan, Traité physique et historique de l'aurore boréale, *Imprimerie Royale*, Paris (1733).

23. A. J. Ångström, Recherches sur le Spectre Solaire, *W. Schultz* 41 (1868).

24. A. L. Broadfoot *et al.*, Extreme ultraviolet observations from Voyager 1 encounter with Jupiter, *Science* **204**, 979 (1979).

25. D. L. Judge, Wu, F.M., Carlson, R.W., Ultraviolet photometer observations of the saturnian system, *Science* **207**, 43 (1980).

26. M. Galand and S. Chakrabarti, Auroral Processes in the Solar System, in Atmospheres in the Solar System: Comparative Aeronomy, *Geophys. Monogr. Ser.* **130** edited by M. Mendillo, A. Nagy, and J. H. Waite, AGU, Washington, D. C. (2002).

27. S.W.H. Cowley, E.J. Bunce, and J.M. ORourke, A simple quantitative model of plasma flows and currents in Saturns polar ionosphere, *J. Geophys. Res.*, **109**, A0521210.1029/2003JA010375 (2004).

28. S. W. H. Cowley and E.J. Bunce, Origin of the main auroral oval in Jupiters coupled magnetosphere-ionosphere system, *Planet. Space Sci.* **49**, 1067 (2001).

29. S. W. H. Cowley and E.J. Bunce, Corotation-driven magnetosphere-ionosphere coupling currents in Saturns magnetosphere and their relation to the auroras, *Ann. Geophysicae* **21**, 1691 (2003).

30. J. Clarke (and 10 others), Ultraviolet emissions from the magnetic footprints of Io, Ganymede and Europa on Jupiter, *Nature* **415**, 998 (2002).

31. D. F. Strobel, Aeronomic systems on planets, moons and comets, in Atmospheres in the Solar System: Comparative Aeronomy, *Geophys. Monogr. Ser.* **130** edited by M. Mendillo, A. Nagy, and J. H. Waite, AGU, Washington, D. C. (2002).

32. J. W. Chamberlain and D. M. Hunten, Theory of Planetary Atmospheres, 2nd Ed., *Academic Press*, London (1987).

33. R. V. Yelle and S. Miller, Jupiter's thermosphere and ionosphere, in Jupiter: Planet, Satellites & Magnetosphere, F. Bagenal, W. McKinnon, and T. Dowling, eds., *Cambridge University Press*, in press (2004).

34. M. H. Stevens, D. F. Strobel, M. E. Summers and R. V. Yelle, On the thermal structure of Triton's thermosphere, *Geophys. Res. Lett* **19**, 669 (1992).

35. D. Charbonneau, M. Brown, R. W. Noyes and R. L. Gilland, Detection of an extra-solar planet atmosphere, *ApJ* **568**, 377 (2002).

36. R. V. Yelle, Aeronomy of extra-solar Giant Planets at Small Orbital Distances, *Icarus*, **170**, 167 (2004).

Ingo Müller-Wodarg

Ingo Müller-Wodarg was born in Copenhagen in 1969 and grew up in Denmark, Sweden, Germany, Egypt and Italy. He studied Physics at the University of Munich, Germany, and moved to London in 1991. At University College London he obtained an M.Sc. in Space Science in 1992 and a PhD in Physics in 1997. For his PhD studies he examined processes in the Earth's upper atmosphere and started using and developing complex computer models. In 1998 he was invited to join a project led by Boston University to develop a similar model for Saturn's largest moon Titan. This move into planetary sciences was to dominate his research ever since, and he not only has made Titan's atmosphere his current focus of attention, but adopted his model to Saturn and Neptune's moon Triton, with other planets to follow. Ingo's work has brought him international reputation in his field and he spends much of his time traveling, collaborating with numerous groups in the U.K., U.S. and rest of Europe. In 2002 Ingo was awarded a Royal Society University Research Fellowship and in 2003 joined the Space and Atmospheric Physics Group at Imperial College London. He also holds an honorary post as Senior Research Associate at Boston University. Since 2002 he is Chair of the UK Planetary Forum and keen on promoting planetary work carried out in the U.K. In his free time Ingo enjoys classical music and loves to spend time in nature, away from the big cities.

CHAPTER 18

THE SOLAR DYNAMO

Steven Tobias

Department of Applied Mathematics, University of Leeds, Leeds LS2 9JT, U.K.
E-mail: smtmaths.leeds.ac.uk

In this article I shall review the fundamentals of solar dynamo theory. I shall describe both historical and contemporary observations of the solar magnetic field before outlining why it is believed that the solar field is maintained by a hydromagnetic dynamo. Having explained the basic dynamo process and applications of the theory to the Sun, I shall conclude by speculating on future directions for the theory.

1. Introduction

It is an exciting time for research into the solar dynamo – the problem of the origin of the Sun's magnetic activity. The beginning of the new Millennium has seen great increases in both theory (made possible by the breathtaking advances in computational power) and high resolution observations (driven by the recent and planned space missions). Taken together, these advances have significantly increased our understanding of the magnetic activity and variability of our nearest star.

Magnetic activity is found in many astrophysical bodies on all length-scales, from planets, through stars to galaxies. This article is concerned with the generation of the magnetic field in our nearest star – the Sun. Explaining the origin of the Sun's magnetic field is the fundamental problem of solar magnetohydrodynamics. This dynamic magnetic field is responsible for all solar magnetic phenomena such as solar flares, coronal mass ejections and the solar wind, and also heats the solar corona to extremely high temperatures. These phenomena all have important terrestrial consequences, causing severe magnetic storms and major disruption to satellites, as well as having a possible impact on the terrestrial climate.

In this article I will outline the complementary roles of observations and theory in aiding our understanding of the generation of the solar field. A more technical and in-depth review of solar and stellar dynamo theory can be found in Weiss (1994) and Weiss & Tobias (2000). In the next section I will review the observations of solar magnetic activity and explain the arguments for believing that the origin of this activity lies in dynamo action. In Section 3, I will explain the basic concepts of dynamo theory, before discussing the application of this theory to the Sun in Section 4, showing how dynamo models are capable of reproducing the basic features of solar magnetic activity. I will conclude by speculating on future developments in solar dynamo theory and discussing possible directions for further research.

2. Observations

Sunspots are the most striking manifestations of the solar magnetic field. These cool dark patches on the solar surface are large enough to be visible with the naked eye (a typical sunspot is comparable in size to the Earth). Although sunspots were systematically observed by Chinese astronomers from the late 1st Century B.C. to the Middle Ages, sunspots were only rediscovered in the west with the advent of telescopes in the 17th century. Galileo and Scheiner made use of the invention of the telescope to conduct programs of observations of sunspots. For over a century observers continued to document the incidence of sunspots, but it was only in 1843 that the systematic properties of sunspots became apparent. Samuel Heinrich Schwabe, who had been meticulously observing the Sun in order to discover the intra-mercurial planets, determined that the average number of sunspots on the solar surface varied cyclically with a period that he estimated to be about 10 years (see Hoyt & Schatten (1997) for this and other historical references). This is the well-known *solar cycle* and is now known to have a mean period of 11 years.

In 1858, the spatial dependence of the incidence of sunspots was linked to their temporal behaviour. Richard Carrington and Gustav Spörer independently demonstrated that the latitude at which sunspots appeared was linked to the phase of the solar cycle. At the beginning of a cycle spots appear on the Sun at latitudes of around $\pm 30°$; as the cycle progresses the location of activity drifts towards the equator and the spots then die out in the next minimum. Carrington also observed that the sunspots rotated more rapidly at lower latitudes, which led him to conclude that the Sun rotated not as a solid body, but differentially – a discovery that will prove important for the solar dynamo. The migration of sunspot activity

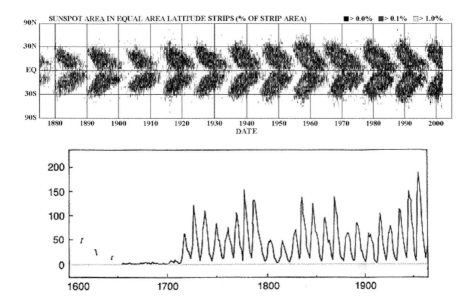

Fig. 1. (a) Solar Butterfly diagram: Location of sunspots as a function of time (horizontal axis) and latitude (vertical axis). As the cycle progresses sunspots migrate from mid-latitudes to the equator (courtesy of D.N. Hathaway). (b) Time-series for yearly sunspot group numbers. Notice the maunder Minimum at the end of the 17th Century.

produces the familiar butterfly diagram first exhibited by Maunder (1913). An up-to-date version showing the incidence of sunspots as a function of latitude is in Figure 1.

Many theories were proposed for the origin of sunspots, but it was not until the early 20th century that they were linked to the solar magnetic field. Hale (1908) used the Zeeman splitting of spectral lines to demonstrate that sunspots were the sites of strong magnetic field, a typical sunspot possessing a magnetic field of about 3000 Gauss. By then it could be seen that sunspots appeared in pairs and Hale demonstrated that the sunspots in the pair have opposite magnetic polarities with the sense of the magnetic polarity being different in the northern and southern hemispheres. Systematic observation then demonstrated that the magnetic polarity of the sunspot pair reversed from one activity cycle to the next and so there is a magnetic cycle with a mean period of 22 years.

It is now widely believed that sunspots are the surface manifestation of a strong azimuthal magnetic field generated deep within the Sun. If this field becomes unstable (to magnetic buoyancy instabilities for example

– see later) it can buckle and buoyantly rise. The two sites where this magnetic field breaks through the visible surface of the Sun can be seen as sunspots. Sunspots appear dark on the solar surface as the magnetic field inhibits energy transport by convection. Active regions (sunspot pairs) are observed at mid-latitudes with loops of strong magnetic field on many scales interacting and twisting as they rise through the solar surface.

Sunspots are not the only signature of the global solar magnetic field. The solar corona, which is clearly visible from the Earth at eclipses shows the radial component of the Sun's field. The coronal field (often termed the poloidal magnetic field by dynamo theorists) has a complicated morphology that changes with the solar cycle. It is now known that the coronal field also reverses, but out-of-phase with the azimuthal (toroidal) sunspot field (Stix 1976) (so that when the sunspot field is at a maximum – often called solar maximum – the coronal field is at a minimum and reversing). The structure of the solar magnetic field is largely consistent with that of an oscillatory dipole field, with the sunspot field antisymmetric and the coronal field symmetric about the solar equator.

In order to construct a reliable theory for the generation of the solar field, it is helpful to make use of historical records. Eddy (1976) and more recently Hoyt & Schatten (1997) have catalogued the sunspot numbers (or alternatively the number of groups of sunspots) to obtain a time-series for the variation in magnetic activity over the past 400 years (since systematic observation began). One such time-series is shown in Figure 1(b). Here the amplitude of the solar activity cycle is seen to be aperiodically modulated. The most striking feature of the time-series is, however, the presence of an episode with a dearth of sunspots at the end of the 17th Century. This interruption is known as the Maunder minimum. There is little doubt that is not due to a lack of observations, but is real phenomenon (see e.g. Ribes & Nesme-Ribes 1993). Careful study of the records also shows that as the Sun emerged from the Maunder minimum, sunspots were almost exclusively observed in the southern hemisphere, but after one asymmetric cycle activity reestablished itself in its "normal mode", with sunspots appearing with little preference for either hemisphere.

Even utilising historical records can only yield information for about twenty magnetic cycles – hardly enough on which to base a theory. Fortunately the record of solar magnetic activity can be extended using terrestrial isotope abundances as proxy data. The cosmic ray flux entering the Earth's atmosphere is modulated by the solar magnetic field (most prominently by the solar wind). These high energy particles are responsible for the production of radioisotopes including ^{10}Be and ^{14}C and so the abundance of these is anti-correlated with solar magnetic activity. The isotope ^{10}Be is preserved in polar icecaps whilst ^{14}C is preserved in tree rings. Analysis of these data demonstrates that the Maunder minimum was not an isolated event, but that several previous minima occurred with a regular mean period of approximately 200 years (see e.g. Beer 2000). Moreover the 11-year activity cycle can be found in the ^{10}Be record stretching back over 4000 years. Also of interest is that although sunspots disappeared during the Maunder minimum the solar magnetic field continued to oscillate at that time, albeit at a much lower amplitude and at a reduced period of about 9–10 years (Beer *et al.* 1998).

Our final pieces of information concerning the origin of the solar magnetic field come not from observations of the Sun, but from observations of stars with similar properties to the Sun (so-called "solar-type" stars). Unfortunately direct observation of magnetic fields in stars is difficult (see e.g. Rosner 2000). However the magnetic activity of a star is known to be correlated with the level of spectral emission found in the Ca^+ H and K lines. In 1966 Wilson set up a systematic programme of observations of the Ca^+ emission for a selected group of nearby stars. As the records of magnetic activity grow longer trends in the activity of the stars are beginning to emerge (Soon *et al.* 1993). It is clear that many of the stars do exhibit magnetic activity with a range of behaviour depending on the rotation rate and spectral type of the star. For a fixed spectral type, the level of activity increases with rotation rate. As the rotation rate increases magnetic activity sets in with periodically varying fields being found for moderate rotators. Further increase in the rotation rate yields increased activity with more complicated time dependence, with the basic stellar cycle being modulated and eventually becoming chaotic. There are examples of stars that appear to be in Maunder Minima type states and even one case where the star appears to be in transit between active and inactive phases. Hence the magnetic activity we see on the Sun is generic and the goal of any theory should be not only to explain the spatio-temporal behaviour of the Sun's magnetic field, but also the systematic changes that are observed in stellar activity as one moves away from solar parameters.

Why is the Sun's magnetic field thought to owe its origin to a hydromagnetic dynamo? For some astrophysical bodies the answer is clear cut. For example the Earth's magnetic field is known from paleomagnetism to have existed for approximately 3.5×10^9 years. One can easily demonstrate that the time it would take a fossil magnetic field (i.e. a magnetic field that was formed at the same time as the Earth) to decay away (the ohmic decay time) is approximately 10^4 years and so the magnetic field must have been sustained against this decay to survive at current levels. Hence dynamo action is invoked to explain the geomagnetic field. If one repeats this argument for the Sun, then one finds that the ohmic decay time of approximately 10^{10} years is comparable to the age of the Sun and so it is possible that the solar field is a fossil field. However it is extremely difficult to explain the wide range of spatio-temporal behaviour of the solar magnetic field if the field were simply a relic field. Why, for example, does the Sun's field reverse every 11 years and why do active regions migrate from mid-latitudes to the equator? Two mechanisms for the generation of the alternating magnetic field in the solar cycle have been proposed. If the poloidal field is fixed and the azimuthal velocity field oscillates with a 22 year period then this angular velocity will generate an azimuthal field that oscillates with the same period. If this were the case the Sun would act like a *Magnetic Oscillator*. However, as stated above, the Sun's poloidal field *does* reverse at sunspot maximum. Observations of flows at the solar surface show that systematic variations in angular velocity occur with a period of 11 years (not 22 years). These torsional oscillations (Howard & Labonte 1980) lead us to a more convincing explanation for the cycle – the generation of the field is due to the action of a self-excited hydrodynamic dynamo limited by the action of the Lorentz force. As we shall see, this force is quadratic in the magnetic field and so fluctuations in angular velocity will depend only on the strength of the field and not on its polarity. The angular velocity would therefore vary with the required period of 11 years. The consensus is therefore that solar magnetic phenomena are due to the existence of a hydromagnetic dynamo operating within the Sun. I will now go on to discuss the basic principles of dynamo theory and the application of this theory to the Sun.

3. Basic Dynamo Theory

Dynamo theory is, by its nature, subtle and contains many elegant pieces of mathematics. I shall, however, for simplicity and ease of presentation concentrate on a physical description of the dynamo process, keeping mathematical details to a minimum. More detailed accounts of aspects of the theory may be found in the books by Moffatt (1978) and Parker (1979), and more recent articles reviewing the theory (Roberts & Soward 1992; Weiss 1994; Chossat *et al.* 2001) indicate just how much progress has been made in different aspects of the subject.

A dynamo process was first tentatively suggested as a possible mechanism for maintaining the solar magnetic field by Larmor (1919) in his brief communication entitled 'How could a rotating body such as the Sun become a magnet?'. The basis of the theory is that the magnetic field is maintained by the motion of an electrically-conducting fluid (such as the highly ionised plasma in the Sun) where the motion of the fluid induces those electric currents needed to sustain the field.

The dynamo process involves many interactions; if the net result of all these interactions is to produce a self-sustaining magnetic field then we have a hydromagnetic dynamo. The dynamo problem is perhaps simplest stated (Roberts 1988) 'How can a magnetic field be maintained by the motion of an electrically-conducting fluid, despite the continual drain of magnetic energy by the resistance of the fluid?'. To solve the full dynamo equations self-consistently requires two steps. It must be demonstrated that there *is* a motion of the fluid **u** that can maintain a non-decaying magnetic field **B** *and* that this motion may be sustained by the forces acting on the fluid.

Mathematically this amounts to finding self-consistent solutions to the equations governing the evolution of the magnetic field (**B**) (the induction equation) and the velocity of the fluid (**u**) (the Navier-Stokes equation). This should in theory be straightforward – many similar equations can easily be solved using modern day computers and powerful numerical techniques. As we shall see it is not that simple; the dynamo process is sufficiently subtle (requiring fully three-dimensional computations which are problematic even for modern computers) and the conditions within the Sun are so extreme (with a large range of spatial and temporal scales needing to be resolved in a successful calculation) that a blind application of computational methods will inevitably lead to confusing (if not simply incorrect) results. What is needed is a more considered approach.

The induction equation is derived by combining the pre-Maxwell equations with Ohm's Law for a moving conductor to yield (see e.g. Mestel 1999)

$$\frac{\partial \mathbf{B}}{\partial t} = \boldsymbol{\nabla} \times (\mathbf{u} \times \mathbf{B}) + \eta \nabla^2 \mathbf{B}, \tag{1}$$

where η is the magnetic diffusivity of the fluid. This equation basically states that there are two processes (given by the terms on the RHS of the equation) that lead to evolution of magnetic field **B**. The second term $\left(\eta \nabla^2 \mathbf{B}\right)$ is a diffusion term and leads to decay of the magnetic field. The first term $\left(\boldsymbol{\nabla} \times (\mathbf{u} \times \mathbf{B})\right)$ represents the *inductive* effects of motions within

the fluid which in general lead to an increase in the magnetic field. The
first part of the dynamo problem described above can now be reformulated
as 'is there a velocity (**u**) for which the inductive term is more efficient at
generating field than the diffusive term is at destroying it?' This sounds like
a simple question to answer, but from the time Larmor posed the question
it took more than fifty years for a convincing demonstration of a dynamo.

The primary reason for this lies in the mathematical complexity of equation 1. In 1934 Cowling demonstrated in his famous antidynamo theorem
that no steady axisymmetric magnetic field could be maintained by dynamo action. Hence any dynamo generated field must be three dimensional.
For the next thirty or so years most of the results in dynamo theory were
negative results; demonstrations of which magnetic fields and velocity fields
could not be generated by or lead to dynamo action. The restrictions placed
on possible solutions of the dynamo equations still have consequences for
researchers in dynamo theory today. Any straightforward numerical calculations of the equations leading to dynamo action must of necessity be
three-dimensional and this leads to severe restrictions on the parameter
regimes that can be investigated.

Despite these mathematical subtleties it became clear that certain properties of the flow were desirable for dynamo action. Progress was made by
decomposing the magnetic field into its poloidal and toroidal parts. (Recall that the toroidal field is the field that is believed to lead to sunspots
whilst the poloidal field corresponds to the large scale coronal field.) The
dynamo problem is now to show that both components of the field may
be sustained by the fluid flow. Dynamo theorists often refer to a dynamo
cycle in which the magnetic field is maintained by the velocity via a continual conversion between poloidal and toroidal fields. One part of this cycle
is easily described. The induction equation shows how any gradients in
the fluid flow (due to the presence of shears or differential rotation) will
lead to the stretching of poloidal field into toroidal field. This process is
often termed the "ω-effect". However there is no comparable term in the
axisymmetric induction equation that leads to the generation of poloidal
field from toroidal. This second process is much more subtle and involves
considering non-axisymmetric components. Parker (1955) and later (on a
more mathematical footing) Steenbeck *et al.* (1966) considered the problem
of a turbulent flow and magnetic field that varied on two scales. The key to
this new approach (dubbed mean field electrodynamics (MFE)) was to consider the interactions of the small-scale magnetic field and velocity (which
would be non-axisymmetric) in producing a large-scale magnetic field. It

was demonstrated that the net effect of the small scale interactions could be so as to produce a large scale poloidal field from a toroidal field *if the small scale flow had an underlying handedness* – this effect is known as the α-effect. The handedness arises naturally in an astrophysical body due to its rotation. The dynamo cycle was therefore complete.

The advent of MFE appeared to kill many birds with one stone. The equations derived for the mean magnetic field could be solved assuming axisymmetry, thus enabling easier computation; all the non-axisymmetry had been relegated to the small-scale field and flow. The beauty (and as we shall see weakness) of MFE is that the small-scale magnetic field is never solved for. The system is closed by postulating how the small-scale magnetic field depends on the large-scale field and small-scale flow. The α-effect is therefore 'modelled' by choosing various parameters in a hopefully physically realistic but nonetheless ad-hoc manner. An advantage is that the mean field equations do not need to be solved for extreme parameter values – all the subtleties and difficulties of MHD turbulence have been avoided once the α-effect is prescribed. For these reasons mean field models have proved very popular – and successful – and they can produce non-decaying magnetic fields that vary cyclically. The application of Mean Field theory to the solar dynamo was a (qualified) success although some outstanding difficulties arise. However, even if one takes mean field theory at face value one has only addressed half the problem. What we have shown is that there exist velocity fields ($\mathbf{u}$) that are capable of sustaining a magnetic field. We have not determined whether the motion may be sustained by the forces acting on the fluid.

One must therefore examine the equation that describes the evolution of the fluid velocity, the Navier-Stokes equation. This is basically a statement of Newton's second law, that the rate of change of momentum of a fluid element is proportional to the forces that act upon it. One such force is the Lorentz Force ($\mathbf{j} \times \mathbf{B}$) – the magnetic field acts back on the fluid and, if it large enough, has a dynamic effect. Hence it is not enough to demonstrate that there exists a velocity field that can sustain a magnetic field; it must also be shown that this continues to occur when the back-reaction of the magnetic field is included in the Navier-Stokes equations. In the next section I will describe how this basic theory has been applied to the solar dynamo before concluding with a discussion of possible future directions for research.

4. The Solar Dynamo

To apply the theory described above to the Sun, we need some information about the velocity field within the Sun. Earlier it was noted that the surface of the Sun rotates not as solid body but differentially and that differential rotation is a crucial ingredient for dynamo action, converting poloidal magnetic field to toroidal field. Of course we are not able directly to observe beneath the visible surface of the Sun and so the behaviour of the interior of the Sun must be inferred either from theory or from indirect observations.

Stellar structure theory provides information on the large scale structure of the solar interior, and a schematic of the results is shown in Figure 2(a). The central region of the Sun is the *inner core* – in this region nuclear reactions (fusion) produce energy. This energy is carried outwards towards the surface of the star initially by radiation. The *radiative interior* extends out to about 210,000km below the solar surface (about 70% by radius). At that point, radiation becomes incapable of transporting the required energy flux towards the solar surface and convection (transport of energy by fluid motions) sets in. This outer *convection zone*, which stretches to the visible surface (photosphere), is therefore a site of highly turbulent fluid flow with many fluid eddies interacting on a wide-range of temporal and spatial scales. The small-scale convection at the solar surface can be seen in the form of granulation.

Until recently our knowledge of the solar interior remained at this basic level. However a major breakthrough came with the new science of helioseismology. This is the study of acoustic waves (pressure fluctuations) within the Sun which can be observed either by ground-based telescopes or from satellites. The frequency of the waves is modified not only by the sound speed of the medium in which they propagate, but also by large scale fluid flows. By measuring the modifications of the frequency not only can the sound speed of the fluid be inferred, but also the large-scale velocities. Helioseismology is capable of providing the most crucial information for dynamo theory – a map of the differential rotation of the interior of the Sun, an example of which is shown in Figure 2(b).

This figure shows contours of constant angular velocity in the outer regions of the Sun. The first thing to note is that the differential rotation observed at the surface is maintained throughout the convection zone (the contours are largely radial in the outer 30% of the Sun). Hence in the convection zone there is little radial differential rotation but some differential rotation with latitude. However this pattern does not continue into the ra-

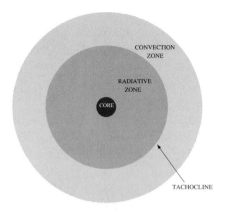 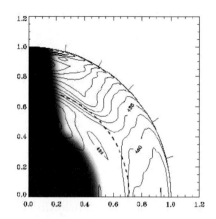

Fig. 2. (a) Schematic showing the internal structure of the Sun and the location of the tachocline. (b) The solar rotation rate from Schou *et al.* (1998). The frequency is in nHz and the shaded area represents the region where there is insufficient data. The base of the convection zone is marked by a dashed line.

diative interior, which rotates largely as a solid body. These two different patterns of differential rotation are joined by a matching layer just below the base of the solar convection zone where the rotation rapidly changes. This layer of extremely strong radial shear is known as the *tachocline*.

We noted earlier that regions of strong differential rotation are ideal sites for the conversion of poloidal magnetic field to toroidal field; one crucial part of the dynamo process. There is now a consensus that the tachocline is responsible for the generation of the strong toroidal field that leads to sunspots. Not only is the tachocline the location of the strongest shear, but, it is able to store even strong magnetic fields for a significant length of time. The presence of magnetic fields in a plasma is known to reduce the local density and so regions of strong magnetic fields are buoyant compared to their surroundings and susceptible to instabilities (just as bubbles of air will rise if surrounded by denser water). As discussed earlier, this *magnetic buoyancy instability* is believed to be the mechanism for the formation of sunspot pairs. However if the magnetic field is too buoyant (as it would be in the solar convection zone) it would rise to the solar surface in about a month (Rosner & Weiss 1992) – not enough time to be amplified by the differential rotation. The tachocline, which is stably stratified, is capable of storing the magnetic field and the differential rotation has time to act so as to strengthen the field. Only very strong magnetic field is buoyant enough to rise from the tachocline to the solar surface and form sunspot pairs (Tobias *et al.* 2001).

So far so good, but we know that to complete the dynamo cycle we must also reconvert the toroidal field into poloidal magnetic field – how is this achieved in the Sun? There are a number of competing theories for the site and mechanism of this reconversion. When little was known about the interior dynamics of the Sun, a dynamo located at the surface of the Sun was proposed (Babcock 1961, Leighton 1969). The Sun was known to rotate differentially at the surface, so the conversion of poloidal to toroidal field was assured. Babcock & Leighton completed the cycle by considering how sunspots behaved as they decayed. They showed that sunspot pairs were swept towards the pole by a meridional flow and decayed in such a way as to create poloidal field. In the language of mean field theory, the decay of active regions leads to the presence of an α-effect at the solar surface. With the advent of helioseismology and the discovery of the tachocline, the preferred location for the ω-effect moved to the base of the solar convection zone. Nevertheless the decay of sunspot fields *did* provide a mechanism for the recreation of poloidal field. The problem is to transport the poloidal field generated at the solar surface down into the depths to be reconverted into toroidal field. This is problematic and relies on the postulation of a "conveyor belt" meridional flow providing the required transport. Recent solar mean field dynamo models (e.g. Dikpati & Charbonneau 1999) utilise such a flow to construct a working dynamo.

However it is much more likely (see e.g. Mason *et al.* 2002) that the site of the generation of poloidal field is closer to the tachocline. If the poloidal and toroidal field are generated in nearby regions then no "conveyor belt" need be postulated. We saw earlier that a turbulent flow in the presence of rotation (cyclonic turbulence) produces an α-effect capable of reconverting the toroidal field into poloidal field. The turbulent convection at the base of the solar convection zone is just such a flow and an ideal candidate for an α-effect. This convection in the lower reaches of the convection zone takes the form of strong downflows which grab and stretch magnetic field, amplifying the field and also recreating a large-scale poloidal field from the toroidal field. These plumes are also believed to overshoot the base of the convection zone into the tachocline and carry the poloidal field with them – a process known as *magnetic pumping*. In this manner the cycle is completed. The differential rotation in the tachocline converts poloidal field into strong toroidal field, which becomes unstable to magnetic buoyancy. As the field rises it is shredded by the convection, and the cyclonic turbulence generates poloidal field. This poloidal field is carried back down by the overshooting plumes into the tachocline and the process begins again. If the toroidal field is locally very strong then it can make it through the turbulence without being shredded. This strongest magnetic field reaches the solar surface to form a sunspot pair. This type of dynamo is termed an *interface dynamo* as it is located at the interface between the solar convection zone and radiative interior. Interface dynamos were first modelled by Parker (1993) who demonstrated that such models could produce oscillating magnetic fields consisting of waves of magnetic activity (dynamo waves), located in the solar tachocline, migrating towards the equator. These mean field interface dynamo models were extended to include the back-reaction of the Lorentz force on the fluid motions by Charbonneau & MacGregor (1996) and Tobias (1997).

Fig. 3. Simulated butterfly diagram from a mean-field model. In the figure time increases horizontally whilst latitude is vertical (as in Figure 1). The field is generally antisymmetric about the equator except when the dynamo emerges from a minimum.

A dynamic magnetic field is capable of producing fluctuations in the differential rotation via the Lorentz force; these correspond to the torsional oscillations described in Section 2. As the Lorentz force is quadratic in the magnetic field the period of these torsional oscillations is half that of the magnetic field, so that if the magnetic field has a cycle period of 22 years the torsional oscillations have one of 11 years. Moreover these fluctuations in the differential rotation can have a significant effect on the generation mechanism for the toroidal field. The net effect of the addition of the dynamic magnetic field to models is to increase their temporal variability. Solutions to the mean field equations may no longer be periodic but modulated on a longer time-scale. Even intervals of reduced magnetic activity can be found. Such behaviour is demonstrated in Figure 3. This shows a simulated butterfly diagram from a mean field interface dynamo model (Beer *et al.* 1998).

The basic magnetic cycle is seen to come and go and periods of reduced activity reminiscent of the Maunder Minimum systematically reoccur. When the dynamo is "working normally" the (toroidal) sunspot field is antisymmetric about the equator and the (poloidal) coronal field (not shown) is symmetric. Hence the field is primarily an oscillating dipole. However as the dynamo emerges from minima in activity the field is largely confined to one hemisphere for a cycle and is asymmetric. This is reminiscent of the behaviour of sunspots as the Sun emerged from the Maunder Minimum. Intriguingly, further results show that it is possible for the Sun to emerge from a minimum with the toroidal field *symmetric* about the equator even if it entered the minimum as an *antisymmetric field*. The minimum has acted as a trigger to change the form of the magnetic field. This is an important result; it is vital to realise that, although the Sun has over the past few hundred years had a magnetic field that is largely dipolar, there is no reason to assume that it possessed such a symmetry in the past or will continue to do so in the future.

We noted earlier two important facts. First a blind application of computational techniques to the problem of the solar dynamo would be counterproductive. Secondly that mean field models (such as those described above) suffer from the ad-hoc nature of the parametrisation of the physical effects that are believed to be important within the Sun. For example it is easy to describe physically how an interface dynamo works, and even to build mean field models with these effects incorporated. Indeed these models are very successful, being able to reproduce not only the basic magnetic cycle but also Grand Minima and the associated symmetry changes. But how robust are these models? How sensitive are they to the assumptions and parametrisations that have been made? It is in this area that numerical simulations on massively parallel computers taken together with a mathematical understanding of the structure of the underlying equations can lead to a greater understanding. I will conclude in the next section by speculating on possible future directions for solar dynamo theory.

5. Conclusions and Speculations

In the past two sections I have sketched an outline of the theory governing the generation of the solar magnetic field. Clearly I have not included all of the facets of the theory; rather I have tried to give a personal view of where dynamo theory stands today. As noted at the start of the article, it is an exciting time for solar dynamo theory. Increases in the resolution of satellite-based observations of the solar magnetic field coupled with the breath-taking increase in computing power mean that theory and observation can both make significant contributions to our understanding of the underlying physics.

On the observational front, the continuing success of the YOHKOH and SOHO satellites will be supplemented by new and improved instruments aboard planned satellite missions. The Solar B mission, due to be launched in September 2005 as the follow-up mission to YOHKOH, will consist of a coordinated set of optical, EUV and X-ray instruments that will investigate the interaction between the Sun's magnetic field and its corona. Moreover NASA will launch the Solar Dynamics Observatory in 2007 as part of its Living with a Star program. These high-resolution observations will provide more information on the temporal and spatial evolution of the solar magnetic field on a wide range of scales. They will also increase our understanding of the fluid flows within the Sun, both the differential rotation so vital for the operation of the solar dynamo and the meridional flows that may lead to transport of magnetic fields. Taken together, these observations of the field and flows will provide vital information to constrain the theory. Sustained observation of the temporal evolution of the differential rotation for example may provide information on the saturation mechanism of the dynamo whilst data about the morphology and twist of the magnetic field in active regions will constrain some of the ad-hoc parametrisations of mean field theory.

Advances in the theory of the solar dynamo are also to be expected in the near future. Mean field theory has proved to be a qualified success, although the qualifications should be recognised. It is now time to move on from considering the dynamics of mean field models as applied to the Sun – we know that a judicious choice of the parameters can yield successful models. Rather we should be concentrating our efforts in trying to understand *why* the mean field models are so successful. We need to examine the mathematical structure of the dynamo equations and use the techniques of nonlinear dynamics to explain their solutions. Moreover we need to understand the physical processes that lead to interaction between the velocity and magnetic fields on small-scales and how these processes can lead to generation of magnetic field on the large-scale. I believe that we are at a point where rapid progress may be made in our understanding of the solar dynamo, but that we must be careful not to assume that the secrets of the generation of the Sun's magnetic field will be found by a naive application of computational power. The solar dynamo is like a jigsaw where each individual piece of physics must be understood before the pieces are assembled in a global model. This sounds simple, yet each piece involves the solution of a problem which may require large numerical simulations. For example – how does the shear flow in the tachocline interact with the strong magnetic field stored there? How can turbulent convection transport and amplify magnetic fields within the convection zone? How do the strongest fields in the tachocline break through the convection to form sunspot pairs at the solar surface. These (and other) pieces of the puzzle need to be understood before a global picture can emerge. That is not to say that global numerical simulations will not be of use; rather that we need to proceed with care. The range of spatial and temporal scales of the solar dynamo is vast – to simulate them would need an increase of 10^9 in computing power. We should therefore not pretend that it is possible to simulate the solar dynamo. However it *is* possible to simulate dynamo action in a rotating sphere (or spherical shell) of electrically conducting fluid, albeit at parameters far removed from those applicable to the Sun. Such simulations, which are now feasible, *can* help us piece together the jigsaw of the solar dynamo by providing information on the global behaviour of magnetic fields. By judicious extrapolation and the derivation of scaling laws we may hope to gain an insight into the generation of magnetic fields in our nearest star.

References

1. Babcock, H.W. 1961 The Topology of the Sun's Magnetic Field and the 22-year Cycle. *Astrophys. J.* **133**, 572–587
2. Beer, J. 2000 Long-term indirect indices of solar variability. *Space Sci. Rev.* **94**, 53–66
3. Beer, J., Tobias, S.M. & Weiss, N.O. 1998 An active Sun throughout the Maunder Minimum. *Solar Phys.* **181**, 237–249
4. Charbonneau, P. & MacGregor, K.B. 1996 On the generation of equipartition-strength magnetic fields by turbulent hydromagnetic dynamos. *Astrophys J.* **473**, L59-L62
5. Chossat, P., Armbruster, D. & Oprea, I. 2001 Dynamo and Dynamics, a Mathematical Challenge (eds) *NATO Science Series II: Mathematics, Physics and Chemistry*, **26**, Kluwer.
6. Cowling, T.G. 1934, The magnetic field of sunspots, *Mon. Not. Roy. Ast. Soc.* **94**, 39–48.
7. Dikpati, M. & Charbonneau, P. 1999 A Babcock-Leighton flux transport dynamo with solar-like differential rotation. *Astrophys. J.* **518**, 508–520
8. Eddy, J.A. 1976 The Maunder minimum. *Science* **192**, 1189–1202
9. Hale, G.E. 1908, On the probable existence of a magnetic field in sun-spots. *Astrophys. J.* **28**, 315–343

10. Howard, R. & Labonte, B.J. 1980, The Sun is observed to be a torsional oscillator with a period of 11 years. *Astrophys. J. Letters* **239**, L33–L36

11. Hoyt, D.V. & Schatten, K.H. *The role of the Sun in Climate Change*, Oxford University Press.

12. Larmor, J. 1919, How could a rotating body such as the Sun become a magnet. *Rep. Brit. Assoc. Adv. Sci. 1919*,159–160

13. Leighton, R.B. 1969 A Magneto-Kinematic Model of the Solar Cycle. *Astrophys. J.* **156**, 1–26

14. Mason, J., Hughes, D.W. & Tobias S.M. 2002 *Astrophys. J* submitted

15. Maunder, E.W. 1913, Note on the distribution of sunspots in heliographic latitude. *Mon. Not. Roy. Ast. Soc.* **64**, 747–761

16. Mestel, L. 1999, *Stellar Magnetism*, Clarendon Press, Oxford.

17. Moffatt, H.K. 1978 *Magnetic Field Generation in Electrically Conducting Fluids*, Cambridge University Press.

18. Parker, E.N. 1955, Hydromagnetic dynamo models. *Astrophys. J.* **122**, 293–314

19. Parker, E.N. 1979 *Cosmical Magnetic Fields: their origin and activity*, Clarendon Press Oxford

20. Parker, E.N. 1993, A solar dynamo surface-wave at the interface between convection and nonuniform rotation. *Astrophys. J.* **408**, 707–719

21. Ribes, J.C. & Nesme-Ribes, E. 1993, The solar sunspot cycle in the Maunder minimum AD 1645 – AD 1715. *Astron. Astrophys.* **276**, 549–563

22. Roberts, P.H. 1988, Future of geodynamo theory. *Geophys. Astrophys. Fluid Dyn.* **44**, 3–31

23. Roberts, P.H. & Soward, A.M. 1992, Dynamo Theory. *Ann. Rev. Fl. Mech.* **24**, 459–512

24. Rosner R. 2000, Magnetic fields of stars: using stars as tools for understanding the origins of cosmic magnetic fields. *Phil. Trans R. Soc. A* **358**, 689-708

25. Rosner, R. & Weiss, N.O. 1992, The origin of the solar cycle. *Asp Conf. Ser.* **127**, 511–531

26. Schou J. *et al.* 1998, Helioseismic studies of differential rotation in the solar envelope by the Solar Oscillations Investigation using the Michelson Doppler Imager. *Astrophys. J.* **505**, 390–417

27. Soon, W.H., Baliunas S.L. & Zhang Q. 1993, An interpretation of cycle periods of stellar chromospheric activity. *Astrophys. J.* **414**, L33-L36

28. Steenbeck, M., Krause, F. & Rädler, K-H 1966, A calculation of the mean electromotive force in an electrically conducting fluid in turbulent motion, under the influence of coriolis forces. *Z. Naturforsch* **21a**, 369–376

29. Stix, M. 1976, Differential Rotation and the Solar Dynamo. *Astron. Astrophys.* **47**, 243–254

30. Tobias S.M. 1997, The solar cycle: Parity interactions and amplitude modulation. *Astron. Astrophys.* **322**, 1007–1017

31. Tobias, S.M., Brummell, N.H., Clune T.L. & Toomre J. 2001, Transport and storage of magnetic field by overshooting turbulent compressible convection. *Astrophys. J.* **549**, 1183-1203

32. Weiss, N.O. 1994, Solar and stellar dynamos. In "Lectures in solar and plan-

etary dynamos" eds Proctor M.R.E. & Gilbert A.D. Cambridge University Press, 59–95

33. Weiss N.O. & Tobias S.M. 2000 Physical causes of solar activity. *Space Sci. Rev.* **94**, 99–112

Steven Tobias

Born in Manchester in 1970, Steve Tobias studied Mathematics at Cambridge where he graduated with first class honours. Having gained a distinction in Part III mathematics, he obtained his PhD in 1996 for research into "Nonlinear Solar and Stellar Dynamos" from Cambridge University. In 1995 he was elected to a Junior Research Fellowship at Trinity College, Cambridge. Between 1996 and 1998 he was a research associate at JILA, University of Colorado, before resuming his fellowship in Cambridge. He joined the Applied Mathematics department at the University of Leeds in 2000 and became a Reader in 2004. He continues to pursue active research in magnetohydrodynamics, with a particular emphasis on dynamo theory, and nonlinear systems with applications to fluid mechanics. Interests include playing football, the guitar and (for his sins) supporting Manchester City.

CHAPTER 19

EXPLOSIONS ON THE SUN

Louise K. Harra

Mullard Space Science Laboratory, University College London,
Holmbury St Mary, Dorking, Surrey, UK, RH5 6NT
E-mail: lkhmssl.ucl.ac.uk

I will describe the most dynamic and highly energetic phenomena in the Solar System - these are the eruptions and flaring that occur on the Sun. They can release as much energy as 10 million volcanoes, and throw out material into the solar system with similar mass to Mount Everest! The theories of what can produce such an explosion are based around the magnetic field that confines the gas. These events can produce emission right across the electromagnetic spectrum. The status of our ability to predict these events is discussed.

1. Introduction

Solar flares are one of the most dynamic and highly energetic phenomena in the Solar System. The physical process to produce so much energy is poorly understood but yet occurs in a variety of contexts in the Universe. Ejections of material from the corona are often related to solar flares. Coronal mass ejections on the other hand hurl 10^{13} kg of material into the solar system in one event. This is approximately the mass of Mount Everest released into the solar system roughly once a day!

The first observation of a solar flare took place in King's Observatory, Kew in England by Carrington in 1869. He was making daily observations of sunspots, and noticed a bright flash in the sunspot. This flash lasted only for a few minutes. It was noted as well, that simultaneously to this flare there were changes in the Earth's magnetic field and large displays of aurora not long afterwards. The link between the two events was unsure at this stage, but it was realised later that this was the first observation of 'space weather' in action! The idea that flaring that happens on the Sun

---

actually has an impact at a distance of 150 million km away on the Earth was initially hard to accept by some! It is now known that space weather effects are seen on planets even further away. For example beautiful aurora are seen on Jupiter, with bright emission even seen in plasma energetic enough to emit in the X-ray energy regime. Recent observations have been carried out by XMM-Newton on the X-ray emission seen by Jupiter led by Branduardi-Raymont.[1]

We are much more aware of space weather effects on Earth now, with our reliance on electricity supplies, satellite communications etc. One of the best known solar storms occurred in 1989 and managed to cause a black-out across Quebec. More frequently spacecraft instrumentation are interrupted through bit-flips or spacecraft de-orbiting early due to increased spacecraft drag. The advent of powerful new space observatories gives us the ability to study solar flares and coronal mass ejections in great detail across a huge range of energies - from gamma-rays to white light. We can study the Sun from under the surface using helioseismology to many solar radii away from the Sun. Understanding, and ultimately predicting the triggering of solar flares and coronal mass ejections is becoming increasingly important. This chapter aims to describe our current understanding of these explosions, along with areas where our knowledge is lacking.

2. The Mysterious Source of the Energy

The Sun is an average middle-aged star. Its main energy source is created in the very dense and very hot core, by the process of Hydrogen being fused to Helium. This energy is transported through the Sun initially by radiation, and then when that process cannot cope any more, by convection. Convection churns the highly ionised gas around, creating with it a magnetic field. This magnetic field can break through the surface of the Sun creating what are known as active regions. These are regions of intense magnetic field, that can be seen in white light images of the Sun. Figure 1 shows a white light image of the Sun - with the dark regions showing sunspots where the intense magnetic field is. The magnetic field strength in these sunspots is a thousand times higher than the magnetic field near household wiring, but with a size many times that of the Earths diameter. Figure 1 also shows what the outer atmosphere of the Sun looks like in X-ray emission. It appears very bright and complex in X-ray emission, and reaches temperature of more than 1 million K. The surface of the Sun is only 6,000K - so why, as we head away from the Sun, does the temperature increase? This has been a problem that scientists have been trying to understand for many years. We do now know that the source of this energy has to be related to the magnetic field. Where the X-ray emission is the brightest, the magnetic field tends to be the strongest. The strongest magnetic fields are also the source of the largest solar flares.

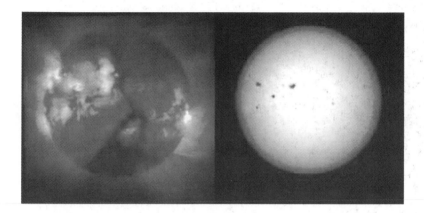

Fig. 1.   The left-hand figure shows a full Sun image in white light, and the right-hand figure shows an X-ray image of the Sun both taken on the same day. The darkest spots on the white-light image show the source of the most intense magnetic field. This corresponds to the brightest regions of X-ray emission. Courtesy of S. Matthews and the Yokkoh team.

For many years, theorists have been working on understanding how to get the energy out of the magnetic fields. This can be achieved by essentially doing two things. Firstly, you can create a situation where the magnetic field collides into another magnetic field line. For this to work the field lines need to be oppositely directed. If they have the same direction the magnetic field essentially just exist along side each other. Magnetic field collisions can occur if the magnetic flux tubes are twisted and sheared in some way, or if you have new magnetic flux emerging from below the surface of the Sun which crashes into the existing magnetic field. Figure 2 shows a cartoon of this process, which is know as magnetic reconnection (from Yokoyama and colleagues[2]). In this case there is a magnetic flux tube that reconnects at the top of the flux tube. When the field lines move close together they push some of the gas out of the flux tubes, and through a simple energy balance this occurs with an inflow of gas. The energy released from the reconnection causes the gas to be heated, and releases energy across the

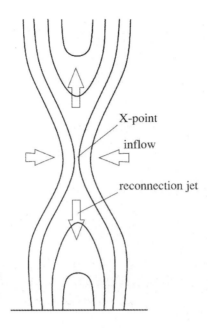

Fig. 2.   A cartoon of magnetic reconnection occurring in a flux tube on the Sun. The reconnection in this case, occurs at the top of of the flux tube (from Yokoyama *et al.*)

full electromagnetic spectrum. The second method for getting energy out of magnetic fields is to shake them to create a wave that travels through flux tube. It is unlikely that this process can produce enough energy for a flare, so magnetic reconnection is the favoured process. Priest and Forbes[3] describe the details of the magnetic reconnection process from both theoretical and observational aspects.

Figure 3 illustrates the appearance of a solar flare in October 2003 in different instruments. The white light image shows how large the sunspot was. It is the sunspot in the south towards the middle of the disk. Further out in the atmosphere bright EUV emission is seen. This flare was associated with a large eruption of material leaving the Sun. This is observed in the coronagraph data. Coronagraphs work by blocking out the central bright disk of the Sun is order to observe the lower density coronal material leave the Sun. Essentially this is replicating an eclipse. Before coronagraphs were invented in the 1930s by Lyot, eclipse observers had often seen complex structures at the edge of the Sun which we now know to be linked with coronal mass ejections. Solar flares and coronal mass ejections throw out

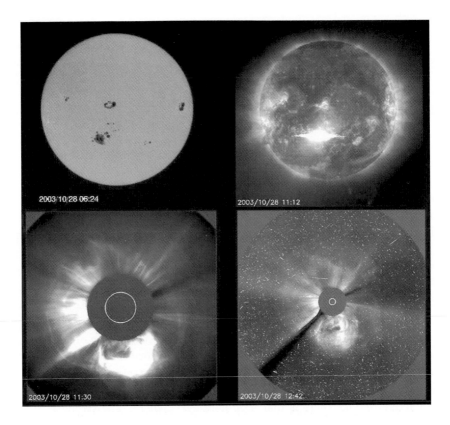

Fig. 3. Active region 10486, which had grown to one of the largest sunspot seen by SOHO, unleashed a spectacular eruption on 28 October 2003. The second largest flare observed by SOHO hurled out a fast-moving Coronal Mass Ejection. The top left figure shows the magnetic field of the active region - seen as the darkest region on the disk. The top right figure shows the flare as seen in EUV - it shows as intense brightness which even saturates the detector. The bottom left figure shows the coronal mass ejection appear in the coronagraph (LASCO) and is heading towards Earth. This coronagraph, LASCO-C2 observes out to 6 solar radii. The bottom right figure shows the ejection hurled further away from the Sun, reaching out to 30 solar radii. By the time it appears in this coronagraph, high energy particles have reached the detectors causing the 'snow' effect seen in this image. The solid white circle shows the size of the Sun's disk in the bottom images.As a result of this storm, aurora were seen as far south as Dublin, Ireland!. Courtesy of the ESA/NASA SOHO team.

high energy particles that hit the Earth within tens of minutes following the flare. This is seen clearly in the lower right hand panel of Figure 3. By the time the ejection reaches the coronagraph C3 on the instrument LASCO on SOHO, the detectors are bombarded with high energy particles that produce a 'snow' type effect on the detectors. The rest of the gas hits the Earth about 12-24 hours later depending on the velocity. In this event, the main ejection reached speeds of over 1000 km/s. As was mentioned earlier, these explosions occur because there is a magnetic field. So it is no great surprise that magnetic fields are also flung into the solar system along with the gas. Whether or not the coronal mass ejection has a big impact on us on Earth is dependent not only on the intensity of the event, but also on the direction of the magnetic field. This all goes back to the process of magnetic reconnection shown in Figure 2. The Earth is protected by its own magnetic field which is generated in the liquid iron core. The field lines emerge from close to the geographic south pole and reenter near the geographic north pole. For the coronal mass ejection to have the biggest impact it must then have a magnetic field that is opposite to that on the Earth - so it must be a southward magnetic field. In this particular flare it was a southward magnetic field, and allows magnetic field to essentially be eroded, and high energy electrons from the Sun to sneak into the Earth's atmosphere. When they reach the atmosphere they can excite the atoms there to produce spectacular auroral displays. The brightest colours seen tend to be red and green which are produced by the excitation of the neutral oxygen atoms in the Earth's atmosphere. The aurora created by this flare were seen at southerly latitudes (or at least where it wasn't cloudy!).

There are flares going on at all spatial scales on the solar disk. Even outside active regions there is magnetic field everywhere. The strength of the field is approximately one thousand times smaller than in an active region. Small brightenings are seen crossing the convective cell and its boundary in the quiet Sun. Figure 3 shows an EUV image of the Sun. The large flare in the active region is the dominant feature, but if you look virtually anywhere else on the disk there is EUV emission. This illustrates that the outer atmosphere has to be hotter than the surface of the Sun and hence must be heated. These small flares have been shown to have some similarities with larger flares, but have a smaller duration, and much produce much smaller energies. These smaller scale events, known as 'nanoflares', are thought to be one of the mechanisms that could possibly heat the corona. This idea was put forward by Eugene Parker.

3. How to Spot a Flare BEFORE It Happens!

The magnetic field plays an important role in understanding the flare trigger process, as described in the previous section. We know that an active region with a high magnetic field is likely to produce flares. We know that if you have a rapidly evolving magnetic field then it is likely to produce flares. We know that if magnetic flux emerges from below that we are likely to have a flare. These characteristics have been known for a while. However predicting exactly WHEN a flare will occur is extrordinarily difficult.

There are now many observations across a wide wavelength range during the impulsive and decay phases of flares. There are however fewer observations during the pre-flare phase and less theoretical understanding of the elusive flare trigger. Previous attempts to understand pre-flare behaviour have not reached strong or consistent conclusions. For example, the fluctuations in the intensity and temperature of solar active regions observed in the pre-flare phase are not vastly different to non-flare behaviour. Current work using the high resolution datasets now available has made possible a big step forward in understanding the pre-flare phase and the flare trigger. We have discovered strong indications that before the main flaring event there are many small scale reconnection events which gradually build up before the main phase of the flare. This oftens occurs away from the main flare core (for example work by Warren and Warshall[4]). It is suggested that these smaller scale reconnection events allow the strong magnetic field lines to open up through the large scale global coronal field.

Other methods are now in use regularly to predict the appearance of large active regions roughly a week before they appear on the disk. The technique, developed by Braun and colleagues,[5] uses seismic holography to probe disturbances in the interior of the Sun, and can be used to observe active region on the far side of the Sun! This information is used on website such as www.spaceweather.com, that attempts to predict space weather for the following days. The important pieces of information they include are magnetic field, the direction of the magnetic field leaving the Sun, and active regions on the far side of the Sun.

4. How to Spot a Coronal Mass Ejection Before It Happens!

There has been debate for many years on whether a CME causes a flare or vice-versa. It is now clearer that neither flares or CMEs cause the other, but that they are caused by changes driven by the same magnetic activity. We do know that if a long duration flare occurs that you are likely to have a coronal mass ejection. Long duration flares and CMEs are also often linked to eruptive prominences. (e.g. Sheeley *et al.*[6]). Prominences are cool dense structures which lie high in the corona. When observed on the disk they are seen as absorption features. Prominences have been shown to be more likely to erupt if they are located near emerging magnetic fields (e.g. Feynman and Martin[7]). There are however, many different types of events that are related to coronal mass ejections that complicate the picture. These are described below.

4.1. *Twisting and shearing!*

S-shaped magnetic structures were first investigated by Rust and Kumar[8] who found that many large transient brightenings in X-rays which were associated with CMEs were S-shaped. They interpreted this as due to an MHD helical kink instability. This work was followed on by a statistical study (Canfield, Hudson and McKenzie[9]) of 117 active regions and Glover et al.[10] Sterling and Hudson[11] studied an S-shaped structure which was related to a coronal mass ejection. Most of the ejected mass comes from two regions which lie close to the ends of the pre-flare S structure. Following the eruption, the sigmoid changes morphology dramatically to a cusp-shaped structure shown in Figure 4. Dimming of the region is long-lived - surviving for several days. It is thought that the cusp shape which develops is due to the reconnection of magnetic field lines following the CME.

Magnetic helicity is a parameter describing the twist, linkage and knottedness in magnetic fields and plasma flows. Magnetic helicity can be measured from observations of the magnetic field. This was carried out for a long-lasting active region. AR 7978 (van Driel-Gesztelyi et al.[12]) survived for six rotations. When the magnetic field first emerged, it produced many large flares, as the new field was colliding into pre-existing field. Many coronal mass ejections were produced during this time also. As the active region progressed through its lifetime, the flaring became weaker and more infrequent. Surprisingly coronal mass ejections continued to be released into the solar system. When the helicity was measured in the active region, it was found to increase with time. This was partly due to the fact that the Sun rotates differentially - the equator rotates faster than the poles. Over several rotations, this adds shear into the existing magnetic field. There is now a strong correlation between the number of coronal mass ejections that occur and the value of helicity. This could be an excellent method of determing liklihood of eruptions in the future (see Démoulin and Berger[13]). Figure 5 shows a model of how a magnetic flux tube can have both twist and writhe added to it. It is clear that a tube that is twisted and sheared in such a way

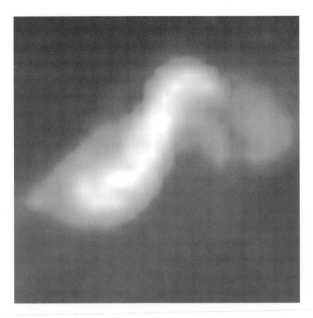

Fig. 4. A soft X-ray image of a coronal loop just before a coronal mass ejection lifted off. The structure shows an 'S-shaped' structure that illustrates twisted and sheared magnetic field. This particular event was studied intensively by Sterling and Hudson, and an 'S-shaped' structure is now used as a proxy for a region that is likely to produce a coronal mass ejection.

Fig. 5. An example of how twist and writhe can be achieved in a magnetic flux tube (Démoulin and Berger)

will at some point want to return to a relaxed state. To do this, a coronal
mass ejection is released.

4.2. *Global coronal waves*

The EUV imaging Telescope (EIT) on SOHO has observed waves in the
corona (see Figure 6 for an example). Shock waves, known as Moreton
waves, were discovered in Hα data (and hence low in the atmosphere) in
the late 1960s. It was expected from theory that there would be a coronal

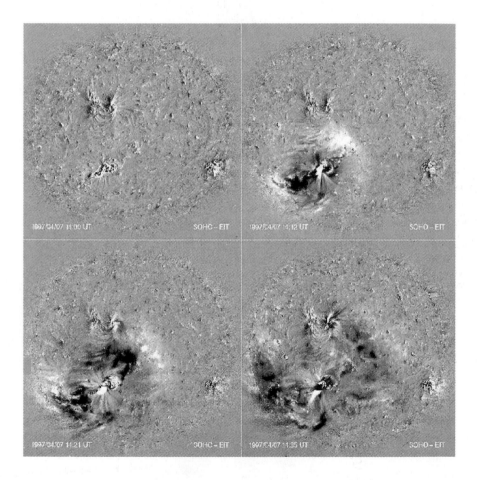

Fig. 6.   An example of a coronal wave. This is a time series of difference images from
the EUV Imaging Telescope on board SOHO. There is a bright front which propgates
across the disk in less than an hour. A strong region of coronal dimming lags the front.

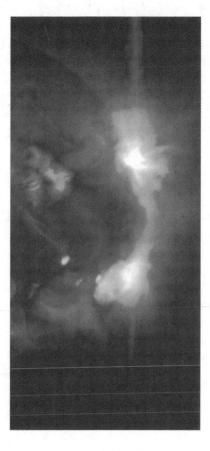

Fig. 7.   The Sun at the west limb in soft X-rays. There are two bright active regions at the limb that are connected by a larger magnetic loop.

counterpart. A comparison of the waves with Hα Moreton waves has shown various similarities. The speeds of the coronal waves are slow (few hundred km/s) but not out of the range of velocities observed in Hα. It is frequently observed that coronal waves are linked to flares and coronal mass ejections. Recent work by Harra and Sterling[14] has described the first spectroscopic observations of a 'coronal' wave, and found that there are outflows of material leaving the Sun which are related to the wave. Are these waves then merely coronal material lifting off which is then seen by coronagraphs as a coronal mass ejection?

4.3. *It's getting dimmer!*

One of the best signatures for CMEs is that of coronal dimming. The early Yohkoh results found dimming related to long duration flares at the limb (Hudson,[15] Hiei, Hundhausen and Sime[16]). Examples of dimming can occur on the disk as well as above the limb (Thompson *et al.*[17]), and this is now one of the most common methods of determining the CME launch location. A good example of dimming related to a halo CME (i.e. one that is approaching earthward) is described in Sterling and Hudson.[11] Dimming occurred following a flare, and an estimate of the mass loss was determined to be 10^{14}g. It was surmised that at least part of the CME mass is detected via the X-ray dimming measurements.

Spectroscopic observations of dimming have been observed with the Coronal Diagnostic Spectrometer on board SOHO (Harrison & Lyons[18]). Observations in six emission lines, with characteristic temperatures ranging from 20,000 K to 2 MK, have been made during a CME. The most significant dimming was found to be the 1MK plasma. Plasmas at other temperatures do not show such a significant change, suggesting that the dimming is due to a decrease in density of the plasma at 1MK. The dimming in this event was consistent with the mass loss of at least 70% of the mass of the CME. However there does seem to be no sudden onset to the dimming and it may begin before the projected onset of the CME as determined from the coronagraph data. The dimming, as observed also by Sterling & Hudson,[11] is a gradual process.

4.4. *Gigantic loops in the corona*

Trans-equatorial loops linking two active regions have been observed many times. An investigation by Tsuneta[19] found that there is evidence for magnetic reconnection between the two active regions, and that the overall structure of the large trans-equatorial loop changes. The formation of these loops is not a random process with some evidence showing that the loops exist well before reconnection takes place.

Recently, a series of homologous disappearances of trans-equatorial active region loops have been observed (Khan & Hudson[20]). Each one of these was associated with a major flare happening before the CME. The CME appears to be above the location of the trans-equatorial loop and not within the flaring arcades. Glover *et al.*[21] studied 18 of these huge loops, and found that not only did some of these loops disappear, but some became brighter.

Harra, Matthews and van Driel-Gesztelyi[21] discovered that these huge loops actually show many similarities with standard flaring!

5. The Future

Most of our knowledge about coronal mass ejections (CMEs) is derived from white light coronagraph data. Pre-Yohkoh it proved difficult to assign any strong connection between the soft X-ray corona and the white-light CMEs. Yohkoh SXT, with its high time cadence and sensitivity, has had the opportunity to study some of the X-ray phenomena related to CMEs and flares in much more detail. The coronagraph LASCO on board SOHO is now providing continuous observations of white-light data providing the ability to observe the outward flows that may be related to soft X-ray features.

Although much progress has been made over the past decade in understanding explosive events, many areas of our understanding is lacking. In particular, we have not found the real source of the trigger of both flares and CMEs in the Sun. Until we do this we will not be able to understand the magnetic energy release that occurs throughout the Universe. Once we understand the trigger, we can then begin making accurate predictions of explosions and their consequences on the Earth's environment. A number of future space missions are currently being developed which will together address one of the major goals in space physics.

5.1. *The follow on to Yohkoh - Solar B*

Solar-B is a joint UK-US-Japan mission which is due to be launched in 2006. It will use a combination of magnetic field, electric current and velocity field measurements in the photosphere along with imaging and spectroscopy in the corona in order to understand what triggers flares and CMEs. The union of such high spatial and temporal resolution will allow us to study the physical process in detail.

5.2. *Stereo - A 3-D look at the Sun*

The NASA Solar Terrestrial Relations Observatory (Stereo) will consist of two identical spacecraft observing the Sun at two different viewpoints providing simultaneous imaging. This will allow for the first time a 3-D image to be built up of a CME, and the complex structures around them. It will be launched in 2006.

5.3. *The Solar Dynamics Observatory*

The Solar Dynamics Observatory (SDO) is a NASA-led cornerstone mission within the International Living with a Star (ILWS) programme. It will be the first space mission to provide full-sun imaging simultaneously above and below the Sun's surface. This will allow us to track the creation of magnetic flux in the convection zone and how this triggers explosive events. SDO will be launched in 2007.

5.4. *Solar orbiter*

The ESA Solar Orbiter mission is planned for launch after 2012. It will reach 0.21 AU to explore the innermost regions of the solar system. In addition it will provide images of the polar regions from heliographic latitudes as high as 38 degrees. It will be able to determine in-situ the properties and dynamics of plasma, fields and particles close to the Sun, and achieve remarkable spatial resolution of up to 30 km.

6. Conclusions

Much has been learnt over the past 10 years about the working of explosive events and their origin. The ultimate goal now is to understand the actual trigger of these spectacular events. The combination of the missions described in the previous section will together provide an entirely new view of the Sun - from very close-up to 3-D to looking below the surface and linking the surface behaviour to the corona. The next decade will unravel even more of the Sun's mysteries.

Acknowledgments

LKH would like to thank PPARC for the award of an Advanced Fellowship.

References

1. G. Branduardi-Raymont, R. F. Elsner, G. R. Gladstone, G. Ramsay, P. Rodriguez, R. Soria, J. H. Waite, Jr, *A&A*, submitted.
2. T. Yokoyama, K. Akita, T. Morimoto, K. Inoue , J. Newmark *ApJ*, **546**, L69 (2001).
3. E. Priest & T. Forbes *Magnetic Reconnection*, Camb Univ press (1999).
4. H.P. Warren and A.D. Warshall *ApJ*, 560, L87 (2001).
5. D.C. Braun, C. Lindsay, Y. Fan and M. Fagan, *ApJ*, **502**, 968-980 (1998).
6. N.R. Sheeley *et al. Solar Phys.* **45** 377 (1975).

7. J. Feynman and S.F. Martin, *J. Geophys. Res.*, **100** 3355, (1995).
8. D.M. Rust and A. Kumar *ApJ* **L199** (1996).
9. R.C. Canfield, H.S. Hudson & D.E. McKenzie, *GRL*, **26**, 627-630 (1999).
10. A. Glover, N.D.R. Ranns, L.K. Harra, J.L. Culhane, *GRL*, **27**, 2161 (2000).
11. A.C. Sterling & H.S. Hudson *ApJL*, **491**, 55-59 (1997).
12. L. van Driel-Gesztelyi *et al.*, *Third Advances in Solar Physics Euroconference: Magnetic Fields and Oscillations, ASP Conference Series* **vol.184** 302 (1999).
13. P. Démoulin and M.A. Berger *Solar Phys.*, *215*, 203-215 (2003).
14. L.K. Harra and A.C. Sterling *ApJ*, *561*, L215 (2001).
15. H.S. Hudson *Magnetodynamic phenomena in the solar atmosphere. Prototypes of stellar magnetic activity* Proceedings of the 153rd colloquium of the International Astronomical Union held in Makuhari; near Tokyo; Japan; May 22-27; 1995; Dordrecht: Kluwer Academic Publishers; edited by Yutaka Uchida, Takeo Kosugi, and Hugh S. Hudson, p.89, 1996.
16. E. Hiei, A.J. Hundhausen & D.G. Sime *GRL*, **20**, 2785-2788 (1993).
17. B.J. Thompson, S.P. Plunkett, J.B. Gurman, A.C. St.Cyr, and D.J. Michels *GRL* **25** 2465 (1998).
18. R.A. Harrison and M. Lyons *Astron. Astrophys.* **358** 1097 (2000).
19. S. Tsuneta *ApJL*, **456**, 63-65 (1996).
20. J.I. Khan and H.S. Hudson *GRL* **27** 1083 (2000).
21. A. Glover, L.K. Harra, S.A. Matthews, C.A. Foley *A&A*, *400*, 759-767 (2003).
22. L.K. Harra, S.A. Matthews and L. van Driel-Gesztelyi *ApJ*, *598*, 59-62 (2003).

Louise K. Harra

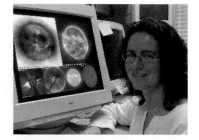 

Louise Harra was born in Co. Armagh, and attended Banbridge Academy. She graduated from Queens University, Belfast with a BSc(Hons) in applied mathematics and physics in 1990, followed by a PhD in physics in 1993. She worked as a resident scientist at the Institute of Space and Astronautical Research in Japan for the Yohkoh mission. Following that she was awarded a PPARC advanced fellowship. She is a Reader at University College London, and was the leader of the Solar and Stellar Physics group at the Mullard Space Science Laboratory. She is a project scientist for the Japan-US-UK mission, Solar-B, which is due to be launched in 2006. In 2004 she received a Philip Leverhulme award for her achievements in solar physics. She enjoys sailing a 22ft keelboat (International Tempest).

CHAPTER 20

SOLAR VARIABILITY, COUPLING BETWEEN ATMOSPHERIC LAYERS AND CLIMATE CHANGE

Neil Arnold

Department of Physics and Astronomy, University of Leicester
University Road, Leicester, LE1 7RH, U.K.
E-mail: nfa1ion.le.ac.uk

One of the enduring puzzles of atmospheric physics is the extent to which changes in the Sun can influence the behaviour of the climate system. Whilst solar–flux changes tend to be relatively modest, a number of observations of atmospheric parameters indicate a disproportionately large response. Global–scale models of the coupled middle and upper atmosphere have provided new insights into some of the mechanisms that may be responsible for the amplification of the solar signal. In particular, modification of the transport of heat and chemicals, such as ozone, by waves during periods of solar activity has been shown to make an important contribution to the climate of the stratosphere and mesosphere. In this paper, a review of some of the recent advances in understanding the coupling between atmospheric layers and how this work relates to Sun–weather relations and climate change will be presented, along with a discussion of some of the challenges that remain.

1. Background

The importance of the Sun to life on Earth has long been recognized and our relationship with our stellar neighbour has featured prominently in religious and philosophical thoughts since ancient times. From a scientific perspective, the Greek notion of an unchanging, perfect celestial sphere has been retained in the concept of the 'solar constant' that has been the paradigm for generations of meteorologists and climatologists. This understanding had to be modified somewhat to accommodate the theory by Milankovitch and others that climatic cycles with periods *ca.* 23,000, 40,000 and 100,000 years matched changes in the precession, obliquity and eccentricity respectively of the Earth's orbit around the Sun (Hays *et al.*[1]). These long–term

---

perturbations coincided with a series of Ice Ages. As the Earth was experiencing a cooling trend in the middle of the 20th century, there were concerns that the current interglacial was coming to a close.

What has yet to be established satisfactorily is how small variations in solar flux related to these orbital changes could bring about temperature swings of greater than 10 K over time–scales of just a few decades (Johnsen et al.[2]). Benzi et al.[3] proposed the theory of stochastic resonance whereby the atmosphere possesses two or more stable states that are separated by an energy barrier that could undergo 'hopping' between these states in harmony with the small changes in the solar flux. The presence of noise in the form of internal climate variability permits 'resonant' interactions that amplify the initial signal considerably. The highly nonlinear nature of the process could help to account for the dramatic reversals in the climate state.

More recently, an upturn in observed global surface temperatures has coincided with anxiety that enhanced levels of man–made greenhouse–gas emissions may be at least in part responsible (IPCC[4]). In addition to increases in these gases, sunspot activity has also been on the rise since the 18th century, a time when sunspots were not observed for several decades (Maunder[5]). Records of geomagnetic activity have revealed a significant rise in the occurrence of magnetic storms (Clilvard et al.[6]). This latter trend has been related to a doubling in the total solar magnetic flux that has been encountered by the Earth over the past century (Lockwood et al.[7]). It has been suggested that the Maunder Minimum in solar activity was responsible for anomalously cold climate conditions in Europe at that time (Eddy[8]). Similar arguments have been employed to account for a Mediaeval warming that could be attributed to a peak in solar activity.

The history of Sun-weather relations goes back a long way (see Hoyt & Schatten[9] for a review). Many interesting connections were reported in the scientific literature, only to be subsequently refuted by the application of more rigorous statistical analyses, or the discovery that the correlation switched signs or broke down over the next solar cycle. One possible explanation for this tendency is that climate is susceptible to a number of other influences. For example, a recent study by Donarummo et al.[10] indicated that sunspot/climate correlations were reversed during periods of high volcanic activity.

Another peak in interest in sun-weather relationships was initiated by the work of Wilcox et al.[11] They found that atmospheric vorticity in the Northern Hemisphere increased a day or two after the reversal of the component of the Interplanetary Magnetic Field that is perpendicular to the ecliptic plane. The IMF is the undulating magnetic structure that radiates outwards from the Sun. Examination of the forecasting skill of weather prediction models indicated that they suffered noticeably at these times.

While the study by Wilcox *et al.*[11] received considerable attention, interest faded when it proved to be difficult to incorporate this discovery into practical forecasting, as the sister field of 'space weather' was (and, to a large extent, still is) very much less mature than its tropospheric counterpart. Even more significantly, the correlations appeared to peter out as observations were acquired into the 1980s. Sceptics have regularly promoted the view that the subject was a classic example of 'auto-suggestion' in action, as researchers subconsciously allowed prejudicial thinking to cloud their objectivity when it came to analyzing and interpreting the data (Pittock[12]). The ideas of Wilcox were given a new lease of life when Tinsley[13] pointed out that the correlation was strongest when the atmospheric aerosol loading was high (there were several strong volcanic eruptions in the early phase of the original analysis) and that the relationship between the IMF and weather was more complex than previously thought.

The dawning of the Space Age had led to two vital pieces of information becoming available that allowed the magnitude of solar variability to be quantified. Firstly, unambiguous evidence of the solar wind was obtained from the first rocket observations. Parker[14] had predicted that the Sun's atmosphere had to extend a long way beyond its photosphere to satisfy the boundary conditions of its atmosphere. The outflow would be continuous and is comprised of a magnetised plasma. Secondly, long–term monitoring of the total solar irradiance by satellites confirmed that there was an 11 year variation in electromagnetic radiation. To date, the magnitude of this perturbation has been observed to be *ca.* $2Wm^{-2}$, which is a few tenths of a percent of the total, *ca.* $1350Wm^{-2}$ (see for example Fröhlich[15]). Short–term variations related to the 27 day rotation period and turbulent solar activity were found to have a similar magnitude. The Sun is clearly a variable star, but it was difficult to see how such a small effect could bring about significant changes to the Earth's weather. A total solar– irradiance change of a few Wm^{-2} has been calculated to cause the global mean surface temperature of the Earth to increase by *ca.* 0.1 K (e.g. Haigh[16]). Attempts to simulate the atmospheric response to the requisite changes in total solar irradiance have been most successful when the variation in solar forcing has been artificially enhanced by a factor of between three and ten (see, for example Wigley & Raper[17]).

Whilst the Sun's visible and infrared radiation has been observed to be stable to within the 0.01 % instrumental accuracy maintained since observations began, at the extremes of the electromagnetic spectrum, the picture is very different. The solar corona is in perpetual flux, as magnetized plasma bursts into the outer atmosphere, generating temperatures of $10^6 - 10^7$. Flares and coronal mass ejections can also release copious amounts of ultraviolet, X–ray and radio emissions. The highly energetic photons cause ionization in the upper atmosphere. The relative contribution of radiation variations, both electromagnetic with wavelengths below 100 nm and corpuscular, accounts for additional heating of *ca.* $0.1 W m^{-2}$.

Particles that stream from the outer corona can reach the Earth's magnetosphere and their associated magnetic fields will perturb the Earth's field, generating enhanced cross–polar–cap potential differences that will drive enhanced convection flows in the ionosphere. The particles themselves can also heat up the ionosphere at high latitudes, forming the Earth's aurora in the process. Above 100 km altitude, the atmospheric temperature can rise by nearly 1000 K between solar minimum and solar maximum and by hundreds of degrees as the plasma ejected from a large solar disturbance passes the Earth's magnetosphere, because the density in this region is extremely low. Below 80 km, the extent of the direct solar perturbation becomes progressively more muted as the pressure increases exponentially, making it increasingly difficult to isolate the solar signature in the observations.

2. Solar Forcing of the Troposphere

Within the troposphere, there is a large number of competing processes that may affect climate. The Intergovernmental Panel on Climate Change,[4,18] identified radiative changes associated with the increase in greenhouse gases such as methane, carbon dioxide, ozone and water vapour in the troposphere and ozone in the stratosphere as being the most significant and the best understood. However, it was recognized that there were also important contributions from changes in sulphate aerosols (due to volcanic activity), fossil fuels and biomass burning and solar forcing. For example, a large eruption such as Mount St. Helens in 1980 can lead to global cooling of a few tenths of a degree for a few years due to enhanced stratospheric aerosol loading reflecting more incoming sunlight. In general, it has proved difficult to identify trends of a few tenths of a degree per decade by anthropogenic influences because of the large natural variability of the climate. Statistical regression techniques have identified solar and aerosol changes as being the most likely agents for change before 1940, whilst a combination of these plus greenhouse gas increases can account for the post Second World War data e.g. Stott *et al.*[19]

High–quality global meteorological records have only been available for the past 40 years or so, and regular satellite monitoring for only half of that time. There remains some, as yet unexplained, discrepancy between measurements made at the surface and those via satellites. Some of the reported global–warming trend may be related to the 'heat island' effect of placing meteorological stations near human activities. For this reason, upper–air measurements may be more reliable, as they have been carefully calibrated against balloon observations and other spacecraft and maintain a high degree of spatial consistency. All data can be fed into a modified version of a global numerical weather–prediction model so that our best estimate of what the weather was doing may be obtained.

Figure 1 illustrates the correlation between monthly mean temperatures on a constant–pressure surface of 500 mbar (corresponding to *ca.* 5.5 km, the altitude where the atmospheric thickness has reduced by approximately one-half) and geomagnetic activity levels (quantified by the planetary Ap index) over the same 40 year interval. Rather than using all of the data available to us, I have made the apparently arbitrary choice of choosing those years where the equatorial winds in the stratosphere were blowing in a particular direction. The reasons for this choice will be made clear in the next section. Statistical significance at the 95 % level was achieved by correlation coefficients of magnitude 0.35 or greater for the number of samples available. Any dark shaded regions fall into this category. Immediately, the eye is drawn to large areas of the Northern Hemisphere where the correlation coefficient exceeds this threshold quite markedly. Much of the winter polar region is positively correlated whilst the mid–latitudes in this hemisphere exhibits what appears to be alternating bands of positive and negative agreement. These bands are related to global–scale normal modes of the atmosphere. Either by coincidence or by design, the atmosphere is resonating with the rhythm of the magnetosphere. Solar–activity–related correlations and anti–correlations with a dividing line down the middle of the North American continent have been identified using the extensive surface observing network (Currie[20]). The pattern was found to be significant but less distinct in the tropical and Southern Hemisphere regions. A physical mechanism was needed to explain any apparent connections.

Svensmark & Friis-Christensen[21] found a strong correlation between

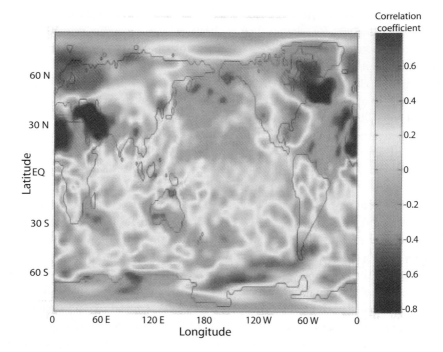

Fig. 1.   Global pattern of statistical significance of solar forcing versus temperature at 500 mb for the eastern phase of the equatorial Quasi Biennial Oscillation during the Northern Hemisphere winter showing clear influence in the North Atlantic storm track region. The data was obtained from the National Center for Environmental Prediction 40 year global climate reanalysis project. (Courtesy of Wesley Ebisuzaki and Chris Calvey).

the amount of cloud cover over the low– latitude oceans (this choice was made because of difficulties in detecting clouds over land and at high latitudes unambiguously) and the level of Galactic Cosmic Ray (GCR) activity. They suggested that these energetic remnants of high energy astrophysical processes outside our Solar System, such as supernovae, would be able to influence the rate of formation of clouds in a manner loosely analogous to that used to study nuclear processes in cloud–chamber experiments. They postulated that the ionized air would provide an efficient centre for water–droplet growth to seed clouds. The link with solar variability comes through the observation that the intensity of cosmic radiation in the atmosphere is anti–correlated with sunspot numbers. Charged particles passing through the heliosphere encounter a deflecting force field, associated with the solar wind, that is more or less intense depending on the phase of the solar cycle (see, for example, Longair[22]).

There were some criticisms of the original paper on the grounds that there was no correlation between individual cloud types and poor agreement at high latitude (Kernthaler *et al.*[23]) and that there was insufficient satellite data (Jorgensen & Hansen[24]) to draw any firm conclusions. Marsh & Svensmark[25] extended the original study using 12 years of infrared radiances from the International Satellite Cloud Climate Project. Cloud–top temperatures and pressures were determined by a combination of modelling and vertical temperature soundings from the Television and Infrared Observations Satellite Observed Vertical Sounder Observed Vertical Sounder. They found that the correlation between cosmic rays and clouds was strongest for low clouds. While the peak ion–production region was $10 - 15$ km, they argued that the low–altitude ionization was responsible for the generation of ultrafine (diameter $< 0.02\mu m$) aerosol. Earlier modelling studies had suggested a stable concentration of several hundred particles per cubic centimetre that was sustained by GCRs in the lower maritime atmosphere (Turco *et al.*[26]).

Alternative hypotheses are available. It has been suggested that solar induced ionization variations can affect the global electric field and associated current flows (see, for example, Tinsley[27]). The presence of space charges on water–ice droplets dramatically accelerates the rate of droplet formation. Changes to radiative transfer properties of clouds and latent–heat release by the phase change of water could potentially modify the climate system. Similarly, Pudovkin & Veretenenko[28] have shown that aerosol loading in the upper troposphere and lower stratosphere will be affected by high–energy solar particles, especially at high latitudes, thereby affecting the albedo and radiative transfer processes within the atmosphere. Considerable work still needs to be done to link these micro–physical processes to the global circulation and then to demonstrate that they can lead to the observed changes within the atmosphere. Fortunately, these challenges coincide with the need to ascertain the full impact of anthropogenic activity on climate and the study of the one will accelerate the progress of the other.

3. Recent Evidence for Enhanced Solar Influences on the Middle Atmosphere

Routine balloon soundings of the stratosphere up to *ca.* 25 km began in the late 1950s, with a fairly comprehensive coverage of the Northern Hemisphere landmasses achieved within the subsequent decade (Peixoto &

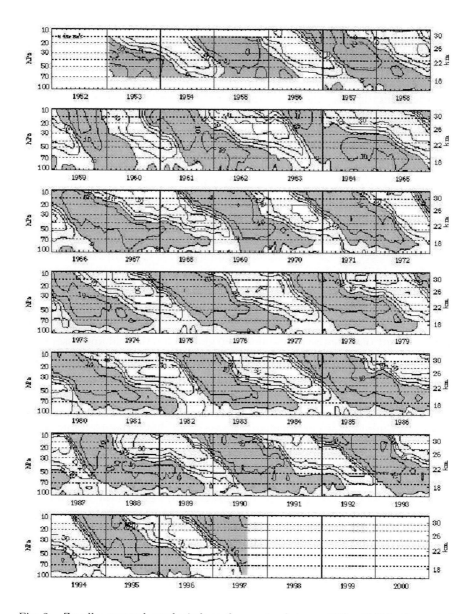

Fig. 2.   Zonally averaged zonal winds at the equator between $5°S - 5°N$ for the period
1958 - 1998 showing the equatorial Quasi-Biennial Oscillation in the stratosphere. The
west winds are positive (shaded) and the east winds are negative. Contour intervals are
$10ms^{-1}$. Updated from Naujokat[61].

Oort[29]). One of the first climate features to be identified by this network was the systematic oscillation of the equatorial winds with an irregular period between 18 - 27 months (Veryard & Ebdon[30]). The data set has now been extended to the present (Figure 2). This phenomenon was explained in terms of waves propagating from the troposphere through to the stratosphere and mesosphere, where they could interact with the background winds, thereby transferring energy and momentum from one region to another. Near the Equator, the Coriolis restoring force is weak and so instabilities may arise. Perturbations in the background flow affect the progress of waves, and it was proposed that a downward–propagating phase front was generated (Holton & Lindzen[31]). The spectrum of waves was altered in such a way that gradually waves of the alternate phase began to dominate at any given altitude and so an oscillation in the winds was set up – referred to as the Quasi Biennial Oscillation (QBO). The time scale is long because the coupling is downward (albeit with the source coming from below) and the mismatch in the respective masses of the layers is substantial (see Baldwin *et al.*[32] for a review). This is an important example of how the lower atmosphere can be significantly influenced by processes that are driven from above (Haynes *et al.*,[33] McIntyre[34]).

From considerations of continuity and the conservation of momentum, changes in one part of the atmosphere must be accompanied by a response elsewhere. Attempts to identify a high–latitude analogue to the QBO met with little success (Holton & Tan[35]) until Labitzke & van Loon[36,37] noticed a 10–12 year modulation in the data in the Northern Hemisphere winter stratospheric polar region. Once the data was divided into East and West components of the QBO, a clear solar signature was evident, especially in the west phase. Instead of the 0.7 K variation that had been calculated to occur in the equatorial region due to solar–induced ozone heating, a change of up to 20 K was observed near the North Pole at a height of *ca.* 25 km. This result was far from self–evident as the polar winter is not exposed to direct sunlight. Clearly, radiative–dynamical coupling processes were involved. More recently, Gray *et al.*[38] have extended the study to include rocket data going up to the stratopause. They found that the evolution of the winds in the upper stratosphere, where solar insolation variations are stronger, was a significant influence on the winter stratospheric circulation.

Simple radiative models predicted that, through an annual cycle, the winter polar stratosphere temperature should drop to 150 K (Fels[39]), while observations put the figure closer to 210 K (Figure 3). Another apparent anomaly is that stratospheric ozone concentrations (halocarbon catalysed

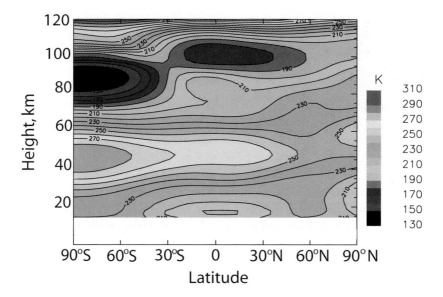

Fig. 3. Climatology of seasonally averaged zonal mean temperatures for the Northern Hemisphere winter for the middle atmosphere and lower thermosphere. The stratosphere is the region between the temperature minimum around 15 km and the maximum at 50 km, while the mesosphere is the region of decreasing temperatures above that. (Taken from the MSIS-90E climate reference model Hedin A. E. 1991 *J. Geophys. Res.* **96** 1159.)

destruction not withstanding) can be the largest in the world in this region, at a location where there is no production in that season. Once again, the only explanation has to be that heat and chemical constituents are being transported there from elsewhere in the atmosphere.

Global–scale waves that are normal modes of the atmosphere provide a means of transporting extra–tropical air masses into the polar region. Land masses in general and mountain ranges in particular, play a role in their generation. When air masses move north or south, they experience a Coriolis force due to the rotation of the Earth, analogous to that experienced by trying to move outwards on a playground roundabout. These waves are associated with the frontal systems in the troposphere that regularly bring bad weather to the west coast of Europe from the Atlantic.

Climatologies of these waves confirmed the theoretical work of Charney and Drazin[40] who found that only the largest of these waves in one favoured direction were able to propagate into the stratosphere, as the motion of the upper–air winds ensured there were no real solutions to the relevant atmospheric equations in the opposite direction (Figure 4 shows prevailing

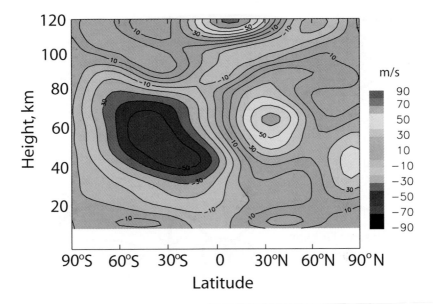

Fig. 4.   Climatology of seasonally averaged zonal winds for the Northern Hemisphere winter. (Taken from the Empirical Wind Model Hedin A. E., E. L. Fleming, A. H. Manson, F. J. Schmidlin, S. K. Avery and S. J. Franke 1993 *NASA Tech. Memo. 104581.*)

winds in the Northern Hemisphere winter). This filters out many medium–scale disturbances and ensures that the winter high–latitude stratosphere is a relatively simple environment to study geophysical fluid dynamics. Also, the extremely cold tropopause freezes out most of the water vapour before it reaches the stratosphere and much of what little remains is due to the breakdown of methane rather than from upward transport.

A conceptual simplification of middle atmosphere dynamics was achieved by Andrews & McIntyre.[41,42] They revisited the superposition theory of waves to demonstrate that steady, linear flows must remain unaffected by their environments and *vice versa*. These well–behaved waves could not account for the systematic transport of heat and chemicals to high latitudes, thereby removing much of the observed planetary wave activity from consideration. The 'residual circulation' that remains directly relates the balance between the net diabatic heating and cooling due to infrared and ultraviolet radiative transfer and transient, nonlinear wave activity, primarily related to planetary waves in the stratosphere.

The combination of differential heating and transient motions generate several global–scale circulations. The first is the Brewer–Dobson circu-

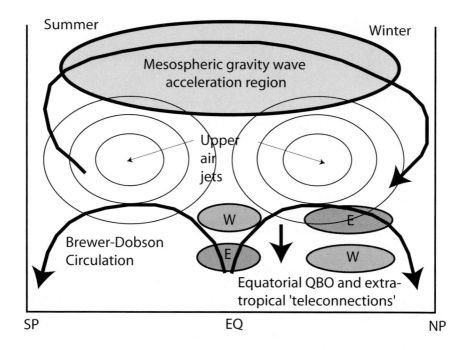

Fig. 5.   Schematic of the global scale circulation in the middle atmosphere at Solstice.
The stratosphere is dominated by the Brewer-Dobson circulation with upflows near the
equator, poleward transport and down-welling at high latitudes. These cells help to
account for the anomalously high ozone values that used to be achieved in these regions before destruction by chlorofluorocarbons became important. The QBO occupies
a narrow zone within 15 degrees of the equator but due to teleconnections, there are
associated flows at higher latitudes. The phase fronts propagate downwards at the rate
of approximately 1 km per month. Gravity wave breaking in the upper mesosphere helps
to close the upper atmosphere jets and provide the mechanical forcing of the summer to
winter transport of air that leads to anomalously low summer temperatures and warm
winter temperatures, far from radiative equilibrium. Increases in solar forcing have been
shown to decrease gravity wave fluxes by 10 %, resulting in a weaker upper atmosphere
circulation and a reduction in the period of the stratospheric QBO.

lation (Brewer[43]; Dobson[44]), where air masses ascend near the Equator,
travel polewards at higher altitudes and then descend towards the poles.
At greater altitudes, air travels from the summer polar mesosphere to the
winter polar mesosphere; the pattern has been named after Murgatroyd &
Singleton[45] who first postulated it 1961 (Figure 5). An important consequence of the work of Andrews & McIntyre was the realization that increases in ultraviolet heating *per se* could not account for an enhancement
of the global meridional circulation (McIntyre[47]), only a change in the east–

west wind strength. For this circulation to be effective, it must be associated with changes to wave activity.

As the Northern Hemisphere polar stratosphere approaches winter conditions, the absence of strong ultraviolet heating results in falling temperatures. The temperature gradient generates an intensification of the zonal winds (due to the Coriolis force being perpendicular to the gradient in meridional temperature in a rotating fluid). This region of strong gradients generates a vortex through which the planetary waves find it difficult to propagate. Without these waves, the polar region gets colder still. There is thus the potential for significant amplification to occur if these conditions are perturbed. At solar maximum, the gradient in temperature is enhanced a little, leading to an enhanced cooling of the polar region. Through the feedback process outlined above, the cooling can then be amplified considerably.

The above picture is complicated by the presence of the QBO. In the eastern phase, this tendency is enhanced, while, during the western phase, the equatorial perturbation is sufficient to disrupt the process. Paradoxically, the influence of solar forcing was found to be strongest at solar maximum in the western phase (Figure 6). In an independent analysis of the origins and nature of inter-annual variability in the winter stratosphere, Kodera[48] suggested that the anomalous patterns in zonal wind exhibit a dipole–like behaviour that extends down to the troposphere. He postulated that the cause was due to internal variability triggered by changes in external forcing, such as solar activity, the equatorial QBO and volcanic activity.

Numerical models of the middle atmosphere were able to reproduce qualitatively the bifurcation in the Northern Hemisphere winter stratosphere with imposed QBO forcing and solar ultraviolet heating perturbations that were significantly larger than those observed (Kodera *et al.*[49]). Introducing solar modifications to the ozone distribution reduced this discrepancy (Haigh[50]). Haigh also found that a consistent change in the tropical circulation pattern near the tropopause resulted (Haigh[16]).

There is another class of waves that are too small in scale to experience the Coriolis force to a significant extent. These are buoyancy waves, also referred to as 'gravity waves', due to the nature of the restoring force. If an air parcel is forced over an object, such as a mountain, then its density will not match its surroundings. Gravitational force will be exerted in the counter direction and an oscillatory motion will be set up. A large proportion of these atmospheric gravity waves are generated in this way by variable topography, but other sources such as tropical cloud convection, instability

N. Arnold

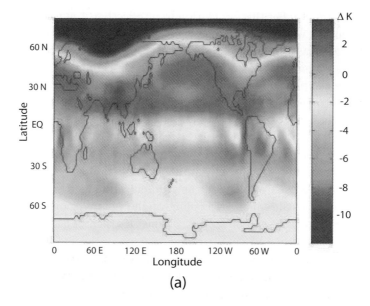

(a)

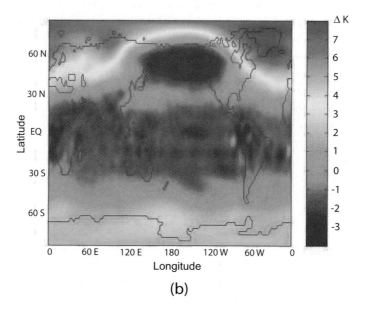

(b)

Fig. 6.   NCEP climate reanalysis air temperature anomalies at 25 km from 40 year mean. (a) 1991 a year with western phase QBO and solar maximum (b) 1974 a year with western phase QBO and solar minimum.

around frontal systems and strong wind shear zones can be important, especially over the oceans. Whilst their amplitudes start off being quite modest, they are able grow exponentially with height in line with the fall in atmospheric density. Those waves that reach the mesosphere can have amplitudes of tens of meters per second and vertical wavelengths of *ca.* 10 km. These waves play a vital role in determining the climatology of the mesosphere. In the summer, observed temperatures are many tens of degrees cooler than radiative equilibrium calculations predict – the opposite of the situation in the winter polar stratosphere – due to mechanical cooling by these waves as they interact with the prevailing winds (see for example Mayr *et al.*[51,52]).

It has been observed experimentally that the gravity–wave flux in the mesosphere is consistently lower at solar maximum than at solar minimum by *ca.* 10 % (Gavrilov[53]). From the work of Andrews & McIntyre, it follows that the global circulation will be similarly reduced and that in turn the QBO will also be affected. A solar influence on the QBO has indeed recently been reported (Salby & Callaghan[54]) and can be seen qualitatively in the data. At solar maximum the QBO has a relatively short period as the phase–front descent rate is relatively strong, while at minimum, the oscillation appears to halt for months at a time before starting up again. Planetary wave propagation into high latitudes will be affected by these changes. A dramatic indication of this is the fact that all of the observed major warming events (defined by achieving a reversal in the wind direction at a particular latitude and altitude where the polar stratospheric vortex normally resides) in the Northern Hemisphere stratospheric winter during the Western phase of the QBO have only ever been observed during high levels of solar activity. In other words, it appears that solar forcing is able to determine some aspects of the large–scale circulation of the winter stratosphere.

Weaker mesospheric gravity waves will lead to stronger winds in the upper atmosphere which in turn will affect the ability of planetary waves to propagate, producing an additional feedback route (Arnold & Robinson[55]). A combination of these radiative and dynamical processes can generate a significant enhancement of the initial forcing (Arnold & Robinson[56]). Figure 7 provides a schematic representation of the role of planetary waves in amplifying the influence of upper atmosphere perturbations on the winter stratosphere. It is important to remember that the upper atmosphere is behaving like the base current in a transistor. The vast bulk of the energy and momentum redistribution is carried out within the stratosphere itself.

While gravity waves are an important component of upper atmosphere accelerations, there are other mechanisms available. Of particular interest

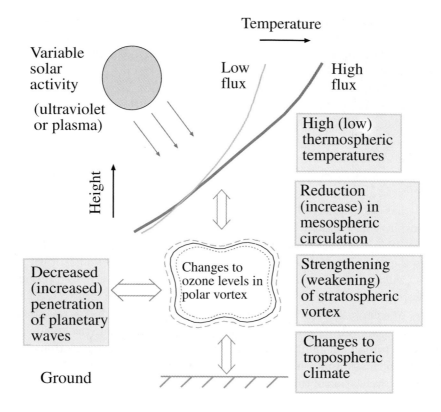

Fig. 7.   Schematic of the influences of solar activity on the (eastern phase) winter stratosphere. Enhanced solar forcing either from ultra violet radiation or space plasma will
warm up the thermosphere and increase the speed of the upper atmosphere jets. Gravity
wave flux is reduced with an associated weakening of the global circulation. Transport
of ozone and heat into the polar regions is reduced as the strengthened polar vortex
diminishes planetary wave penetration.

is the acceleration of the lower thermosphere associated with the enhancement of ion flows that have been initiated by the transfer of plasma from
the solar wind into the ionosphere. The conditions required to bring such
processes about are beyond the scope of this paper, so the reader is referred
to Cowley.[57] In the mid–thermosphere, speeds in excess of 1000 ms^{-1} can
be achieved. The resultant changes to the mean circulation at a height of
110 km will couple down through continuity to the mesosphere where interaction between the background air and the gravity waves will be affected
and then into the stratosphere where planetary wave transport will become
involved. Modelling studies by Arnold & Robinson[58] have indicated that

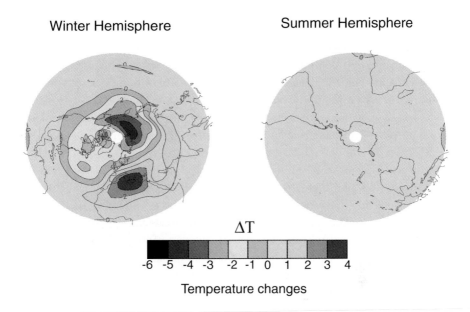

Fig. 8. 8 Model simulation of temperature difference between (a) solar maximum and (b) solar minimum conditions in the Northern Hemisphere stratospheric winter. The plot is a cross-section of the stratosphere at *ca.* 40 km altitude. The polar vortex has cooled by up to 6 K, as mixing due to planetary wave activity has been suppressed.

these space plasma effects can be as effective as more direct ultraviolet heating variations in modifying the middle atmosphere (Figure 8) and could go some way to accounting for the effect first observed by Wilcox *et al.*[11]

We can begin to understand why many of the early reports of Sun–weather relationships were destined to be contradictory. In the lower atmosphere, there is a strong interplay between radiative forcing and atmospheric dynamics. At a given stage in the annual cycle, the atmosphere can occupy a number of states, especially in the winter months and the energy difference between them is very small. Just like electron states in an excited heavy atom, these states are quasi-degenerate. Over time, the states that are available to the circulation will also change, for example due to the eruption of a volcano with its attendant stratospheric ash or the onset of an El Nino event in the tropospheric circulation. Using models, it is possible to isolate the solar contribution and effectively control the natural variability in the accompanying conditions and thereby demonstrate that the signal is indeed present and often quite strong.

4. Future Challenges

With the exception of radiative processes such as ultraviolet heating of ozone and infrared cooling of enhanced levels of carbon dioxide, many of the physical mechanisms within the atmosphere are only moderately understood. Quite often micro–physical processes, such as the catalytic destruction of ozone by the daughter products of chlorofluorocarbons on cloud droplets, play a crucial role in large scale processes. Representing all of these scales in models that purport to simulate the climate of the distant future appears to be almost beyond our abilities. Considerable efforts have to go into 'parameterizing' small scale processes onto courser model grids. Success has been achieved with gravity waves, allowing us to simulate a number of fundamental atmospheric processes for the first time (e.g. Mengel *et al.*[59]). Fortunately, detailed knowledge of individual phenomena is often less important than the accumulated statistical characteristics of the driving mechanism. For example, the spectrum of atmospheric gravity waves obeys a power law.

Also, in the upper atmosphere, transport processes redistribute energy and momentum and chemistry affects constituent distributions on timescales shorter than those that are believed to be important in the lower atmosphere, so approximations can often be made. A new generation of three–dimensional computer models that include the neutral middle and upper atmospheres along with processes relevant to the ionosphere are being developed that overcome the tradition boundaries between climate and solar–terrestrial research (e.g. Harris *et al.*[60]).

The transition from relatively simple mechanistic models of the role of solar variability to weather prediction has yet to be attempted in a systematic fashion. Atmospheric data is only routinely available up to an altitude of 45 - 55 km. One reason for this is that more sophisticated instrumentation is often required to obtain measurements of similar quality at greater heights. The first step is to demonstrate the efficacy of such addition information to forecasting skill, by attempting to assimilate research satellite data into modified numerical weather prediction models that have been extended up to the lower thermosphere. Whilst the output of solar ultraviolet radiation can be estimated quite well over an 11-year cycle, space plasma effects are inherently more unpredictable. Useful information can still be obtained by using an ensemble of forecasting simulations with variable solar forcing to determine potential regions of susceptibility to solar activity, such as the winter North Atlantic storm tracks. Climate reanalysis would not be as problematic, as there are a number of space weather observing systems in operation at any given time providing past and current conditions.

In addition to the Sun's output varying, the Earth's magnetic field is also in a state of flux. Over the past century, the field strength has diminished by approximately 10 percent (Clilverd *et al.*[6]). Assuming this trend continues, we are due to experience a field reversal in the next few thousand years. This will have important consequences for the ability of the magnetosphere to shield the atmosphere from high–energy particles from the Sun and cosmic rays from outside the solar system. A coupled magnetosphere–ionosphere–atmosphere model would allow us to estimate what sort of a climate regime we would experience under these conditions. Certainly, the location of the auroral zones would change and the thermosphere would be much more disturbed, but what of the ozone layer and the North Atlantic storm tracks?

A micro-satellite mission has been proposed to investigate the links between solar variations and climate change, to overcome the inherent difficulties of bringing together data from a range of heterogeneous sources(Arnold [61]). Coordinated observations will contribute to treating the Sun-Earth system as a whole, rather than dividing it into isolated regions as has been the tendency in the past. This multi-disciplinary approach has gained ground over recent years and is simply following in the footsteps of pioneers such as Sidney Chapman, who was interested in the ozone layer, the upper atmosphere and the magnetosphere, to name but a few. We are rediscovering the curiosity that our ancestors had for the subtle influences that the Sun exerts on human affairs.

Acknowledgments

The author wishes to recognize the many invaluable contributions to this work made by Terry Robinson, Chris Calvey and the assistance of the National Center for Environmental Prediction for providing the climate database, especially Wesley Ebisuzaki. I am grateful to the referee for a number of constructive comments. Much of the research by the author has been made possible by the award of an Advanced Fellowship from the UK Particle Physics and Astronomy Research Council.

References

1. J.D. Hays, J. Imbrie and N. J. Shackleton, *Science* **194**, 1121 (1976).
2. S. Johnsen, D. Dahl-Jensen, W. Dansgaard and N. Gundestrop, *Tellus* **47B**, 624 (1992).

3. R. Benzi, G. Parisi, A. Sutera and A. Vulpiani, *Tellus* **34**, 10(1982).
4. Intergovernmental Panel on Climate Change, *Climate Change 1995: The Science of Climate Change*, Eds. J. T. Houghton, L. G. Meira Filho, B. A. Callander, N. Harris, A. Kattenberg and K. Maskell, (Cambridge University Press, 1996).
5. E.E. Maunder, *J. Br. Astr. Soc.* **32**, 140 (1922).
6. M. A. Clilverd, T. D. G. Clark and H. Rishbeth, *J. Atmos. Solar-Terr. Phys.* **60**, 149 (1998).
7. M. Lockwood M., R. Stamper and M. N. Wild M.N., *Nature* **399**, 437 (1999).
8. J. A. Eddy, *Science* **192**, 1189 (1976).
9. D. V. Hoyt D.V. and K. Schatten K., *The role of the Sun in climate change*, (Oxford University Press, 1997).
10. J. am M. Donarummo and M. Stolz, *Geophys. Res. Lett.* **29**, 1361 (2002).
11. J. M. Wilcox, P.H. Scherrer, L. Svalgaard, W. O. Roberts and R. H. Olson, *Science* **180**, 185 (1973).
12. A. B. Pittock, In *Solar Terrestrial influences on weather and climate* Eds. B. M. McCormac and T. A. Seliga, (P. 181 Reidel, 1979).
13. B. A. Tinsley, *EOS Trans. Am. Geophys. Un.* **75**, 369 (1984).
14. E. N. Parker, *Astrophys. J.* **128**, 664, (1958).
15. C. Fröhlich, *Space Sci. Revs.* **94**, 15 (2000).
16. J.D. Haigh, *Q.J.R. Meteorol. Soc.* **125**, 871 (1999).
17. T. M. L. Wigley and S. C. B. Raper *Geophys. Res. Lett.* **17** 2169 (1990).
18. Intergovernmental Panel on Climate Change, *Climate Change: The IPCC Assessment*, Eds. J. T. Houghton, G. J. Jenkins and J. J. Ephraums, (Cambridge University Press, 1990).
19. P. A. Stott, S. F. B. Tett, G. S. Jones, M. R. Allen, J. F. B. Mitchell and G. J. Jenkins, *Science* **290**, 2133 (2000).
20. R. G. Currie, *Int. J. Clim.* **13**, 31 (1993).
21. H. Svensmark H. and E. Friis-Christensen E., *J. Atmos. Solar Terr. Phys.* **59**, 1225 (1997).
22. M.S. Longair, *High energy astrophysics*, (Cambridge University Press, 1981).
23. S.C. Kernthaler, R. Toumi R. and J. D. Haigh, *Geophys. Res. Lett.* **26**, 863 (1999).
24. T. S. Jorgensen and A. W. Hansen, *J. Atmos. Solar Terr. Phys.* **62**, 73 (2000).
25. N. D. Marsh and H. Svensmark, *Phys. Rev. Lett.* **85**, 5004 (2000).
26. R. P. Turco, J. X. Zhao and F. Yu, *Geophys. Res. Lett.* **23**, 635 (1998).
27. B. A. Tinsley, *Space Sci. Revs.* **94**, 231 (2000).
28. M. I. Pudovkin and S. V. Veretenenko, *J. Atmos. Terr. Phys.* **57**, 1349 (1995).
29. J. P. Peixoto and A. H. Oort *Physics of Climate*, (Am. Inst. Physics, 1992).
30. R. G. Veryard and R. A. Ebdon, *Meteor. Mag.* **90** 125 (1961).
31. J. R. Holton and R. S. Lindzen, *J. Atmos. Sci.* **29**, 1076 (1972).
32. M. P. Baldwin, L. J. Gray, T.J. Dunkerton, K. Hamilton, P.H. Haynes, J.R. Holton, M.J. Alexander, I. Hirota, T. Horinouchi, D.B.A. Jones, J.S. Kinnersley, C. Marquartdt, K. Sato and Takahashi, *Rev. Geophys.* **39**, 179 (2001).
33. P.H. Haynes, C.J. Marks, M.E. McIntyre, T.G. Shepherd and K.P. Shine, *J. Atmos. Sci.* **48**, 651 (1991).

34. M.E. McIntyre, In *NATO ASI Seris C: Transport processes in the middle atmosphere,* Eds. G. Visconti and R. Garcia, (p. 267 Springer-Verlag, 1987).

35. J. R. Holton and H. -C. Tan, *J. Atmos. Sci.* **37**, 2200 (1980).

36. K. Labitzke and H. van Loon, *J. Atmos. Terr. Phys.* **50**, 197 (1988).

37. K. Labitzke K. and H. van Loon H., *Ann. Geophysicae* **11**, 1084(1993).

38. L.J. Gray, S. J. Phipps, T.J. Dunkerton, M.P. Baldwin, E.F. Drysdale and M.R. Allen, *Q.J.R. Meteorol. Soc.* **125**, 1985 (2001).

39. S.B. Fels, *Adv. Geophys.* **28A**, 277 (1985).

40. J.G. Charney and P.G. Drazin, *J. Geophys. Res.* , **66** 83 (1961).

41. D. G. Andrews and M.E. McIntyre, *J. Atmos. Sci.* **33**, 2031 (1976).

42. D. G. Andrews and M. E. McIntyre, *J. Atmos. Sci.* **35**, 175 (1978).

43. A. W. Brewer, *Quart. J. R. Met. Soc.* **75**, 351 (1949).

44. G. M. B. Dobson, *Phil. Trans. Roy. Soc. Lon.* **A236**, 187 (1956).

45. R. J. Murgatroyd and F. Singleton, *Quart. J. R. Met. Soc.* **87**, 125 (1961).

46. M. E. McIntyre, In *Transport processes in the middle atmosphere* Eds. G. Visconti & R. Garcia (P 267 Springer, 1987)).

47. K. Kodera K., *J. Geophys. Res.* **100**, 14077 (1995).

48. K. Kodera K., M. Chiba and K. Shibata, *Geophys. Res. Lett.* **18**, 1209(1991).

49. J.D. Haigh, *Nature* **370**, 544 (1994).

50. H. G. Mayr, J. G. Mengel, C. O. Hines, K. L. Chan, N. F. Arnold, C. A. Reddy and H. S. Porter *J. Geophys. Res.* **102**, 26077 (1997a).

51. H.G. Mayr, J.G. Mengel, C.O. Hines, K. L. Chan, N. F. Arnold, C. A. Reddy and H. S. Porter **J. Geophys. Res. 102**, 26093 (1997b).

52. N. M. Gavrilov, in *NATO ASI Series I: Global environmental change, 50, Gravity wave processes: their parameterization in global climate models,* Ed. K. Hamilton (p. 45 Springer–Verlag, Berlin, 1997).

53. M. Salby M. and P. Callaghan, *J. Clim.* **13**, 2652 (2000).

54. N. F. Arnold and T. R. Robinson, *Ann. Geophysicae* **16**, 69 (1998).

55. N. F. Arnold and T. R. Robinson, *Space Sci. Revs.* **94**, 279 (2000).

56. S. W. H. Cowley, *Earth in Space* **8**, 9 (1996).

57. N. F. Arnold and T. R. Robinson, *Geophys. Res. Lett.* **28**, 2381 (2001).

58. J. G. Mengel, H. G. Mayr, K. L. Chan, C. O. Hines, C. A. Reddy, N. F. Arnold and H. S. Porter *Geophys. Res. Lett.* **22**, 3029 (1995).

59. M. J. Harris, N. F. Arnold and A.D. Aylward, *Ann. Geophysicae* **20**, 225 (2002).

60. N. F. Arnold, *Phil. Trans. Roy. Soc.* **A361**, 127 (2003).

61. B. Naujokat, *J. Atmos. Sci.* **43**, 1873 (1986).

Neil F. Arnold

Born in Urmston, Lancashire, Neil Arnold graduated with a first-class honours degree in physics with astrophysics from the University of Kent at Canterbury in 1986 and obtained a DPhil in atmospheric physics from the University of Oxford in 1990. For the majority of the past ten years he has been a researcher at the University of Leicester, but he also spent a couple of years at the Goddard Space Flight Center in Maryland, USA as a Visiting Scientist. He was awarded a PPARC Advanced Fellowship in 1999 and is due to become a lecturer in October 2004, aged 39. Scientific interests focus on the influence of space on the Earth's climate system and the use of satellites, radars and computer models to gain insight into the underlying processes. Recreations include playing musical instruments and spending time with his two young sons.

INDEX